A HISTORY OF
BRITISH EARTHQUAKES

A HISTORY

OF

BRITISH EARTHQUAKES

BY

CHARLES DAVISON, Sc.D., F.G.S.

CAMBRIDGE

AT THE UNIVERSITY PRESS

1924

CAMBRIDGE UNIVERSITY PRESS
Cambridge, New York, Melbourne, Madrid, Cape Town, Singapore, São Paulo, Delhi

Cambridge University Press
The Edinburgh Building, Cambridge CB2 8RU, UK

Published in the United States of America by Cambridge University Press, New York

www.cambridge.org
Information on this title: www.cambridge.org/9780521140997

First published 1924
This digitally printed version 2009

A catalogue record for this publication is available from the British Library

ISBN 978-0-521-14099-7 paperback

PREFACE

IN a study that has occupied the greater part of my leisure time during the last thirty-five years, in one moreover that depends so greatly on the co-operation of others, I have received much and valued help from friends and strangers, and I take this opportunity of expressing my debt to them and of offering them my sincere and hearty thanks. If I have been led in this way to realise how many careful observers there are in this country, I have also experienced the ready courtesy with which they have placed their records at my disposal.

Among the many thousands who have assisted me, I should mention especially the late Dr Tempest Anderson, F.G.S., Dr H. H. Bemrose, F.G.S., Mr E. S. Cobbold, F.G.S., the late Prof. Grenville A. J. Cole, F.R.S., Mr E. Greenly, F.G.S., the late Prof. J. Milne, F.R.S., the late Mr H. Cecil Moore, the Rev. W. M. Morris, Mr R. C. Mossman, F.R.S.E., and Mr J. T. Stobbs, F.G.S. For numerous observations of the Menstrie earthquakes, I have to thank Mr J. Dempster, Mr T. J. H. Drysdale, Mr T. B. Johnston, Mr W. H. Lindsay, Mr E. J. Sim and others, who by their interested care have preserved an important series of earthquakes from oblivion. I have received much help in connexion with the geology of the central districts of various earthquakes from Dr J. Horne, F.R.S., the late Prof. C. Lapworth, F.R.S., and many other geologists.

All who have worked in the Cambridge University Library will realise how much I owe to the various officials, to their courtesy and helpfulness.

For permission to make use of various papers published by them I have pleasure in thanking the Councils of the Geological Society and of the Royal Society of Edinburgh, and Dr R. H. Rastall, F.G.S., the Editor of the *Geological Magazine.*

In writing the following pages, I have had three main objects in view—to compile a catalogue of all known British earthquakes, to trace the zones in which crust-changes have recently occurred, in which the faults are yet alive, and to discover some of the laws that rule the growth of faults. How slight has been the advance in each direction, few perhaps can know more fully than myself. But "if," with the materials at my command, "I have done well, and as is fitting the story, it is that which I desired; but if slenderly and meanly, it is that which I could attain unto[1]."

CHARLES DAVISON.

70 CAVENDISH AVENUE, CAMBRIDGE
July 1924

[1] II Maccabees xv. 38.

CONTENTS

CHAPTER   PAGE

I. INTRODUCTION 1

 The Study of British Earthquakes. 1
 Methods of Investigation 7
 Discrimination of Spurious Earthquakes . . . 10

II. CATALOGUE OF BRITISH EARTHQUAKES . . . 12

III. EARTHQUAKES OF INVERNESS AND OF OTHER
  CENTRES NEAR THE GREAT GLEN FAULT . . . 35

 Inverness Earthquakes 36
 Fort William Earthquakes 56
 Ballachulish Earthquake 58
 Oban Earthquakes 58
 Torosay (Mull) Earthquake 61
 Phladda Earthquake 61
 Dhu Heartach Earthquake 61

IV. EARTHQUAKES OF COMRIE AND OF OTHER CENTRES
  NEAR THE HIGHLAND BORDER FAULT . . . 62

 Dunkeld Earthquakes 62
 Comrie Earthquakes 62
 Dunoon Earthquakes 119
 Rothesay Earthquakes 121
 Kintyre Earthquakes 121

V. MENSTRIE EARTHQUAKES 123

VI. EARTHQUAKES OF MISCELLANEOUS AND UNKNOWN
  CENTRES IN SCOTLAND 152

 Ullapool Earthquakes 152
 Invergarry Earthquakes. 153
 Strontian Earthquakes 157
 Kinloch Rannoch Earthquake 159
 Pitlochry Earthquake 159
 Aberfeldy Earthquake 159
 Amulree Earthquake 159
 Ardvoirlich Earthquakes 159
 Perth Earthquakes 160
 Dunning Earthquakes 160
 Glasgow Earthquakes 160
 Edinburgh Earthquakes 163
 Leadhills Earthquakes 165
 Penpont Earthquakes 166
 Ecclefechan Earthquakes 166
 Earthquakes of Unknown Epicentres in Scotland . . 168

CHAPTER   PAGE

VII. EARTHQUAKES OF NORTH WALES 172

Carnarvon Earthquakes. 172
Beddgelert Earthquake 180
Barmouth Earthquakes 180
Bala Earthquake 181
Rhyl Earthquake 181
Earthquakes of Unknown Epicentres in North Wales . 181

VIII. EARTHQUAKES OF SOUTH WALES 183

Pembroke Earthquakes 183
Carmarthen Earthquakes 191
Swansea Earthquakes 196
Earthquakes of Unknown Epicentres in South Wales . 203

IX. EARTHQUAKES OF THE NORTH-WEST OF ENGLAND . 204

Carlisle Earthquakes 204
Grasmere Earthquakes 210
Kendal Earthquakes 212
Rochdale Earthquakes 216
Bolton Earthquakes 219
Shap Earthquake 221
Dalton-in-Furness Earthquakes 221
Settle Earthquake 221
Clitheroe Earthquakes 222
Halifax Earthquakes 222
Everton (Liverpool) Earthquake 222
Earthquakes of Unknown Epicentres in the North-west
of England 223

X. EARTHQUAKES OF THE NORTH-EAST OF ENGLAND . 224

York Earthquakes 224
Wetherby Earthquakes 226
Earthquakes of Unknown Epicentres in Yorkshire . . 226

XI. EARTHQUAKES OF THE WEST OF ENGLAND . . 227

Cheshire Earthquake 227
Shropshire Earthquakes 228
Worcestershire Earthquakes 229
Malvern Earthquake 229
Gloucestershire Earthquakes 231
Herefordshire Earthquake 231

XII. EARTHQUAKES OF THE MIDLAND COUNTIES OF
ENGLAND 233
Staffordshire Earthquakes 233
Derbyshire Earthquakes 233

CHAPTER   PAGE

    Nottinghamshire Earthquakes 235
    Mansfield Earthquakes 236
    Leicestershire Earthquake 237
    Oakham Earthquake 237
    Birmingham Earthquakes 238
    Northamptonshire Earthquakes 239
    Stamford Earthquakes 239
    Huntingdonshire Earthquakes 241
    Oxfordshire Earthquakes 241
    Banbury Earthquakes 241
    Oxford Earthquakes 242

XIII. TWIN EARTHQUAKES OF THE WEST AND MIDLAND
     COUNTIES OF ENGLAND 244

    Hereford Earthquakes 244
    Stafford Earthquakes 259
    Derby Earthquakes 263
    Leicester Earthquakes 278
    Northampton Earthquakes 283
    Doncaster Earthquakes 284

XIV. EARTHQUAKES OF THE EAST OF ENGLAND . . . 290

    Gainsborough Earthquake 290
    Lincoln Earthquakes 291
    Grantham Earthquakes 292
    Coningsby Earthquakes 293
    Cambridgeshire Earthquakes 293
    Norfolk Earthquakes 293
    Bungay Earthquake 294
    Suffolk Earthquakes 294
    Ixworth Earthquake 295
    Framlingham Earthquakes 295

XV. EARTHQUAKES OF THE SOUTH-WEST OF ENGLAND . 296

    Penzance Earthquakes 296
    Helston Earthquakes 299
    Falmouth Earthquake 305
    Truro Earthquakes 305
    St Agnes Earthquake 306
    St Austell Earthquake 307
    Camelford Earthquakes 307
    Altarnon Earthquakes 308
    Liskeard Earthquakes 309
    Launceston Earthquakes 309

Earthquakes of Unknown Epicentres in Cornwall . . 311
Okehampton Earthquakes 312
Dartmouth Earthquake 313
Exmouth Earthquakes 314
Barnstaple Earthquakes 314
Earthquakes of Unknown Epicentres in Devon . . 317
Taunton Earthquakes 318
Wells Earthquakes 319
Earthquakes of Unknown Epicentres in Somerset . . 321

XVI. EARTHQUAKES OF THE SOUTH OF ENGLAND . . 323
Dorsetshire Earthquakes 323
Wiltshire Earthquakes 324
Hampshire Earthquakes 324

XVII. EARTHQUAKES OF THE SOUTH-EAST OF ENGLAND . 325
Chichester Earthquakes 325
Lewes Earthquakes 329
Canterbury Earthquakes 330
Reigate Earthquakes 331
Maidstone Earthquake 332
London Earthquakes 332
St Albans Earthquake 336
Colchester Earthquakes 337

XVIII. EARTHQUAKES OF UNKNOWN EPICENTRES IN ENG-
LAND 345

XIX. EARTH-SHAKES IN MINING AND LIMESTONE DIS-
TRICTS 353
Earth-Shakes in Mining Districts 353
Earth-Shakes in Cornwall 354
Earth-Shakes in the Rhondda Valleys (Glamorganshire) 356
Pendleton Earth-Shakes 359
Barnsley Earth-Shake 361
Kilsyth Earth-Shake 362
Characteristics of Earth-Shakes in Mining Districts . 362
Origin of Earth-Shakes in Mining Districts . . . 364
Earth-Shakes in Limestone Districts 365

XX. EXTRA-BRITISH EARTHQUAKES 367
Earthquakes in Ireland 367
Earthquakes in the Channel Islands 368
Earthquakes in the Shetland Islands 370
The Lisbon Earthquake of 1755 Nov. 1 . . . 371

CHAPTER		PAGE
XXI.	SOUND-PHENOMENA OF BRITISH EARTHQUAKES .	377
	Nature of the Earthquake-Sound	377
	Inaudibility of the Sound to some Observers. . .	378
	Variation throughout the Sound-Area	379
	Relations between the Sound-Area and Disturbed Area.	380
	Time-Relations of the Sound and Shock . . .	381
XXII.	DISTRIBUTION OF BRITISH EARTHQUAKES IN SPACE	
	AND TIME	383
	Relations between the Earthquakes of England, Scotland	
	and Wales	383
	Distribution of British Earthquakes in Space. . .	385
	Distribution of British Earthquakes in Time . . .	389
	Periodicity of British Earthquakes	390
XXIII.	ORIGIN OF BRITISH EARTHQUAKES	394
	Origin of Simple Earthquakes	395
	Nature of Twin Earthquakes	398
	Twin Earthquakes connected with a Twin-Focus . .	400
	Classification of Twin Earthquakes	402
	Origin of Twin Earthquakes	403
	INDEX	410

LIST OF ILLUSTRATIONS

FIG.   PAGE

1. Inverness Earthquakes of 1888 Feb. 2 and 1890 Nov. 15 . 39
2. Inverness Earthquakes of 1890: After-Shocks . . . 41
3. Inverness Earthquake of 1901 Sept. 18 44
4. Inverness Earthquakes of 1901: Minor Shocks . . . 47
5. Inverness Earthquakes: Distribution of Epicentres . . 53
6. Fort William Earthquake of 1894 Jan. 12 57
7. Oban Earthquakes of 1880 Nov. 28 and 1907 Jan. 17 . . 60
8. Comrie Earthquakes of 1789 Nov. 5, 1895 July 12 and 1898 Aug. 22 67
9. Comrie Earthquake of 1839 Oct. 23 80
10. Dunoon Earthquakes of 1904 Sept. 18 and 1908 July 3 . . 120
11. Kintyre Earthquake of 1889 July 15 122
12. Menstrie Earthquakes of 1900 Sept. 17 and 22 . . . 125
13. Menstrie Earthquake of 1905 July 23 126
14. Menstrie Earthquake of 1905 Sept. 21 128
15. Menstrie Earthquakes of 1906 Oct. 8, Dec. 28 and 30 . . 130
16. Menstrie Earthquakes of 1908 Oct. 20 134
17. Menstrie Earthquake of 1912 May 3 144
18. Menstrie Earthquakes: Distribution of Epicentres . . . 150
19. Ullapool Earthquake of 1892 Mar. 4 152
20. Invergarry Earthquakes: Places of Observation . . . 153
21. Strontian Earthquake of 1902 Oct. 14 158
22. Glasgow Earthquake of 1910 Dec. 14 162
23. Edinburgh Earthquake of 1889 Jan. 18 164
24. Penpont Earthquake of 1873 Apr. 16 and Ecclefechan Earthquake of 1896 May 29 167
25. Carnarvon Earthquake of 1903 June 19 173
26. Carnarvon Earthquake of 1903 June 19: Seismographic Record at Birmingham 175
27. Carnarvon Earthquakes of 1903 June 19 and 1906 June 29 . 178
28. Pembroke Earthquakes of 1892 Aug. 18 184
29. Pembroke Earthquakes of 1892 Aug. 18 and 22 . . . 188
30. Carmarthen Earthquake of 1893 Nov. 2 192
31. Carmarthen Earthquakes of 1893 Nov. 2 194
32. Swansea Earthquake of 1906 June 27 197
33. Swansea Earthquakes of 1906 June 27 and 1907 July 3 . . 201
34. Carlisle Earthquake of 1786 Aug. 11 205
35. Carlisle Earthquakes of 1901 July 9 207
36. Grasmere Earthquakes of 1885 June 30 and 1911 May 16 . 211
37. Kendal Earthquake of 1843 Mar. 17 and Lancashire Earthquake of 1750 Apr. 13 213
38. Kendal Earthquakes of 1871 Mar. 17 214
39. Rochdale Earthquake of 1777 Sept. 14 216
40. Rochdale Earthquakes of 1869 Mar. 15 and 25 . . . 218

FIG.   PAGE

41. Bolton Earthquake of 1889 Feb. 10 220
42. York Earthquakes of 1822 Mar. 20 and 1885 June 18 and Wetherby Earthquake of 1890 June 25 225
43. Malvern Earthquake of 1907 Sept. 27 230
44. Herefordshire Earthquake of 1775 Sept. 8 232
45. Derbyshire Earthquake of 1795 Nov. 18 234
46. Mansfield Earthquake of 1816 Mar. 17 236
47. Oakham Earthquake of 1898 Jan. 28 237
48. Stamford Earthquakes of 1813 Sept. 24 and 1844 June 12 . 240
49. Banbury Earthquake of 1731 Oct. 21 and Oxford Earthquakes of 1666 Jan. 18 and 1683 Sept. 17 242
50. Hereford Earthquakes of 1863 Oct. 6 and 1868 Oct. 30 . . 245
51. Hereford Earthquake of 1896 Dec. 17 250
52. Hereford Earthquake of 1896 Dec. 17: Coseismal Lines . 255
53. Hereford Earthquakes of 1853 Mar. 27, 1896 Dec. 17 and 1924 Jan. 26 256
54. Stafford Earthquake of 1916 Jan. 14 260
55. Stafford Earthquake of 1916 Jan. 14: Central District . . 261
56. Derby Earthquake of 1903 Mar. 24 264
57. Derby Earthquake of 1903 Mar. 24: Seismographic Record at Birmingham 267
58. Derby Earthquake of 1903 May 3 270
59. Derby Earthquake of 1904 July 3, 3.21 p.m. 271
60. Derby Earthquake of 1904 July 3, 11.8 p.m. 274
61. Derby Earthquake of 1906 Aug. 27 275
62. Leicester Earthquake of 1893 Aug. 4 279
63. Leicester Earthquake of 1904 June 21 280
64. Northampton Earthquake of 1750 Oct. 11 283
65. Doncaster Earthquake of 1905 Apr. 23 286
66. Doncaster Earthquakes of 1902 Apr. 13 and 1905 Apr. 23 . 287
67. Gainsborough Earthquake of 1703 Dec. 28 . . . 291
68. Grantham Earthquakes of 1750 Sept. 3 and 1792 Feb. 25 . 292
69. Bungay Earthquake of 1757 Jan. 10, Ixworth Earthquake of 1869 Jan. 9 and Framlingham Earthquake of 1901 Oct. 19. 294
70. Penzance Earthquake of 1904 Mar. 3 and St Agnes Earthquake of 1905 Jan. 20 298
71. Helston Earthquake of 1842 Feb. 17 299
72. Helston Earthquakes of 1892 May 16 and 17 . . . 301
73. Helston Earthquakes of 1898 Mar. 29 and Apr. 1 . . . 302
74. Altarnon Earthquake of 1889 Oct. 7 and Camelford Earthquake of 1891 Mar. 26 308
75. Launceston Earthquake of 1883 June 25 309
76. Launceston Earthquake of 1896 Jan. 26 310
77. Okehampton Earthquake of 1858 Sept. 28 312
78. Dartmouth Earthquake of 1886 Jan. 4 313

FIG.   PAGE

79.   Barnstaple Earthquakes of 1894 Jan. 23 and 1920 Sept. 9 .   315
80.   Taunton Earthquakes of 1868 Jan. 4 and 1885 Jan. 22 .   .   318
81.   Wells Earthquakes of 1893 Dec. 30 and 31   320
82.   Chichester Earthquake of 1750 Mar. 29   .   .   .   326
83.   Chichester Earthquakes of 1811 Nov. 30 and 1834 Jan. 23,
      Lewes Earthquake of 1864 Apr. 30, Canterbury Earthquake
      of 1776 Nov. 27, and Reigate Earthquake of 1551 May 25   328
84.   London Earthquakes of 1750 Feb. 19 and Mar. 19   .   .   334
85.   Colchester Earthquake of 1884 Apr. 22   .   .   .   338
86.   Colchester Earthquake of 1884 Apr. 22: Central District   .   340
87.   Colchester Earthquake of 1884 Apr. 22: Level of Water in
      Well at Bocking   343
88.   Earthquake of 1852 Nov. 9   351
89.   Blisland Earth-Shake of 1895 Aug. 25   354
90.   Camborne Earth-Shake of 1902 June 4   355
91.   Rhondda Valleys Earth-Shakes of 1889 June 22, 1894 Apr. 11,
      1896 Oct. 16, 1907 May 17 and 1910 Feb. 16 .   .   .   357
92.   Pendleton Earth-Shakes of 1899 Feb. 27 and 1905 Nov. 25 .   359
93.   Barnsley Earth-Shake of 1903 Oct. 25   361
94.   Distribution of British Earthquakes in Space   .   .   .   386
95.   Monthly Variation in Frequency of British Earthquakes   .   390
96.   Annual Periodicity of British Earthquakes   391
97.   Hourly Variation in Frequency of British Earthquakes   .   392
98.   Diurnal Periodicity of British Earthquakes   392
99.   Origin of Twin Earthquakes   404
100.   Connexion between Twin Earthquakes and the Structure of
      the Earth's Crust   408

AUTHORITIES

THE following list contains the names and editions of the principal authorities on which this history depends. They are arranged under three headings—Chronicles and Histories, Scientific and other Journals, and Catalogues of Earthquakes. Each work is preceded by the abbreviation used for reference in the text. Newspapers and local histories and various books to each of which but a few references are made are not included in this list.

CHRONICLES AND HISTORIES

Anglo-Saxon Chron.—The Anglo-Saxon Chronicle. Rolls Series, 2 vols. 1861.

Annal. Cambriae.—Annales Cambriae. Rolls Series, 1860.

Annal. Monast.—Annales Monastici. Rolls Series, 4 vols. 1864–1869.

Annals of Loch Cé.—A Chronicle of Irish Affairs. Rolls Series, 2 vols. 1871.

Baker, Sir R.—A Chronicle of the Kings of England, etc. 1674.

B. de Cotton.—Bartholomaei de Cotton monachi Norwicensis, Historia Anglicana. Rolls Series, 1859.

Brut y Tywysogion—or, The Chronicle of the Princes. Rolls Series, 1860.

Calderwood.—The History of the Kirk of Scotland, by David Calderwood. Woodrow Society, 1842–1849.

Camden.—The History of the Most Renowned and Victorious Princess Elizabeth, late Queen of England, etc., by William Camden, 1688.

Chron. Angliae.—Chronicon Angliae, auctore monacho quodam Sancti Albani. Rolls Series, 1874.

Chron. Monast. de Melsa.—Chronica Monasterii de Melsa, auctore Thoma de Burton, Abbate, etc. Rolls Series, 3 vols. 1866–1868.

Chron. Monast. S. Albani.—Chronica Monasterii S. Albani Johannis de Trokelowe et Henrici de Blaneforde, Monachorum S. Albani. Rolls Series, 1866.

Chron. of Melrose.—The Chronicles of Melrose. Trans. by Rev. J. Stevenson (The Church Historians of England, vol. 4, pt. 1, 1856).

Chron. Scotorum.—Chronicon Scotorum: a Chronicle of Irish Affairs. Rolls Series, 1866.

Eulog. Hist.—Eulogium (Historiarum, sive Temporis), a Monacho quodam Malmesburiensi exaratum. Rolls Series, 3 vols. 1858–1863.

Evelyn's Diary.—Diary of John Evelyn, Esq., F.R.S., edited by H. B. Wheatley, 4 vols. 1879.

Fascic. Zizan.—Fasciculi Zizaniorum Magistri Johannis Wyclif cum Tritico, ascribed to Thomas Netter of Walden. Rolls Series, 1858.

Florence of Worcester.—The Chronicle of Florence of Worcester. Trans. by T. Forester, 1854.

Flor. Hist.—Flores Historiarum. Rolls Series, 3 vols. 1890.

Gervase of Cant.—The Historical Works of Gervase of Canterbury. Rolls Series, 2 vols. 1879, 1880.

Henry of Huntingdon.—The History of the English, by Henry, Archdeacon of Huntingdon. Rolls Series, 1879.

Henry of Knyghton.—Sir R. Twysden, Historiae Anglicanae Scriptores X. 1652.

Hist. Monast. S. August. Cant.—Historia Monasterii S. Augustini Cantuariensis, by Thomas of Elmham, formerly monk and treasurer of that foundation. Rolls Series, 1858.

Holinshed.—Chronicles of England, Scotland and Ireland, by R. Holinshed, 6 vols. 1807–1808.

John Hardyng's Chron.—The Chronicle of John Hardyng,...with the continuation by Richard Grafton, 1812.

John de Oxenedes.—Chronica Johannis de Oxenedes. Rolls Series, 1859.

John of Brompton.—Sir R. Twysden, Historiae Anglicanae Scriptores X. 1652.

John of Hexham.—The Chronicle of John, Prior of Hexham. Trans. by Rev. J. Stevenson (The Church Historians of England, vol. 4, pt. 1, 1856).

Le Livere de Reis de Engleterre.—Le Livere de Reis de Brittania e le Livere de Reis de Engleterre. Rolls Series, 1865.

Materials for the Hist. of Thomas Becket.—Rolls Series, 7 vols. 1875–1885.

Matthew Paris.—Matthaei Parisiensis, Monachi Sancti Albani, Historia Anglorum, etc. Rolls Series, 3 vols. 1866–1869.

Monum. Franc.—Monumenta Franciscana. Rolls Series, 2 vols. 1858, 1882.

N. Triveti.—Annales Sex Regum Angliae, etc., by F. Nicolai Triveti. 1845.

Ralph de Diceto.—The Historical Works of Master Ralph de Diceto, Dean of London. Rolls Series, 2 vols. 1876.

Robert Fabyan.—The New Chronicles of England and France, etc., by Robert Fabyan. Named by himself the Concordance of Histories. 1811.

Roger de Hoveden.—Chronica Magistri Rogeri de Hovedene. Rolls Series, 4 vols. 1868–1871.

Roger of Wendover.—The Flowers of History, by Roger de Wendover. Rolls Series, 3 vols. 1886–1889.

Speed.—The History of Great Britaine, etc., by John Speed, 2nd edit. 1627.

Stow.—The Annales of England,...from the first inhabitation ontill this present yeere 1601, by John Stow.

Strype.—Ecclesiastical Memorials, etc., by John Strype, 3 vols. 1721.

Symeon of Durham.—Symeonis Monachi Opera Omnia, Historia Ecclesiae Dunhelmensis, Historia Regum, etc. Rolls Series, 2 vols. 1882, 1885.

Thomas Walsingham.—Thomae Walsingham, quondam Monachi S. Albani, Historia Anglicana. Rolls Series, 2 vols. 1863.

William of Malmesbury.—The History of the Kings of England and of his own Times, trans. by Rev. J. Sharpe (The Church Historians of England, vol. 3, 1854).

William Rishanger.—Chronica Monasterii S. Albani Willelmi Rishanger, quondam monachi S. Albani. Rolls Series, 1865.

William Thorn.—Sir R. Twysden, Historiae Anglicanae Scriptores X. 1652.

SCIENTIFIC AND OTHER JOURNALS

Ann. de Chim. et de Phys.—Annales de Chimie et de Physique (1816–1837, when the annual lists of earthquakes ceased to be published).

Ann. of Phil.—Annals of Philosophy, etc., by Thomas Thomson (1813–1826).

Ann. Reg.—Annual Register (1758–).

Brit. Ass. Rep.—Reports of the British Association (1831–).

Brit. Met. Soc. Proc.—Proceedings of the British Meteorological Society (1861–1871).
Brit. Rainfall.—Symons's British Rainfall (1866–).
Cornwall Geol. Soc. Trans.—Transactions of the Royal Geological Society of Cornwall (1818–).
Edin. Journ. Sci.—Edinburgh Journal of Science (1824–1832).
Edin. New Phil. Journ.—Edinburgh New Philosophical Journal (1826–1864).
Edin. Roy. Soc. Proc.—Proceedings of the Royal Society of Edinburgh (1845–).
Edin. Roy. Soc. Trans.—Transactions of the Royal Society of Edinburgh (1788–).
Gent. Mag.—The Gentleman's Magazine: or, Trader's Monthly Intelligencer (1731–1868).
Geol. Ass. Proc.—Proceedings of the Geologists' Association (1859–).
Geol. Mag.—Geological Magazine (1864–).
Irish Acad. Trans.—Transactions of the Royal Irish Academy (1787–).
Journ. Sci.—Journal of Science (1879–1885).
London Mag.—London Magazine or Gentleman's Monthly Intelligencer (1732–1755).
Mon. Met. Mag.—Symons's Monthly Meteorological Magazine (1866–).
Nature (1870–).
Notes and Queries (1849–).
Phil. Mag.—The Philosophical Magazine (1798–).
Phil. Trans.—Philosophical Transactions of the Royal Society (1665–).
Quart. Journ. Geol. Soc.—Quarterly Journal of the Geological Society (1845–).
Woolhope N. F. C. Trans.—Transactions of the Woolhope Naturalists' Field Club (1852–).

CATALOGUES OF EARTHQUAKES

Drummond.—A table of shocks of earthquakes, from September 1839 to the end of 1841, observed at Comrie, near Crieff. By James Drummond, Phil. Mag. vol. 20, 1842, pp. 240–247.
Fuchs.—Statistik der Erdbeben, von 1865–1884: England. By C. W. C. Fuchs, Sitzungsb. der k. Akad. der Wissensch., Wien, 1886, pp. 134–139.
Howard.—The Climate of London, etc. By Luke Howard, 3 vols. 1833.
Lauder.—Account of the late earthquake [1816 Aug. 13] in Scotland. By T. L. Dick (afterwards T. D. Lauder), Ann. of Phil. vol. 8, 1816, pp.364–377.
Lowe.—Natural Phenomena and Chronology of the Seasons. By E. J. Lowe, 1870.
Mallet.—Catalogue of recorded earthquakes from 1606 B.C. to A.D. 1850. By R. Mallet, Brit. Ass. Rep. 1852, pp. 1–176; 1853, pp. 118–212; 1854, pp. 2–326.
Milne.—Notices of earthquake-shocks felt in Great Britain, and especially in Scotland, etc. By David Milne (afterwards Milne Home). Edin. New Phil. Journ. vol. 31, 1841, pp. 92–122, 259–309; vol. 32, 1842, pp. 106–127, 362–378; vol. 33, 1842, pp. 372–388; vol. 34, 1843, pp. 85–107; vol. 35, 1843, pp. 137–160; vol. 36, 1844, pp. 72–86, 362–377. Reprinted,

with additions, in 1887. Referred to thus: Milne, vol. 31, p. 95, and Milne, reprint, respectively.

O'Reilly.—Catalogue of the earthquakes having occurred in Great Britain and Ireland during historical times, etc. By J. P. O'Reilly, Irish Acad. Trans. vol. 28, 1884, pp. 285–316.

Parfitt.—On earthquakes in Devonshire. By E. Parfitt, Devon Ass. Trans. vol. 16, 1884, pp. 641–661; vol. 17, 1885, pp. 281–284; vol. 19, 1887, pp. 547–554. Referred to as Parfitt, I, II, and III.

Perrey.—Sur les tremblements de terre dans les Iles Britanniques. By A. Perrey, Lyon, Soc. Agric. Annal. vol. 1, 1849, pp. 115–177. Perrey's annual catalogues were published in Dijon, Mém. Acad. des Sci. from 1845 to 1854; Bruxelles, Bull. Acad. Roy. des Sci. from 1848 to 1856, and Mém. Cour. from 1859 to 1875. Referred to thus: Perrey; Perrey, Dijon Acad.; Perrey, Bull.; and Perrey, Mém. Cour. respectively.

Prestwich.—On earthquakes. By Sir John Prestwich, Bart. Geol. Mag. vol. 7, 1870, pp. 541–544.

Roper.—A list of the more remarkable earthquakes in Great Britain and Ireland during the Christian Era. By W. Roper, Lancaster, 1889.

Short.—A General Chronological History of the Air, Weather, Seasons, Meteors, etc., in Sundry Places and different Times, more particularly for the Space of 250 years, 2 vols. 1749.

Taylor.—An account of repeated shocks of earthquakes felt at Comrie in Perthshire. By Rev. Ralph Taylor, Edin. Roy. Soc. Trans. vol. 3, 1794, pp. 240–246.

Walford.—The Famines of the World, etc. By C. Walford, 1879, pp. 52–68.

CHAPTER I

INTRODUCTION

THE STUDY OF BRITISH EARTHQUAKES

THE earliest known catalogue including British earthquakes is contained in an anonymous and almost inaccessible work attributed to Dr Thomas Short (1690?–1772), a physician of Sheffield. In 1749, he published in two volumes *A General Chronological History of the Air, Weather, Seasons, Meteors, etc., in Sundry Places and different Times, more particularly for the Space of 250 years*[1]. Earthquakes, comets, storms and pestilences are all included in this history, which, so far as British earthquakes are concerned, begins with the year 10 and ends with 1748. The great defect of the catalogue is the almost entire want of references, especially for the first half of the Christian era. Though Dr Short's statements have been accepted without question by several writers, such as Meldola, Roper and Milne, it is difficult to understand what sources of information can have been accessible to him in the middle of the eighteenth century that are closed to us now. Thus, while quoting some of Dr Short's records in their appropriate places, I have omitted them from the catalogue of British earthquakes in the following chapter, unless they are of earthquakes near his own time or are supported by other evidence, that is to say, I have not regarded the shocks as undoubted earthquakes.

The study of British earthquakes may be said to date from the year 1750. During this "year of earthquakes," as Dr Stukeley called it, there were five notable shocks in England. London and the home counties were shaken on Feb. 19 and Mar. 19 (N.S.), Portsmouth and the Isle of Wight on Mar. 29, the north-west of England and north-east of Wales on Apr. 13, Northamptonshire and the surrounding counties on Oct. 11. The interest and alarm created by them, and especially by the early pair of shocks in London, were intense. From Feb. 19 until about the end of the year, nearly fifty articles and letters were read before the Royal Society. Dr W. Stukeley (1687–1765) and Dr S. Hales (1677–1761) contributed others on the causes and philosophy of earthquakes, and the Royal Society adopted the unique expedient of publishing these papers out of the usual chronological

---

[1] The only copy of this book to which I have been able to obtain access is in the library of the Royal Meteorological Society, and I take this opportunity of expressing my thanks for the permission, courteously given me, to consult it.

order as an appendix to the *Philosophical Transactions* for the year
1750[1]. Though most of these articles are merely personal narratives,
they form a collection that is of the highest value to the student of
British earthquakes.

To the disturbances of 1750, we are also indebted for our earliest
catalogues of purely British earthquakes. Two appeared at the same
time, one in the *London Magazine* for March 1750 (pp. 102–103), the
other in the *Gentleman's Magazine* for that month (vol. 20, p. 56),
both being the work of anonymous writers. The former contains
notices of 33 earthquakes in England "as recorded in our ancient historians" from Symeon of Durham to Stow. The dates range from
A.D. 974 to 1692. To this, the editor added a supplementary list of
15 other earthquakes from that of 1665 to the latest shock on 1750
Feb. 19. The second catalogue gives the dates, and little more than
the dates, of 24 earthquakes between the years 1048 and 1750.

Towards the close of the same year, a third and more important
catalogue was issued in a pamphlet—*A Chronological and Historical
Account of the most memorable Earthquakes that have happened in the
World, from the beginning of the Christian Period to the present year
1750; with an Appendix, containing a distinct series of those which have
been felt in England, and a Preface, seriously addressed to all Christians
of every Denomination*. Though it was published anonymously as
"By a Gentleman of the University of Cambridge," its author is
known to have been the Rev. Zachary Grey (1688–1766), an antiquary
of wide reading and an eager controversialist. The appendix contains
notices of 41 earthquakes from 974 to 1750. Though the *London
Magazine* and the Grey catalogues depend to a great extent on the
same authorities, the former notices 12 earthquakes which are not
referred to in the latter, while this contains 8 records which are not
to be found in the former. The *London Magazine* catalogue cannot
therefore be an earlier version of the Grey catalogue.

During the latter half of the eighteenth century, interest in British
earthquakes was intermittent and local[2]. On 1757 July 15, a shock,
felt over the greater part of Cornwall and in the Scilly Isles, was the

---

[1] *Philosophical Transactions: being an Appendix to those for the year* 1750. Consisting of a Collection of several Papers laid before the Royal Society, concerning
several Earthquakes felt in England, and some neighbouring Countries, in the
year 1750. No. 497, pp. 601–750.

[2] Towards the end of the year 1755, many communications were made to the
Royal Society on the effects of the great Lisbon earthquake of Nov. 1, especially
on the seiches observed in the lakes and pools of the British Isles and of the great
sea-waves which, a few hours later, reached our southern coasts.

subject of the first scientific study of a British earthquake, for which we are indebted to the Rev. W. Borlase, D.D. (1695-1772)[1]. Again, on 1795 Nov. 18, an earthquake widely felt throughout the midland counties was carefully investigated by Dr E. W. Gray (1748-1806)[2]. About the close of the century (1789-1801), a remarkable series of shocks felt at and near Comrie in Perthshire attracted the close attention of the Rev. S. Gilfillan (1762-1826) at Comrie and the Rev. R. Taylor at Ochtertyre (4 miles east of Comrie). The results of their valuable work will be noticed in Chapter IV. The strongest of Scottish earthquakes occurred in the neighbourhood of Inverness on 1816 Aug. 13. A brief description of this earthquake was published by Mr (afterwards Sir) T. Dick Lauder (1784-1846), and was prefaced by a list of 20 earthquakes in England and Scotland[3], which seems to have formed the foundation of several later catalogues.

In the autumn of 1833, a centre near Chichester, which had been in action from time to time for nearly two centuries, gave rise to a series of moderately strong earthquakes that continued for nearly two years. These earthquakes were studied by an unofficial committee in Chichester, of which Mr J. P. Gruggen was secretary. The report which he drew up was communicated to the Royal Society, but only a very brief summary was published[4]. Of considerable value in itself, it is interesting as the first report of a committee instituted for the study of British earthquakes.

After 1801, the Comrie earthquakes diminished in strength and frequency until the autumn of 1839, when a second series began early in October, culminating in the strong shock of Oct. 23 that was felt over nearly all Scotland. This in its turn was followed by a large number of after-shocks. At the ensuing meeting of the British Association, a committee was appointed with Mr David Milne (afterwards Milne Home, 1805-1890) as secretary, and charged with the registration of earthquakes in Scotland and Ireland. The work of this committee (1841-1844)[5] was confined almost entirely to the Comrie earthquakes and will be described in Chapter IV. During the same years, Mr Milne published a valuable series of papers on British earthquakes[6],

---

[1] *Phil. Trans.* vol. 50, 1759, pp. 499-506.   [2] *Phil. Trans.* 1796, pp. 353-381.
[3] *Annals of Phil.* vol. 8, 1816, pp. 364-377.
[4] *Roy. Soc. Proc.* vol. 3, 1837, p. 338. The original report is, however, preserved in the Archives of the Society, and I am indebted to the courtesy of the Council for permission to make extracts from it.
[5] *Brit. Ass. Rep.* 1841, pp. 46-50; 1842, pp. 92-98; 1843, pp. 120-127; 1844, pp. 85-90.
[6] See List of Authorities: Catalogues of Earthquakes.

copies of which with some additions were bound together and re-
issued in 1887. The first part contains a register of 238 earthquake
shocks felt in Great Britain from 1608 to 1839 Oct., founded on the
Rev. S. Gilfillan's diary of the earlier Comrie shocks and on papers
in various scientific journals. The second and longer part deals with
the Comrie earthquakes from 1839 Oct. 3 to the end of 1844, and
includes a full account of the strong shock of 1839 Oct. 23—the
most detailed report up to that time published on any British
earthquake.

About the middle of the last century, the study of earthquakes in
general was advanced by the labours of Prof. Alexis Perrey (1807–1882)
in France and Mr Robert Mallet (1810–1881) in this country. Perrey's
work was almost confined to the cataloguing of earthquakes. In 1849,
his memoir "Sur les tremblements de terre dans les Iles Britanniques"
was published[1], in which about 325 earthquakes between the years
1043 and 1847 are described. In addition to this valuable memoir,
Perrey published catalogues of earthquakes for the whole world for
each year from 1843 to 1871, which include occasional references to
British earthquakes[2].

Mallet's great catalogue of recorded earthquakes occupies many
pages in the Reports of the British Association for the years 1852–1854[3].
It was intended to cover the period from 1606 B.C. to A.D. 1850. The
completeness of Perrey's annual catalogues, however, rendered its
continuation after 1842 unnecessary. Of the total number of earthquakes (nearly seven thousand), 458 are British earthquakes, the first
of which occurred in 974 and the last in 1842. To the pioneer work of
Perrey and Mallet, no historian of British earthquakes can look back
without admiration and gratitude.

After 1862 Mallet took no active interest in earthquakes. His place,
so far as British earthquakes are concerned, was filled to some extent
by Mr E. J. Lowe (1825–1900), who, in his observatory at Beeston
(near Nottingham), erected a pendulum 30 feet in length, with which
he recorded a few local shocks. We are indebted to Mr Lowe for a
study of the Hereford earthquake of 1863, for notices of several minor
shocks, and for a work on *The Natural Phenomena and Chronology of
the Seasons* (1870), in which, among remarkable frosts, droughts,

[1] Lyon, *Soc. Agric. Annal.* vol. 1, 1849, pp. 115–177.
[2] Perrey's annual catalogues are published in Dijon *Mém. Acad. des Sci.* from
1845 to 1854; Bruxelles *Bull. Acad. Roy. des Sci.* 1848 to 1856, and *Mém. Cour*
from 1859 to 1875
[3] *Brit. Ass. Rep.* 1852, pp. 1–176; 1853, pp. 118–212; 1854, pp. 2–326.

storms, plagues, famines, etc., he records the occurrence of 134 earthquakes between the years 369 and 1750.

The Committee appointed by the British Association in 1840 to study Scottish earthquakes lapsed after the meeting of 1844. The Comrie earthquakes were then becoming slight and infrequent. Their observation was, however, continued by Mr P. Macfarlane, the village postmaster, whose records were communicated to Prof. Perrey and published in his annual lists. In 1869, probably on the initiative of Dr James Bryce (1806–1877), a new committee was appointed to investigate the earthquakes of Scotland. Mr Milne Home was a member of this committee, of which Dr Bryce, who had also belonged to the earlier body, was secretary. Seven reports, chiefly relating to the few slight shocks at Comrie, were presented, the work of the committee being terminated abruptly by the untimely death of its secretary in 1877[1]. Mr Macfarlane, who had registered the earthquakes since 1839, died a few years before, in June 1874, and, from this time onwards, the Comrie earthquakes ceased to have any regular observers.

In a statistical work on *The Famines of the World: Past and Present* (1879), Mr C. Walford (1827–1885) gives a list of "Comets, cyclones, earthquakes, hailstorms, hurricanes, and violent storms generally, chronologically arranged" (Table v, pp. 52–68), in which there are brief notices of 19 British earthquakes from 1048 to 1871. Another table (VII, pp. 71–85) on the literature of the same subjects contains references to many early books and pamphlets relating to the earthquakes of this country.

Five years later, 1884, Prof. J. P. O'Reilly (1829–1905) communicated to the Royal Irish Academy a paper on the earthquakes of Great Britain and Ireland during historical times[2]. He compiled two lists, one of 53 earthquakes which disturbed areas of some extent between the years 1426 and 1880 inclusive; the other of the places at which the earthquakes were felt with the number of shocks experienced at each and the years in which they occurred. On the map which accompanies the paper, these places are denoted by marks corresponding to the numbers of shocks felt in them, districts are tinted, the depth of colour indicating the frequency of the earthquakes within them, and the boundaries of the disturbed areas of seven earthquakes are

---

[1] *Brit. Ass. Rep.* 1870, pp. 48–49; 1871, pp. 197–198; 1872, pp. 240–241; 1873, pp. 194–197; 1874, p. 241; 1875, pp. 64–65; 1876, p. 74.

[2] *R. Irish Acad. Trans.* vol. 28, 1884, pp. 285–316.

represented approximately by circles. Rough as it is, this map is of interest as the first attempt to depict graphically the distribution of British earthquakes.

In the following year, Prof. C. W. C. Fuchs gave a list of 76 British earthquakes during the twenty years 1865–1884 in his memoir "Statistik der Erdbeben[1]"; and Messrs R. Meldola (1849–1915) and W. White presented their detailed "Report on the East Anglian Earthquake of April 22nd, 1884[2]." In the Historical Introduction to this important work, the authors give a list of 58 British earthquakes which have caused structural damage, 21 of which before the end of 1842 are not included in Mallet's catalogue. For most of these exceptions, they are indebted to Dr Short's history, the notices in which appear to me to be of somewhat doubtful value. The appendix contains a further list of 16 destructive British earthquakes obtained from the proofs of the work next to be mentioned.

Several of the catalogues described above, including those of the year 1750, are the work of antiquarians rather than of scientific writers, and the same may be said of the latest and most complete yet published, namely, that compiled by Mr W. Roper and after his death issued as a pamphlet by his son[3]. Mr Roper's list contains notices (for the most part brief) of 597 earthquakes from the year 10 to 1889 Feb. 10[4]. Of this large number, however, 89, depending on Dr Short's history, must, I think, be regarded as doubtful. Several others seem to be of spurious origin, but, when all of these are deducted, the catalogue remains of great value and is clearly the result of wide reading and considerable labour.

In his great seismic survey of the world, M. F. de Montessus de Ballore (1851–1923) included the British Isles in his memoir on the "Seismic Phenomena of the British Empire[5]." The number of earthquakes studied by him was 1023; they are grouped about 221 places, which may be regarded as close to the epicentres, and these places are spread over the following ten districts: Scottish Lowlands, Perth-

---

[1] *Sitzungsb. der k. Akad. der Wissensch.* vol. 92, 1886, pp. 134–139.

[2] Essex Nat. Field-Club, *Special Memoirs*, vol. 1, 1885, pp. 1–223.

[3] *A List of the more Remarkable Earthquakes in Great Britain and Ireland during the Christian Era*, 1889, pp. 1–46.

[4] The late Prof. Milne's great catalogue of destructive earthquakes (*Brit. Ass. Rep.* 1911, pp. 649–740) contains notices of 60 British shocks from 103 to 1896. With the exception of two earthquakes, these notices are based on the catalogues of Mallet, Meldola and White, and Roper.

[5] *Quart. Journ. Geol. Soc.* vol. 52, 1896, pp. 651–668. See also the same author's *Les Tremblements de Terre. Géographie Seismologique*, 1906, pp. 43–55.

shire and north-eastern coasts of Scotland, Northern and Central England, English coast of the Channel, Caledonian Canal, South-eastern Ireland, Wales, East Anglia, North-eastern Scotland, and Shetland Isles. The distribution of earthquakes throughout the country is represented in two maps. In one, the places of observation are indicated by circular spots, the diameters of which depend on the numbers of earthquakes originating near them. In the other map, the regions into which de Montessus de Ballore divides the country are outlined, each region being crossed by two series of equidistant perpendicular lines forming small squares, the sides of which are inversely proportional to $\sqrt{pA/n}$, where p is the number of shocks recorded in a region of area A within an interval of n years.

Lastly, I may refer to my studies of British earthquakes, begun in 1889, which form the foundation of a large part of the present volume[1].

METHODS OF INVESTIGATION

The methods of investigation which I have used in studying British earthquakes[2] are based on the theory that earthquakes are the results of successive steps in the growth of faults. No recorded British earthquake has ever been accompanied by perceptible crust-displacements at the surface. We may therefore, on this theory, regard our earthquakes as due to waves of vibration generated as one rock-mass adjoining a fault slides over and against the other. The seismic focus is then a portion of the fault-surface. For our present purpose, it may be regarded as a limited plane surface inclined to the horizon, longer as a rule in the horizontal, than in the vertical, direction. In the simplest case, the relative displacement of the rock-masses adjoining the fault is greatest in the central region of the focus and decreases gradually towards the margins until it vanishes. From the central portion come the principal vibrations which cause the perceptible shock; from the margins, and especially from the upper and lateral

[1] *Quart. Journ. Geol. Soc.* vol. 47, 1891, pp. 618–632; vol. 53, 1897, pp. 157–175; vol. 56, 1900, pp. 1–7; vol. 58, 1902, pp. 371–397; vol. 60, 1904, pp. 215–242; vol. 61, 1905, pp. 1–17; vol. 62, 1906, pp. 5–12; vol. 63, 1907, pp. 351–374. *Geol. Mag.* vol. 8, 1891, pp. 57–67, 306–316, 364–372, 450–455; vol. 9, 1892, pp. 299–305; vol. 10, 1893, pp. 291–302; vol. 3, 1896, pp. 75–78, 553–556; vol. 7, 1900, pp. 106–115, 164–177; vol. 8, 1901, pp. 358–361; vol. 1, 1904, pp. 487–490, 535–542; vol. 2, 1905, pp. 219–223; vol. 3, 1906, pp. 171–178; vol. 5, 1908, pp. 296–309; vol. 7, 1910, pp. 315–320, 410–419; vol. 6, 1919, pp. 302–312. *Edin. Roy. Soc. Proc.* vol. 36, 1917, pp. 256–287. *The Hereford Earthquake of Dec.* 17, 1896 (1899), pp. 1–298.
[2] A full account of these methods is given in Gerland's *Beiträge zur Geophysik*, vol. 9, 1908, pp. 201–236.

margins, proceed the minute and rapid vibrations which are sensible to us as sound.

It is evident that a focus of such a form and position must affect the distribution of intensity throughout the disturbed area. Thus, from the forms, magnitudes and relative positions of the isoseismal lines, it should be possible to infer the direction and hade of the originating fault, the approximate course of the fault-line and, roughly, the horizontal length of the focus. These important elements of the originating fault may be determined as follows:

(i) When the isoseismal lines are elongated, the direction of their longer axes is approximately that of the fault.

(ii) On the side towards which the fault hades, the isoseismal lines of a weak earthquake are farther apart than on the other side; those of a strong earthquake are farther apart at a short distance from the epicentre and closer together at a great distance.

(iii) In slight earthquakes, the sound-area overlaps the disturbed area on the side from which the fault hades.

(iv) The fault-line must pass at a short distance from the centre of the innermost isoseismal and on the side from which the fault hades.

(v) The length of the seismic focus is approximately equal to, though somewhat greater than, the difference between the length and width of the innermost isoseismals.

In some earthquakes, known as "twin earthquakes," there are two foci, which are practically or entirely detached, and within which impulses occur either simultaneously or at a very short interval apart. Their origin will be considered in Chapter XXIII. They give rise in most places to two series of vibrations separated by an interval of rest and quiet, lasting at most for two or three seconds. When they occur nearly or quite simultaneously, the two series coalesce and form a single shock along a narrow band, which I have called the "synkinetic band." It crosses the disturbed area between the two epicentres in a direction at right angles to the longer axes of the isoseismal lines. The form of the synkinetic band enables us to determine whether the impulses occurred simultaneously or, if otherwise, which focus was first in action. Thus, in a twin earthquake the chief points to be determined, in addition to the elements of the originating fault, are the positions of the two epicentres, the order of occurrence of the two impulses, and at which focus the stronger impulse occurred.

The methods described above can be applied to few earthquakes but those after the year 1889. For earthquakes before that year, it is seldom possible to draw any isoseismal lines with the exception of the boundary of the disturbed area. The centre of this area is assumed to coincide with the epicentre, but it should be remembered that the coincidence, at any rate in strong earthquakes, can seldom be more than approximate.

Scale of Intensity.

In drawing the isoseismal lines, I have used the following scale of intensity—a modification of the well-known Rossi-Forel scale, but containing one test only for each degree of the scale:

1. Recorded only by instruments.
2. Felt only by a few persons lying down and sensitive to weak tremors.
3. Felt by ordinary persons at rest; not strong enough to disturb loose objects.
4. Window, doors, fire-irons, etc., made to rattle.
5. The observer's seat perceptibly raised or moved.
6. Chandeliers, pictures, etc., made to swing.
7. Ornaments, vases, etc., overthrown.
8. Chimneys thrown down and cracks made in the walls of some, but not many, houses in one place.
9. Chimneys thrown down and cracks made in the walls of about one-half the houses in one place.

Types of Earthquake-Sound.

Descriptions of the earthquake-sound are given in terms of the following scale:

1. Waggons, carriages, traction-engines or trains passing, generally very rapidly, on hard ground, over a bridge or through a tunnel; the dragging of heavy boxes or furniture over the floor.
2. Thunder, a loud clap or heavy peal, but most often distant thunder.
3. Wind, a moaning, roaring or rough strong wind; the rising of the wind, a heavy wind pressing against the house, the howling of wind in a gap or chimney, a chimney on fire, etc.
4. Loads of stones, etc., falling; such as the tipping of a load of coals or bricks.
5. Fall of heavy bodies, the banging of a door, the blow of a wave on the seashore.
6. Explosions, distant blasting, the boom of a distant heavy gun.

7. Miscellaneous, such as the trampling of many animals, an immense covey of partridges on the wing, the roar of a waterfall, a low pedal note on the organ, and the rending or settling together of huge masses of rock.

DISCRIMINATION OF SPURIOUS EARTHQUAKES

In Great Britain, spurious earthquakes are usually due to the firing of heavy guns or the bursting of meteorites. Explosions also give rise to shocks which at a distance may be mistaken for earthquakes and perceptible tremors are sometimes caused in buildings by unusually loud peals of thunder.

With the exception of the shocks due to explosions, the vibrations are propagated chiefly through the air, and, when the observations are detailed, this fact usually leads to their detection. At some distance from the source, however, the resemblance to earthquakes may be so close as to mislead observers who have lived in seismic countries.

Reported earthquakes have been traced to the firing of heavy guns on 1890 Jan. 7 in south-west Essex, 1893 May 5–6 in the Isle of Man, 1900 July 18 along the south coast of England, 1902 June 3 in west Essex and adjoining counties, and 1903 June 6 in North Wales and Co. Dublin. The following disturbances, which have found admittance to earthquake catalogues, are also probably due to the same cause: 1859 Aug. 13 in Sussex and Norfolk, 1866 Sept. 13 in Devon, 1871 Apr. 10 in the Shetland Islands, 1871 Aug. 28 in Sussex, and 1873 Apr. 14 in Devon. Others have been traced to the bursting of meteorites on 1887 Nov. 20 in the midland counties and on 1894 Jan. 25 in the west of England; and to explosions on 1899 May 12 in Lancashire and 1904 Jan. 5 in Cornwall and Devon.

The following are the tests which have been used in discriminating reported earthquakes due to gun-firing and the bursting of meteorites. Those due to other causes can as a rule be detected without difficulty.

For gun-firing: (i) when several shocks are observed on one day, they are nearly equal in strength and tend to occur at regular intervals; (ii) the sound is a deep boom and is usually compared to thunder (23 per cent. of the records) or explosions (55 per cent.); (iii) the disturbance is obviously transmitted through the air, the sound gives rise to a slight deafening of the ears, and windows are shaken without any accompanying tremor of the ground; and (iv) the area disturbed may lie not far from well-known practising or manœuvring grounds.

For the bursting of meteorites: (i) the disturbed area is very elongated, the length being unusually great for the slightness of the vibration; (ii) when there are two or more explosions, the disturbed area may consist of detached portions with a linear arrangement; (iii) the sound is far more prominent than the shock, the sound-area overlapping the disturbed area in all directions; (iv) the sound is usually compared to thunder (31 per cent.) or explosions (53 per cent.); and (v) it is obviously transmitted through the air[1].

[1] *Nature*, vol. 60, 1889, pp. 139–141.

CHAPTER II

CATALOGUE OF BRITISH EARTHQUAKES

IN the following catalogue of British earthquakes, I have included only those disturbances which seem to me to have an undoubted right to the name. This limitation has led to the exclusion of a large number of supposed earthquakes which have hitherto found a place in British lists. Many of those omitted may have been true earthquakes, but the evidence at our disposal is insufficient to place their nature beyond doubt. I have also excluded all those with origins in Ireland, in or near the Channel Islands, or in Norway and other parts of Europe. Lists of such shocks will, however, be found in Chapter xx.

If space had been a matter of no consideration, the reasons for rejecting what have hitherto been regarded as earthquakes ought to have been given in each case. To have done so would have added many pages to the length of this book. Every known list of British earthquakes has, however, been examined, and I have excluded no shock except for what seem to me good reasons.

My chief difficulty has been to determine the epoch with which the catalogue should begin. For the reality of the earthquake of the year 974, there seems to be abundant evidence. The accounts of the earthquakes before that year are mostly legendary, and none of them appears to me to be based on trustworthy records. The first date in the catalogue is thus 974. But, even after this year, the monastic annalists and many later writers were prone to include as earthquakes landslips, subsidences and other movements for the seismic origin of which there is no evidence whatever. To detect the character of these disturbances is a simple matter. There are, however, other shocks —such as those due to the bursting of meteorites—which are less easy to recognise, and I cannot feel sure that a few, but probably a very few, of those given in the catalogue may not have been connected with such a cause.

A second difficulty arises from the inaccurate chronology of some of the older annalists. Though there is evidently much copying by one monastic chronicler from another, there is occasionally some variation in the date given for an earthquake—usually, though not always, in the year. For instance, the church of St Michael's outside Glastonbury is said to have been thrown down by an earthquake on 1274 Dec. 5, on 1275 Sep. 11, and in 1276. As the older authorities agree on the second of these dates and others record without details a great earthquake in the same year and not in the other years, there can be little doubt that 1275 is the correct date. Again, earthquakes are said to have occurred on Mar. 27 in each of the years 1076, 1077 and 1081, leading to a suspicion that the three records refer to one and the same earthquake. The first year is given in Matthew Paris, B. de Cotton, the *Annales Monastici*, and the *Flores Historiarum*; the second in Holinshed, who refers, but erroneously, to Matthew Paris; and the third in Roger of Wendover, Matthew Paris, B. de Cotton, and the *Flores Historiarum*. It may be inferred, I think, that there were earthquakes on Mar. 27 in both the years 1076 and 1081, but that the earthquake attributed to 1077 is identical with one of the others, probably with that in 1076.

The catalogue is arranged in six columns. The first contains the number assigned to the earthquake, the second the year and (when known) the month and day, New Style being used from the beginning of 1750 though it was not legally introduced until 1752; the third gives the hour in Greenwich mean time; in the fourth will be found the centre or district in which the earthquake originated, in the fifth the intensity according to the scale on p. 9, and in the sixth the magnitude of the disturbed area in square miles. For the early earthquakes, the part of the country chiefly affected is rarely known; later, the county of origin may become defined; as a rule, it is only during the last two centuries that an earthquake can be assigned to a known centre. From and after the year 1839, two figures are occasionally given in the last column. In such cases, the left-hand, and usually smaller, figure refers to the area within the isoseismal 4 in square miles. When the two figures are the same, the boundary of the disturbed area is an isoseismal of that intensity.

The catalogue should perhaps have ended with the year 1916. The three succeeding years are blank, so far as the history of British earthquakes is concerned, and the lists for 1920, 1921 and 1924 are probably incomplete. I have, however, added six shocks in these three years, owing to their connexion with important earthquake centres.

NO.	DATE	HOUR	CENTRE	INT.	DIST. AREA, ETC.
1	974	—	England	—	—
2	1048 May 1	—	Worcestershire	—	—
3	1060 July 4	—	England	—	—
4	1067	—	,,	—	—
5	1076 Mar. 27	—	,,	—	—
6	1081 ,, 27	—	Hampshire	—	—
7	Dec. 25	—	England	—	—
8	1088	—	,,	—	—
9	1089 Aug. 11	—	,,	—	—
10	1099 Nov. 3	—	,,	—	—
11	1110	—	Shrewsbury	—	—
12	1116 Dec. 13	—	England	—	—
13	1119 Sept. 29	—	Worcestershire	—	—
14	1120 ,, 28	—	England	—	—
15	1122 July 25	—	Somerset	—	—
16	1129	—	England	—	—
17	1132	—	,,	—	—
18	1133 Aug. 4	—	,,	—	—
19	1142 Dec.	—	Lincoln	—	—
20	1158	—	England	—	—
21	1165 Jan. 26	—	Cambridgeshire	—	—
22	1180 Apr. 25	—	Nottinghamshire	8	—
23	1184 Jan. 16	—	England	—	—
24	1185 Apr. 15	—	Lincoln	8	—
25	1201 May 22	—	Somerset	—	—
26	Jan. 8	—	England	—	—
27	1228 Apr. 23	—	,,	—	—
28	1233 Nov.	—	Huntingdonshire	—	—
29	1246 June 1	—	Canterbury	8	—
30	Mar. 11	—	England	—	—
31	1247 Feb. 13	—	London	—	—
32	1248 Dec. 21	—	Wells	8	—
33	Feb. 19	—	S. Wales	8	—
34	1250 Dec. 13	—	St Albans	—	—
35	1255 Sept.	—	Wales	—	—
36	1275 Sept. 11	—	Somerset	8	—
37	1298 Jan. 5	—	England	—	—
38	1318 Nov. 14	—	,,	—	—
39	1319 Dec. 1	—	,,	—	—
40	1349	—	Yorkshire	—	—
41	1382 May 21	—	Canterbury	8	—
42	,, 24	—	England	—	—
43	1385	—	,,	—	—
44	1480 Dec. 28	—	Norfolk	8	—
45	1487 ,, 21	—	,,	—	—
46	1508 Sept. 19	—	England&Scotland	—	—
47	1551 May 25	c. noon	Reigate	—	c. 190
48	1553	—	Chichester	—	—
49	1563 Sept.	—	Northamptonshire	—	—
50	1564 Sept.–Nov.	—	,,	—	—
51	1573	—	York	—	—
52	1575 Feb. 26	5–6 p.	England	8	—
53	1580 Apr. 6	c. 6 p.	London	8	—
54	May 1	—	Canterbury	—	—
55	1581 Apr.	c. 6 p.	York	—	—
56	1597 July 23	8–9 a.	N. Scotland	—	—
57	1600	—	York	—	—
58	1601 Dec. 24	—	London	—	—
59	1608 Nov. 8	c. 9 p.	Scotland	—	>11000
60	1621	—	,,	—	—

NO.	DATE	HOUR	CENTRE	INT.	DIST. AREA, ETC.
61	1638	—	Chichester	—	—
62	1656 Aug. 17	4 a.	Glasgow	—	—
63	1657 July 8	—	Cheshire	—	—
64	1661	—	England	—	—
65	1666 Jan. 18	c. 6 p.	Oxford	—	c. 105
66	1668 Apr. 3	—	Shropshire	—	—
67	1678 Nov. 4–5	—	Staffordshire	—	—
68	1680 Jan. 4	—	Somerset	—	—
69	1683 Sept. 17	c. 7 a.	Oxford	—	c. 770
70	Oct. 9	c. 11 p.	Staffordshire	—	—
71	1686 Feb. 18	—	York	—	—
72	1687 May 12	—	England	—	—
73	1690 Aug. 27	c. 8 p.	Carmarthen	—	—
74	Oct. 7	c. 7.30–	England	—	—
75	1692 Sept. 7	c. 2 p.	„	—	—
76	1696	—	Falmouth	—	—
77	1703 Nov.	—	Lincolnshire	—	—
78	Dec. 28	5.3 p.	Gainsborough	5	c. 3000
79	1707 Oct. 25	3.30 p.	Chichester	5	—
80	1712	—	Shropshire	—	—
81	1726 Oct. 25	—	London	—	—
82	1727 July 19	4–5 a.	Devon	—	—
83	Mar. 1	4 a.	S. Scotland	—	—
84	—	—	England	—	—
85	1731 Oct. 21	c. 4 a.	Banbury	—	c. 50
86	1732 July 11	2–3 p.	Glasgow	—	—
87	1734 Oct. 25	1 a.	Chichester	—	—
88	„ 25	c. 3.50 a.	„	5	—
89	„ 31	—	Derbyshire	—	—
90	1736 Apr. 30	noon	Menstrie	c. 8	—
91	May 1	1 a.	„	c. 8	—
92	1737 May 13	—	Shropshire	—	—
93	1738 July 10	3–4 a.	Derbyshire	—	—
94	Dec. 30	—	Halifax	—	—
95	1744 Feb. 5	—	Merionethshire	—	—
96	1747 July 1	10–11 p.	Taunton	5	—
97	1748 Oct.	—	Leadhills (Dumfriesshire)	—	—
98	1749 Feb. 14 N.S.	8–9 a.	„	5	—
99	1750 Feb. 19	c. 0.40 p.	London	7	c. 240
100	„ 19	c. 1.10 p.	„	—	—
101	„ 20	c. 1 a.	Devon	—	—
102	Mar. 19	1–2 a.	London	—	—
103	„ 19	c. 5.30 a.	„	7	c. 1200
104	„ 20	2 a.	„	—	—
105	„ 29	c. 5.45 p.	Chichester	7	c. 18000
106	„ 30	3.30 a.	„	—	—
107	Apr. 6	—	„	—	—
108	„ 13	10 p.	Lancashire	5	—
109	May 4	c. 10 a.	Dorsetshire	4	—
110	Sept. 3	c. 6.45 a.	Grantham	—	c. 2500
111	„ 18	6 p.	Chichester	—	—
112	Oct. 11	c. 0.30 p.	Northampton	7	c. 4500
113	1752 Feb. 23	—	Okehampton	—	—
114	Mar. 31	—	Somerset	—	—
115	Apr. 16	—	„	—	—
116	July 15	—	Cornwall	—	—
117	1753 June 8	11–12 p.	Rochdale	5	—
118	„ 22	11.40 p.	Bolton	—	—

NO.	DATE	HOUR	CENTRE	INT.	DIST. AREA, ETC.
119	1754 Apr. 19	c. 11 p.	Yorkshire	4	—
120	1755 July 31	6–7 a.	Northamptonshire	—	—
121	Aug. 1	—	Stamford	—	—
122	Nov. 17	—	Cumberland	—	—
123	Dec. 31	c. 1 a.	Glasgow	—	—
124	1756 Feb. 27	6 p.	Barnstaple	—	—
125	June 1	—	Canterbury	—	—
126	Oct. 10	—	Scotland	—	—
127	Nov. 17	11.50 p.	Dunoon	c. 5	—
128	Dec. 26	—	Cornwall	—	—
129	1757 Jan. 10	—	Bungay	—	c. 680
130	May 17	—	Halifax	5	—
131	July 15	c. 6.30 p.	Penzance	c. 6	—
132	1758 Jan. 24	2 a.	Reigate	4	—
133	Dec. 20	—	London	—	—
134	1759 Feb. 24	10 p.	Liskeard	—	—
135	1761 June 9	11.45 a.	Dorsetshire	—	—
136	1764 Nov. 6	c. 4.15 a.	Oxford	5	—
137	1766 Feb. 10	—	Swansea?	—	—
138	1767 Apr. 20	9.0–	Menstrie	—	—
139	,, 20	9.15–	,,	—	—
140	1768 Jan. 3	1 a.	Northampton	—	—
141	,, 18	—	Rhyl ?	—	—
142	May 15	4.15 p.	Lancashire	—	—
143	Oct. 24	—	Inverness ?	—	—
144	Dec. 21	5–6 p.	Worcestershire	—	—
145	1769 June 15	—	Barmouth	—	—
146	Nov. c. 14	—	Inverness	8	—
147	Nov. 23	4 p.	Birmingham	—	—
148	1771 July	—	Cornwall	—	—
149	1772 Nov. 15	4 a.	Birmingham	—	—
150	1775 Sept. 8	c. 9.45 p.	Herefordshire	—	c. 28000
151	1776 Oct. 27	c. 9.45 p.	Northamptonshire	—	—
152	Nov. 27	8.15 a.	Canterbury	6	c. 300
153	1777 Sept. 14	c. 10.55 a.	Rochdale	7	c. 9000
154	1780 Aug. 28	c. 8.45 a.	N. Wales	—	—
155	Dec. 9	4–5 p.	,,	—	—
156	1782 Oct. 5	8.40 p.	Carnarvon ?	—	—
157	1783 Apr. 23	c. 1.15 p.	Shropshire	—	—
158	Aug. 9	—	Launceston	—	—
159	1786 June 16	—	Unknown	—	—
160	Aug. 11	c. 2.20 a.	Carlisle	7	c. 27000
161	1787 Jan. 6	10–11 a.	Glasgow	4	—
162	July 6	—	Carlisle	—	—
163	1788 July 8	—	Isle of Man	—	—
164	Nov. 11	—	Comrie	—	—
165	1789 May 5	3.15 a.	Barnstaple	—	—
166	Sept. 2	c. 11 p.	Comrie	—	—
167	Nov. 5	5.57 p.	,,	5	c. 250
168	,, 10	c. 3 p.	,,	5	—
169	,, 11	—	,,	—	—
170	Dec. 29	1 p.	,,	—	—
171	,, 30	—	nr. Edinburgh	—	—
172	1791 Sept. 2	5.5 p.	Comrie	—	—
173	1792 Feb. 25	c. 8.30 p.	Grantham	5	c. 5200
174	Mar. 2	c. 8.45 p.	Stamford	5	c. 4000
175	Oct. 10	—	Comrie	—	—
176	bet. Oct. 12 and Nov. 18	7.30 p.	,,	—	—
177	Nov. 10	—	Kinloch Rannoch (Perthshire)	—	—

NO.	DATE	HOUR	CENTRE	INT.	DIST. AREA, ETC.
178	1792 Nov. 18	11 a.	Comrie	—	—
179	1793 Feb. 3	—	,,	—	—
180	,, 25	10.30 p.	,,	—	—
181	May	—	,,	—	—
182	Sept. 28	imm. after 4 p.	Wiltshire	—	—
183	1794 May 2	4 p.	Comrie	—	—
184	Sept. 28	3 p.	,,	—	—
185	Oct. 1	5.30 p.	,,	—	—
186	,, 2	11 p.	,,	—	—
187	,, 18	1 a.	,,	—	—
188	Dec. 3	5.30 p.	,,	—	—
189	,, 4	10 p.	,,	—	—
190	,, 25	1.15 p.	,,	—	—
191	,, 30	8 p.	,,	—	—
192	1795 Jan. 2	1.50 a.	,,	—	—
193	,, 22	2.40 p.	,,	—	—
194	Mar. 12	11 p.	,,	—	—
195	,, 13	—	,,	—	—
196	,, 16	—	,,	—	—
197	,, 21	—	,,	—	—
198	,, 23	—	,,	—	—
199	,, 27	8 p.	,,	—	—
200	Apr. 8	3 p.	,,	—	—
201	,, 25	c. 6 p.	,,	—	—
202	June 19	—	,,	—	—
203	July 14	—	,,	—	—
204	,, 15	—	,,	—	—
205	,, 25	6.30 p.	,,	—	—
206	Sept. 1	—	,,	—	—
207	,, 4	2–3 p.	,,	—	—
208	Oct. 4	—	,,	—	—
209	Nov. 18	c. 11.10 p.	Derbyshire	8	c. 30000
210	1796 Jan.	—	Comrie	—	—
211	Mar. 16	—	,,	—	—
212	1797 Feb. 8	c. 7 p.	,,	—	—
213	,, 10	0.20 a.	,,	—	—
214	,, 10	6 a.	,,	—	—
215	,, 17	—	,,	—	—
216	May 12	—	,,	—	—
217	Aug. 24	—	,,	—	—
218	Nov. 19	11 a.	,,	—	—
219	Dec. 19	5 a.	,,	—	—
220	1798 Apr. 19	—	,,	—	—
221	May 6	10 a.	,,	—	—
222	1799 Jan. 17	—	,,	—	—
223	Feb. 24	1.50 p.	,,	—	—
224	bet. Feb. 24 and Mar. 3	—	,,	—	—
225	1800 Dec. 8	9 a.	,,	—	—
226	1801 Jan. 11	7 a.	,,	—	—
227	,, 11	after 11 a.	,,	—	—
228	June 1	1.30 a.	Cheshire	—	—
229	Sept. 6	1.15 p.	Comrie	—	—
230	,, 7	4 a.	,,	—	—
231	,, 7	6 a.	,,	7	c. 3600
232	,, 18	—	,,	—	—
233	,, 25	10 a.	,,	—	—
234	1802 June 10	11 p.	,,	—	—
235	Aug.	5 a.	Menstrie	—	—
236	,,	6 a.	,,	—	—

NO.	DATE	HOUR	CENTRE	INT.	DIST. AREA, ETC.
237	1802 Oct. 3–8	—	Comrie	—	—
238	,, 21	c. 8.15 p.	Carmarthen	—	—
239	1804 Mar. 4	c. 4 p.	Comrie	—	—
240	,, 11	c. 11 a.	,,	—	—
241	,, 14	c. 2 a.	,,	—	—
242	1805 Jan. 12	c. 7 p.	Rhyl	—	—
243	1806 May 29	—	Comrie	—	—
244	1809 Jan. 9	c. 5.30 a.	,,	—	—
245	,, 17	2–3 a.	Menstrie	4	—
246	,, 18	c. 1 a.	Dunning (Perthshire)	—	—
247	,, 18	c. 2 a.	,,	5	—
248	,, 31	—	Strontian	—	—
249	Feb. 1	—	,,	—	—
250	,, 4	—	,,	—	—
251	,, 5	—	,,	—	—
252	,, 6	—	,,	—	—
253	May 3	—	Barnstaple	—	—
254	1810 Nov. 14–15	—	Comrie	—	—
255	,, 18	—	Oxfordshire	—	—
256	1811 Nov. 30	c. 2.40 a.	Chichester	5	c. 470
257	1812 Jan. 18	—	Oxford	4	—
258	May 1	—	Gloucestershire	—	—
259	Sept. 3	8–9 p.	Nottinghamshire	—	—
260	,, 10	—	Comrie	—	—
261	1813 Mar. 21	6 a.	Exmouth	—	—
262	Sept. 24	c. 3.15 p.	Stamford	—	350
263	1814 Nov. 20–25	—	Comrie	—	—
264	Dec. 6	a few mins. before 2 a.	Chichester	6	—
265	1816 Mar. 17	0.30–1 p.	Mansfield	7	3100
266	Aug. 6	10.45 p.	Dunkeld (Perthshire)	—	—
267	,, 13	10.45 p.	Inverness	8	c. 50000
268	,, 13	c. 11.15 p.	,,	—	—
269	,, 19	—	,,	—	—
270	,, 20	—	,,	—	—
271	1817 Jan. 27	11 p.	Mansfield	—	—
272	Apr. 26	6.30 a.	Scotland	—	—
273	June 10	1.20 p.	Inverness	—	—
274	,, 16	5 a.	,,	—	—
275	,, 30	—	,,	—	—
276	Aug. 17	8.20 a.	,,	—	—
277	,, 31	—	,,	—	—
278	Sept. 7	c. 3.30 a.	,,	—	—
279	Oct. 31	—	Dalton-in-Furness	—	—
280	Nov. 9	—	Kendal	—	—
281	1818 Feb. 6	—	Coningsby (Lincs.)	4	—
282	,, 19	c. 2.20 p.	Scotland	4	—
283	,, 20	1.20–	Inverness	—	—
284	,, 20	c. 3 p.	Coningsby (Lincs.)	—	—
285	Apr. 30	—	Lincolnshire	—	—
286	June 9	c. 2.20 p.	Sutherland	—	—
287	,, 19	—	Comrie	—	—
288	Sept. 1	—	Inverness	—	—
289	Nov. 9	9 p.	,,	—	—
290	,, 9–10	c. midn.	,,	—	—
291	,, 10	0.20 a.	,,	6	—
292	,, 10	4 a.	,,	—	—

NO.	DATE	HOUR	CENTRE	INT.	DIST. AREA, ETC.
293	1818 Dec. 7	c. 9 a.	Carnarvon	—	—
294	1819 Feb. 11	—	Aberfeldy (Perthshire)	—	—
295	Nov. 28	1.30 a.	Comrie	4	—
296	Dec. 4	7.30 p.	Amulree (Perthshire)	4	—
297	1820 Feb. 22	8.30 a.	Glasgow	5	—
298	May 20	—	Leadhills (Dumfriesshire)	—	—
299	Sept. 27	9 p.	Barmouth	—	—
300	Nov. 28	c. 8 a.	Leadhills (Dumfriesshire)	—	—
301	,, 28	c. 11 p.	,,	—	—
302	Dec. 20	a little after 9 p.	Barmouth	—	—
303	,, 25	—	Inverness-shire	—	—
304	1821 Feb. 1	c. 2.30–	Chichester	—	—
305	Oct. 15	—	Rothesay	—	—
306	,, 22	—	Scotland	—	—
307	,, 23	c. 3 p.	Menstrie?	—	—
308	,, 29	—	Glasgow	—	—
309	,, end	—	Rothesay	—	—
310	1822 Mar. 20	—	York	—	—
311	Apr. 13	c. 9.30 a.	Comrie	—	—
312	,, 22	9.30 a.	Dunkeld (Perthshire)	—	—
313	May 18	9–10 a.	Comrie	6	—
314	1824 May 31	4 p.	Rochdale	—	—
315	Aug. 8	—	Comrie	4	—
316	,, 10	—	Perth	—	—
317	Dec. 6	1.45 p.	Chichester	5	c. 330
318	1825, about	—	Mansfield	4	—
319	1826 Aug. 30	1 a.	Cornwall	—	—
320	Nov. 26	c. 4 p.	Rothesay	4	—
321	Dec. 25	2 a.	Ardvoirlich (Perthshire)	—	—
322	,, 25	2 p.	Leadhills (Dumfriesshire)	—	—
323	1827 Feb. 9	7 p.	Carnarvon?	—	—
324	1828 May 20	—	Leadhills (Dumfriesshire)	—	—
325	Dec. 9	—	Comrie	—	—
326	1831 Mar. 1	11 p.	Ardvoirlich (Perthshire)	4	—
327	,, 2	—	Canterbury	—	—
328	,, 17	—	Carnarvon?	—	—
329	1832 Dec. 30	8.20–	Swansea?	—	—
330	1833 Sept. 18	c. 10 a.	Chichester	—	—
331	Nov. 13	2.40 a.	,,	—	—
332	,, 13	6 a.	,,	—	—
333	1834 Jan. 23	2.45 a.	,,	5	c. 780
334	Feb. 20	2 a.	,,	—	—
335	Aug. 25	—	Comrie	—	—
336	,, 27	10.25 p.	Chichester	—	c. 1100
337	Sept. 21	11.20 a.	,,	—	—
338	1835 Jan. 12	8 a.	,,	—	—
339	Aug. 3	—	,,	—	—
340	,, 19–20	midn.	Clitheroe	—	—
341	,, 20	c. 3.33 a.	,,	5	c. 450

NO.	DATE		HOUR	CENTRE	INT.	DIST. AREA, ETC.
342	1837 Oct.	20	c. 2 p.	Liskeard	4	—
343	,,	27	—	Camelford	—	—
344	Nov.	24	—	,,	—	—
345	Dec.	8	11.15 p.	Stamford	—	—
346	1838 Mar.	17	1 p.	Shropshire	—	—
347	,,	27	1 p.	,,	—	—
348	Sept.	14	—	Banbury	—	—
349	Nov.		—	Comrie	—	—
350	1839 Mar.	20	c. 3 a.	Inverness-shire	—	—
351	May	24	2 a.	Comrie	—	—
352	,,	24	—	Glasgow	—	—
353	Sept.	1	1 a.	Gloucestershire	—	—
354	,,	2	—	England	—	—
355	Oct.	3	3.30 p.	Comrie	—	—
356	,,	7	4.30 a.	,,	—	—
357	,,	7	11.30 a.	,,	—	—
358	,,	7	3 p.	,,	—	—
359	,,	7	4 p.	,,	—	—
360	,,	7	5.30 p.	,,	—	—
361	,,	9	2.15 a.	,,	—	—
362	,,	10	4.15 a.	,,	—	c. 80
363	,,	11	10 p.	,,	—	—
364	,,	12	1 p.	,,	5	—
365	,,	12	2.30 p.	,,	—	—
366	,,	12	3 p.	,,	7	c. 700
367	,,	12	4 p.	,,	—	—
368	,,	12	c. 4.30 p.	,,	—	—
369	,,	13	9 a.	,,	—	—
370	,,	13	11 a.	,,	—	—
371	,,	13	11.30 a.	,,	—	—
372	,,	14	2.45 p.	,,	5	c. 150
373	,,	15	3 p.	,,	—	—
374	,,	16	2.30 a.	,,	5	—
375	,,	16	6 a.	,,	—	—
376	,,	16	5.45 p.	,,	—	—
377	,,	17	—	,,	—	—
378	,,	18	—	,,	—	—
379	,,	19-22	—	,,	—	—
380	,,	23	10.15 p.	,,	8	c.26500; c.26500
381	,,	23	c. 10.45 p.	,,	—	c. 700
382	,,	23-24	c. midn.	,,	—	—
383	,,	24	5 a.	,,	—	—
384	,,	25	7 p.	,,	—	—
385	,,	26	7 p.	,,	—	—
386	,,	26	7-11.30 p.	,,	—	—
387	,,	26	11.30 p.	,,	—	—
388	,,	27	—	,,	—	—
389	,,	28	11.30 p.	,,	—	—
390	,,	29	2.45 a.	,,	—	—
391	,,	30	—	,,	—	—
392	,,	31	—	,,	—	—
393	Nov.	1	2.30 p.	,,	—	—
394	,,	7	4 a.	,,	—	—
395	,,	9	—	,,	—	—
396	,,	19-28	4 a.	,,	—	—
397	,,	29	—	,,	—	—
398	,,	30	—	,,	—	—
399	Dec.	2	—	,,	—	—
400	,,	3	—	,,	—	—
401	,,	4	—	,,	—	—

NO.	DATE	HOUR	CENTRE	INT.	DIST. AREA, ETC.
402	1839 Dec. 5	—	Comrie	—	—
403	,, 6	2 a.	,,	—	—
404	,, 7	—	,,	—	—
405	,, 8	—	,,	—	—
406	,, 11	9.30 p.	,,	—	—
407	,, 12	3 a.	,,	—	—
408	,, 13–18	—	,,	—	—
409	,, 20	3.15 a.	,,	—	—
410	,, 23	c. 4 p.	Leicester	—	—
411	,, 24	—	Comrie	—	—
412	,, 28	3 a.	,,	—	—
413	,, 28	noon	,,	—	—
414	,, 31	2 a.	,,	—	—
415	1840 Jan. 2	1.30 a.	,,	—	—
416	,, 4	11.45 p.	,,	—	—
417	,, 8	0.30 a.	,,	—	—
418	,, 8	1 a.	,,	—	—
419	,, 11	2 a.	,,	—	—
420	,, 11	noon	,,	—	—
421	,, 18	9.45 p.	,,	4	—
422	,, 18	10 p.	,,	—	—
423	,, 19	5 a.	,,	—	—
424	,, 19	9 a.	,,	—	—
425	,, 20	9 a.	,,	—	—
426	,, 27	6 a.	,,	—	—
427	Feb. 6	8.30 a.	,,	—	—
428	,, 9	9 a.	,,	—	—
429	,, 10	4 a.	,,	—	—
430	,, 14	1 a.	,,	—	—
431	,, 25	2 p.	,,	—	—
432	,, 26	1 a.	,,	—	—
433	Mar. 8	6 a.	,,	—	—
434	,, 8	4 p.	,,	—	—
435	,, 9	5.30 a.	,,	—	—
436	,, 11	5 a.	,,	—	—
437	,, 11	6 a.	,,	—	—
438	,, 11	—	,,	—	—
439	,, 13	8 p.	,,	—	—
440	,, 14	9.30 p.	,,	—	—
441	,, 21	0.30 a.	,,	—	—
442	,, 24	8.30 a.	,,	—	—
443	,, 24	11 a.	,,	—	—
444	,, 25	1 p.	,,	—	—
445	,, 27	1.30 p.	,,	—	—
446	,, 27	2.30 p.	,,	—	—
447	Apr. 1	1 p.	,,	—	—
448	,, 1	1.30 p.	,,	—	—
449	,, 7	4 p.	,,	—	—
450	,, 7	4.15 p.	,,	—	—
451	,, 7	5 p.	,,	—	—
452	,, 11	2.45 a.	,,	—	—
453	,, 11	—	Kendal	—	—
454	,, 12	1 p.	Comrie	—	—
455	,, 13	1 p.	,,	—	—
456	May 19	1.45 p.	,,	—	—
457	,, 19	2 p.	,,	—	—
458	,, 19	2.15 p.	,,	—	—
459	,, 22	1.30 a.	,,	—	—
460	,, 22	5.30 a.	,,	—	—
461	July 3	11 a.	,,	—	—

NO.	DATE	HOUR	CENTRE	INT.	DIST. AREA, ETC.
462	1840 July 11	11 a.	Comrie	—	—
463	,, 16	3.30 a.	,,	—	—
464	,, 17	8.30 a.	,,	—	—
465	,, 23	1 a.	,,	—	—
466	Aug. 5	6 p.	,,	—	—
467	,, 6	2 a.	,,	—	—
468	Sept. 19	3.30 a.	,,	—	—
469	,, 21	3 p.	,,	—	—
470	,, 26	3 a.	,,	—	—
471	Oct. 4	9 p.	,,	—	—
472	,, 20	11 a.	,,	—	—
473	,, 26	6.45 p.	,,	—	—
474	Nov. 12	2.30 p.	,,	—	—
475	,, 13	2.30 a.	,,	—	—
476	,, 13	4 a.	,,	—	—
477	,, 13	7 a.	,,	—	—
478	,, 16	3 a.	,,	—	—
479	,, 16	6 a.	,,	—	—
480	,, 24	4 a.	,,	—	—
481	Dec. 6	7 p.	,,	—	—
482	,, 7	—	,,	—	—
483	,, 8	—	,,	—	—
484	,, 10	2 a.	,,	—	—
485	,, 18	2 a.	,,	—	—
486	1841 Jan. 6	—	,,	—	—
487	,, 18	—	,,	—	—
488	,, 24	3–4 a.	Carmarthen	—	—
489	,, 31	2 a.	Comrie	—	—
490	,, 31	9 p.	,,	—	—
491	,, 31	11.30 p.	,,	—	—
492	Feb. 1	7.30 p.	,,	—	—
493	,, 14	9.45 a.	,,	—	—
494	,, 16	—	,,	—	—
495	Mar. 6	9 a.	,,	—	—
496	,, 10	4.45 p.	,,	—	—
497	,, 11	1 a.	,,	—	—
498	,, 22	1 a.	,,	—	—
499	,, 22	6.15 a.	,,	—	—
500	,, 23	1 a.	,,	—	—
501	Apr. 3	8 a.	,,	—	—
502	,, 9	7.45 a.	,,	—	—
503	,, 12	5 a.	,,	—	—
504	,, 14	5 a.	,,	—	—
505	,, 17	—	,,	—	—
506	,, 19	5.30 a.	Oban	—	—
507	,, 19	11 a.	,,	—	—
508	,, 19	2.30 p.	,,	—	—
509	,, 21	1.35 a.	,,	—	—
510	,, 24	1.45 p.	Comrie	—	—
511	,, 24	9.45 p.	,,	—	—
512	,, 24	a little after 9.45 p.	,,	—	—
513	,, 25	1 a.	,,	—	—
514	May 5	9 a.	,,	—	—
515	,, 8	noon	,,	—	—
516	,, 22	noon	,,	—	—
517	,, 26	—	,,	—	—
518	,, 27	8 a.	,,	—	—
519	,, 28	7 a.	,,	—	—
520	,, 30	6.45 a.	,,	—	—

NO.	DATE	HOUR	CENTRE	INT.	DIST. AREA, ETC.
521	1841 June 29	—	Comrie	—	—
522	July 2	5.45 a.	,,	—	—
523	,, 4	c. 9.30 p.	Inverness-shire	—	—
524	,, 23	1 a.	Comrie	—	—
525	,, 23	4.15 p.	,,	—	—
526	,, 25	4.45 p.	,,	—	—
527	,, 26	3 a.	,,	—	—
528	,, 30	8 a.	,,	—	—
529	,, 30	2.30 p.	,,	8	<1100
530	,, 30	3.30 p.	,,	—	—
531	,, 31	8 a.	,,	—	—
532	Aug. 1	—	,,	—	—
533	,, 10	—	,,	—	—
534	,, 12	10 a.	,,	—	—
535	,, 30	—	,,	—	—
536	Sept. 8	3 a.	,,	—	—
537	,, 9	11.45 p.	,,	—	—
538	,, 10	2.30 a.	,,	—	—
539	,, 10	4.30 a.	,,	—	—
540	,, 10	11.15 a.	,,	—	—
541	,, 11	2.15 a.	,,	—	—
542	,, 16	—	,,	—	—
543	,, 16	9.30 p.	,,	—	—
544	,, 17	1 a.	,,	—	—
545	,, 17	4.30 a.	,,	—	—
546	,, 22	11.30 p.	,,	—	—
547	,, 23	1.15 a.	,,	—	—
548	,, 23	2.30 a.	,,	—	—
549	,, 25	9 p.	,,	—	—
550	,, 29	—	,,	—	—
551	,, 29	9.15 a.	,,	—	—
552	Oct. 5	—	,,	—	—
553	,, 13	midn.	,,	—	—
554	,, 23	midn.	,,	—	—
555	Nov. 3	midn.	,,	—	—
556	,, 5	1 a.	,,	—	—
557	,, 6	8 a.	,,	—	—
558	,, 7	—	,,	—	—
559	,, 8	8 a.	,,	—	—
560	,, 18	8 a.	,,	—	—
561	,, 26	c. 1 p.	,,	—	—
562	Dec. 3	8.30 a,	,,	—	—
563	,, 6	3 a.	,,	—	—
564	,, 7	3 a.	,,	—	—
565	,, 19	1.30 a.	,,	—	—
566	,, 20	4 p.	Ross-shire	—	—
567	1842 Jan. 2	—	Comrie	—	—
568	,, 7	—	,,	—	—
569	Feb. 17	c. 8.30 a.	Helston	4	c. 49; c. 160
570	Mar. 10	1 p.	Comrie	—	—
571	Apr. 21	3 p.	,,	—	—
572	,, 22	10.30 p.	,,	—	—
573	,, 22	11 p.	,,	—	—
574	May	6–7 a.	Carnarvon	—	—
575	June 1	—	Comrie	—	—
576	,, 2	1 a.	,,	—	—
577	,, 8	1.30 a.	,,	—	—
578	,, 8	1.45 a.	,,	—	—
579	,, 21	10–11 p.	Dalton-in-Furness	4	—
580	July 1	2.30 p.	Comrie	—	—

NO.	DATE		HOUR	CENTRE	INT.	DIST. AREA, ETC.
581	1842 July	1	4.30 p.	Comrie	—	—
582	,,	1–10	9 a.	,,	—	—
583	Aug.	19	—	Pitlochry (Perthshire)	—	—
584	,,	22	—	Carnarvon	—	—
585	,,	27	11.45 p.	Comrie	—	—
586	Sept.	2	3 a.	,,	—	—
587	,,	24	5.53 a.	,,	—	—
588	,,	24	6.59 p.	,,	—	—
589	,,	25	—	,,	—	—
590	Nov.	18	midn.	,,	—	—
591	,,	29	11 p.	,,	—	—
592	Dec.	4	2 a.	,,	—	—
593	,,	17	8 a.	,,	—	—
594	,,	17	—	,,	—	—
595	1843 Feb.	25	c. 8.12 p.	Argyllshire	5	c. 1450
596	Mar.	3	8.40 p.	,,	4	—
597	,,	10	c. 8 a.	Rochdale	—	—
598	,,	16	c. 10.30 p.	Kendal	—	—
599	,,	17	0.55 a.	,,	pr. 7	c. 24000
600	,,	23	8.30 p.	Comrie	—	—
601	,,	23	11 p.	,,	—	—
602	May	14	1 a.	,,	—	—
603	,,	14	2 a.	,,	—	—
604	,,	28	0.20 p.	,,	—	—
605	June	4	0.15 p.	,,	—	—
606	,,	4	3 p.	,,	—	—
607	,,	10	2.30 a.	,,	—	—
608	,,	10	3 a.	,,	—	—
609	,,	15	—	,,	—	—
610	,,	17	4 a.	,,	—	—
611	Aug.	25	10.40 a.	,,	—	—
612	,,	25	11 a.	,,	—	—
613	,,	25	5 p.	,,	—	—
614	Sept.	1	5.15 a.	,,	—	—
615	,,	1–2	midn.	,,	—	—
616	,,	2	6 a.	,,	—	—
617	,,	16	11 a.	,,	—	—
618	Nov.	1	11.15 p.	Torosay (Mull)	—	—
619	,,	25	2 a.	Comrie	—	—
620	,,	26	4 a.	,,	—	—
621	,,	30	—	,,	—	—
622	1844 Jan.	1	2 a.	,,	—	—
623	,,	3	3 a.	,,	—	—
624	,,	9	5 a.	,,	—	—
625	,,	14	1 p.	,,	—	≮ 525
626	,,	14	1.10 p.	,,	—	—
627	,,	14	1.20 p.	,,	—	—
628	,,	14	1.37 p.	,,	—	—
629	,,	14	2 p.	,,	—	—
630	,,	14	9 p.	,,	—	—
631	,,	14	11 p.	,,	—	—
632	,,	14	11.45 p.	,,	—	—
633	,,	23	2 a.	,,	—	—
634	Feb.	6	4 a.	,,	—	—
635	,,	16	—	,,	—	—
636	,,	20	6.30 a.	,,	—	—
637	Mar.	3	7 a.	,,	—	—
638	,,	15	c. 11.20 p.	Galashiels	—	—
639	,,	19	1 p.	Comrie	—	—

NO.	DATE		HOUR	CENTRE	INT.	DIST. AREA, ETC.
640	1844 Apr.	1	10.15 a.	Comrie	—	—
641	May	11	10 p.	,,	—	—
642	,,	14	4 a.	,,	—	—
643	,,	22	7 a.	,,	—	—
644	,,	25	5.30 a.	,,	—	—
645	June	12	c. 7 p.	Stamford	—	c. 45
646	July	9	3 a.	Comrie	—	—
647	Aug.	25	3 a.	,,	—	—
648	Sept.	1	11.15 p.	,,	—	—
649	,,	4	6.15 p.	,,	—	—
650	,,	13	5.30 a.	,,	—	—
651	,,	22	8.50 p.	,,	—	—
652	Oct.	8	5.45 a.	,,	—	—
653	,,	22	9.15 a.	,,	—	—
654	,,	24	8.15 p.	,,	—	—
655	,,	27	8.15 p.	,,	—	—
656	1845 Jan.	31	1 a.	,,	—	—
657	Mar.	9	—	Glasgow	—	—
658	,,	26	soon after 9 p.	Huntingdon	—	—
659	May	12	1.30 a.	Comrie	—	—
660	Aug.	7	1.15 p.	,,	—	—
661	Sept.	7	7 p.	,,	—	—
662	,,	22	2 p.	,,	—	—
663	,,	22	2.30 p.	,,	—	—
664	,,	22	4 p.	,,	—	—
665	Oct.	5	7.15 p.	,,	—	—
666	,,	13	2 a.	,,	—	—
667	,,	13	9.15 a.	,,	—	—
668	,,	18	9.45 p.	,,	—	—
669	,,	19	1–2 a.	,,	—	—
670	Nov.	12	8.15 p.	,,	—	—
671	Dec.	21	10 a.	,,	—	—
672	1846 Feb.	3	8 p.	,,	—	—
673	,,	4	7 a.	,,	—	—
674	,,	16	10.45 a.	,,	—	—
675	Mar.	2	c. 11 a.	,,	—	—
676	Nov.	24	11–12 p.	Menstrie?	—	—
677	,,	25	1 a.	Comrie	—	—
678	,,	25	1.15 a.	,,	—	—
679	,,	25	4 p.	,,	—	—
680	Dec.	17	11.45 p.	,,	—	—
681	,,	18	3 a.	,,	—	—
682	1847 Jan.	20	3.30 a.	,,	—	—
683	,,	20	9.30 a.	,,	—	—
684	,,	28	3 p.	,,	—	—
685	,,	28	9 p.	,,	—	—
686	Feb.	1	—	,,	—	—
687	,,	19	11.15 p.	,,	—	—
688	May	13	9 p.	,,	—	—
689	,,	13	10.0–10.15 p.	,,	—	—
690	,,	20	11.30 p.	,,	—	—
691	Oct.	7	7 p.	,,	—	—
692	,,	17	7.30 p.	,,	—	—
693	1848 May	7	7 p.	,,	—	—
694	July	7	noon	,,	—	—
695	,,	19	6.30 p.	,,	—	—
696	Oct.	19	7 a.	Canterbury	—	—
697	Nov.	13	8.30 a.	Comrie	—	—
698	,,	13–15	—	,,	—	—
699	,,	15	midn.	,,	—	—

NO.	DATE	HOUR	CENTRE	INT.	DIST. AREA, ETC.
700	1848 Nov. 23	9.30 a.	Comrie	—	—
701	1851 Feb. 15	1–2 a.	,,	—	—
702	May 17	11.30 p.	,,	—	—
703	Nov. —	—	Barmouth	5	—
704	1852 Apr. 1	5.30 a.	Wells	—	—
705	,, 3	3 a.	Somerset	—	—
706	,, 3	5.35 a.	,,	4	—
707	June 1	c. 7.45 a.	Swansea	—	—
708	Aug. 12	c. 8 a.	Altarnon	—	—
709	Nov. 9	4.25 a.	Unknown	—	c. 56000
710	,, 17	—	Penzance	—	—
711	1853 Feb. 19	c. 2.30 p.	Inverness	—	—
712	Mar. 27	c. 11.30 p.	Hereford	5	—
713	1855 May 7	c. 2 a.	Perth	—	—
714	,, 30	3 p.	Penzance	4	—
715	Aug. 7	—	Skye	—	—
716	1857 Jan. 25	3.20 p.	Nottinghamshire	—	—
717	1858 Apr. 2	—	Cornwall	—	—
718	Sept. (beg.)	—	Okehampton	—	—
719	,, 28	c. 7.45 p.	,,	—	95
720	1859 Oct. 21	c. 6.45 p.	Truro	4	—
721	,, 21	c. 7.45 p.	,,	—	—
722	Dec. 15	—	Settle	—	—
723	1860 Jan. 13	10.30 p.	Truro	5	—
724	,, 13	11.30 p.	,,	—	—
725	Sept. 3	3.30 p.	Maidstone	—	c. 120
726	1863 Oct. 5	11 p.	Everton (Liverpool)	—	—
727	,, 6	1.30 a.	Staffordshire	—	—
728	,, 6	c. 2.30 a.	Hereford	—	—
729	,, 6	3.10 a.	,,	—	—
730	,, 6	3.22 a.	,,	8	c. 85000
731	1864 Apr. 30	11.6 p.	Lewes	5	c. 80
732	July 3	—	Comrie	—	—
733	Aug. 21	1.27 a.	Lewes	5	—
734	Sept. 26	0.35 a.	Rochdale	5 or 6	—
735	1865 Jan. 2	1–2 a.	Nottinghamshire	—	—
736	,, 15	11 a.	Dalton-in-Furness	—	—
737	May 7	8–9 p.	Comrie	4	—
738	,, 10	—	,,	—	—
739	June 14	1.20 a.	Exmouth	—	—
740	1867 Feb. 22	—	Dalton-in-Furness	—	—
741	,, 23	c. 1.15 a.	Grasmere	4	—
742	Aug. 20	1–2 a.	Ullapool	5	—
743	1868 Jan. 4	5.10 a.	Taunton	4	c. 200
744	Oct. 30	10.35 p.	Hereford	—	c. 10000
745	Nov. 19	7.30 p.	Leicestershire	—	—
746	1869 Jan. 9	11.17 a.	Ixworth (Suffolk)	4	c. 280
747	Feb. 15	—	Swansea?	—	—
748	Mar. 9	c. 8 p.	Fort William?	—	—
749	,, 15	6.6 p.	Rochdale	6	c. 580
750	,, 16	c. 3 p.	York	—	—
751	,, 25	8.25 p.	Rochdale	—	—
752	July 7	5 p.	Comrie	—	—
753	1870 Apr. 29	c. 11 p.	,,	—	—
754	Nov. (end)	—	Fort William	—	—
755	Dec. 3	c. 2 a.	,,	—	—
756	1871 Mar. 17	6.20 p.	Kendal	4	c. 1180
757	,, 17	10.56 p.	,,	4	—

NO.	DATE	HOUR	CENTRE	INT.	DIST. AREA, ETC.
758	1871 Mar. 17	11.5 p.	Kendal	7	c. 23400
759	,, 17–18	—	,,	—	—
760	,, 18	c. 3 a.	,,	—	—
761	,, 19	c. 10.20 a.	Halifax	—	—
762	,, 20	c. 9.50 p.	Hereford	—	—
763	Apr. 15	7.55 p.	Dunoon	—	—
764	1872 Aug. 8	4.10 p.	Menstrie	5	c. 700
765	Dec. 24	—	Leadhills (Dumfriesshire)	—	—
766	1873 Feb. 1	7 a.	Pembroke?	—	—
767	Apr. 16	9.55 p.	Penpont (Dumfriesshire)	4	19
768	,, 29	c. 2.40 p.	Doncaster	4	—
769	1874 Jan. 6	8.13 p.	Dhu Heartach (Argyllshire)	—	—
770	Nov. 15	2 a.	Carnarvon?	—	—
771	1875 Sept. 23	9.57 p.	Kendal	4	—
772	1876 Jan. 14	—	Comrie	—	—
773	,, 16	c. 3 a.	,,	—	—
774	,, 16	—	,,	—	—
775	Aug. 11	11–12 p.	Penpont (Dumfriesshire)	—	—
776	,, 12	3 a.	,,	—	19
777	1877 Mar. 11	11.30 a.	Oban	4	—
778	Apr. 15–21	—	,,	—	—
779	,, 23	c. 3.40 a.	,,	4	—
780	May 11	2–3 a.	Comrie	—	—
781	1878 Dec. 3	5 a.	Ross-shire	—	—
782	1879 Apr. 8	c. 8.35 p.	Denbighshire	4	—
783	June 16	—	Mull	—	—
784	Oct. 25	5.30 a.	Cumberland	—	—
785	1880 Nov. 28	c. 5.52 p.	Oban	5 or 6	c. 50000
786	Dec. 12	—	Shap (Westmorland)	—	—
787	1881 Jan. 12	c. 7 a.	Menstrie	—	—
788	1882 Apr. 8	7.37 p.	Phladda (Argyllshire)	—	—
789	Oct. 14	c. 3 p.	Comrie	—	—
790	,, 14	c. 7.30 p.	,,	—	—
791	1883 Jan. 16	c. 5 p.	Gloucestershire	—	—
792	June 25	1.39 p.	Launceston	5	c. 3020; c. 3020
793	,, 25	2.7 p.	,,	4	c. 2080; c. 2080
794	1884 Feb. 18	c. 1.20 a.	Colchester	—	—
795	Apr. 22	9.18 a.	,,	9	c.53000; 100000
796	Nov. 14	5.10 p.	Clitheroe	—	—
797	1885 Jan. 22	8.30–9 p.	Taunton	4	c. 210
798	June 18	c. 1 a.	Ballachulish	—	—
799	,, 18	c. 10.50 a.	York	5	c. 2000; c. 2000
800	,, 30	5.45 a.	Grasmere	—	c. 165
801	1886 Jan. 4	10.19 a.	Dartmouth	5	c. 100
802	,, 20	c. 7 a.	St Austell	4 or 5	—
803	Apr. 25	c. 0.30 a.	Comrie	—	—
804	Sept. 15	—	Kendal	—	—
805	1887 Dec. 1	c. 6.50 a.	Bolton	—	—
806	,, 18	5–6 p.	Ullapool	—	—
807	1888 Jan. 5	5.30 a.	Invergarry	4	—
808	,, 12	—	,,	—	—
809	,, 31	c. 2 a.	Birmingham	—	—
810	Feb. 2	5.1 a.	Inverness	pr. 7	c. 20000
811	,, 29	8.10 p.	Invergarry	—	—

NO.	DATE	HOUR	CENTRE	INT.	DIST. AREA, ETC.
812	1888 Mar. 1	9.15 a.	Invergarry	—	—
813	Apr. 4	9 a.	,,	—	—
814	,, 4	11 a.	—	—	—
815	,, 11	—	N. Wales	—	—
816	May 20	6.10 p.	Invergarry	—	—
817	July 3	2.30 p.	,,	—	—
818	,, 19	shortly before 4 a.	Ecclefechan	5	—
819	Oct. 22	1.25 p.	Invergarry	—	—
820	1889 Jan. 18	c. 4.10 a.	Edinburgh	—	—
821	,, 18	6.53 a.	,,	6	c. 830
822	Feb. 10	10.36 p.	Bolton	6	c. 1900; c. 2480
823	May 22	1.58 p.	Fort William	c. 4	—
824	June 19	7.40 a.	Invergarry	—	—
825	July 15	c. 6 p.	Kintyre	5	c. 350
826	Oct. 7	c. 1.45 p.	Altarnon	4	c. 400; c. 400
827	1890 Jan. 5	2.30 a.	Invergarry	4	—
828	,, 5	4.35 p.	,,	—	—
829	,, 5	4.40 p.	,,	—	—
830	,, 5	4.47 p.	,,	—	—
831	,, 19	4.55 p.	,,	—	—
832	Mar. 15	8.45 a.	,,	—	—
833	May 29	4.45 p.	,,	—	—
834	June 25	c. 10.30 p.	Wetherby	4	c. 60; c. 60
835	,, 26	c. 1 a.	,,	—	—
836	July 24	11.37 a.	Kintyre	5	c. 350
837	Aug. 8	3.35 p.	Invergarry	—	—
838	Nov. 15	5.50 p.	Inverness	7	5700; c. 7500
839	,, 15	6.15 p.	,,	4	c. 370
840	,, 16	9.15 a.	,,	4	—
841	,, 16	8.30 p.	,,	5	c. 910
842	,, 16	8.30 p.	Invergarry	—	—
843	,, 18	2.20 a.	Inverness	—	—
844	,, 19	1.33 a.	Invergarry	—	—
845	,, 19	1.40 a.	Inverness	—	—
846	,, 19	2.10 a.	,,	—	—
847	,, 19	4.10 a.	,,	4	—
848	Dec. 1	0.15 a.	,,	4	—
849	,, 1	10.10 a.	Invergarry	—	—
850	,, 8	4.10 a.	Inverness	5	—
851	,, 14	3.30 a.	,,	6	c. 820
852	,, 26	6.10 p.	Invergarry	—	—
853	1891 Feb. 24	10.55 p.	,,	—	—
854	,, 24	11.20 p.	,,	—	—
855	,, 25	1.15 a.	,,	—	—
856	Mar. 1	3.15 p.	,,	—	—
857	,, 1	9.25 p.	,,	—	—
858	,, 2	10.15 p.	,,	—	—
859	,, 26	c. 11.30 a.	Camelford	4	c. 170
860	Apr. 24	2.30 p.	Invergarry	4	—
861	Aug. 27	4.30 a.	,,	—	—
862	,, 27	6 a.	,,	—	—
863	,, 30	4.15 p.	,,	—	—
864	Nov. 16	10.25 a.	,,	—	—
865	,, 16	2.15 p.	,,	—	—
866	,, 16	8.45 p.	,,	—	—
867	Dec. 6	9.55 a.	,,	—	—
868	,, 26	1.20 a.	,,	—	—
869	,, 28	8.20 p.	,,	—	—
870	,, 28	9.24 p.	,,	—	—

NO.	DATE	HOUR	CENTRE	INT.	DIST. AREA, ETC.
871	1891 Dec. 30	9.45 a.	Invergarry	—	—
872	1892 Feb. 29	10.35 p.	,,	—	—
873	,, 29	11.15 p.	,,	4	—
874	Mar. 4	c. 7.30 a.	Ullapool	5	c. 364; c. 364
875	Apr. 3	7.35 a.	Invergarry	—	—
876	May 16	c. 10.30 p.	Helston	3	c. 120
877	,, 17	c. 1.30 a.	,,	4	c. 224; c. 480
878	Aug. 17	c. 11.30 p.	Pembroke	3	—
879	,, 18	0.22 a.	,,	—	—
880	,, 18	0.24 a.	,,	7	c.44860;c.44860
881	,, 18	0.37 a.	,,	c. 4	c. 4800
882	,, 18	c. 1.5 a.	,,	3	—
883	,, 18	c. 1.40 a.	,,	c. 5	c. 29000
884	,, 18	2.50 a.	,,	3	—
885	,, 18	c. 4 a.	,,	2 or 3	c. 127
886	,, 22	c. 11.55 a.	,,	4	c. 112
887	,, 23	4.30 a.	,,	—	—
888	Sept. 25	8.13 a.	Invergarry	—	—
889	Oct. 24	10.10 a.	,,	4	—
890	Nov. 18	2.17–	,,	—	—
891	1893 Jan. 2	7.20 p.	,,	—	—
892	Aug. 4	6.41 p.	Leicester	5	1170; 2200
893	Nov. 2	5.45 p.	Carmarthen	7	c. 35900; 63600
894	,, 2	6.1 p.	,,	4	c. 1000
895	,, 2	6.15–6.30 p.	,,	—	—
896	,, 3	c. 1 a.	,,	—	—
897	Dec. 11	c. 3 p.	Invergarry	—	—
898	,, 30	11.20 p.	Wells	4	38; 159
899	,, 31	0.28 a.	,,	4	43; 180
900	,, 31	c. 4 a.	,,	c. 3	—
901	1894 Jan. 12	c. 11.50 p.	Fort William	4	140
902	,, 23	c. 9 a.	Barnstaple	4	228; 389
903	,, 25	1.7 p.	Invergarry	4	—
904	Mar. 8	c. noon	Ecclefechan	—	—
905	May 14	—	,,	4	—
906	July 12	c. 11 p.	Comrie	—	—
907	Sept. 18	10.10 a.	Invergarry	—	—
908	1895 Jan. 9	5.45–5.50 a.	Fort William	—	—
909	July 12	c. 7.40 a.	Comrie	3	c. 40
910	1896 Jan. 26	6.50 a.	Launceston	3	86
911	May 29	4.47 a.	Ecclefechan	4	33
912	June 5	—	Fort William	—	—
913	Dec. 16	c.11.0 or 11.30p.	Hereford	pr. 4	c. 6300
914	,, 17	c. 1 a.	,,	3	—
915	,, 17	c. 1.30 or 1.45a.	,,	3	—
916	,, 17	c. 2 a.	,,	4	—
917	,, 17	c. 3 a.	,,	pr. 4	c. 6400
918	,, 17	c. 3.30 a.	,,	pr. 3	—
919	,, 17	c. 4 a.	,,	4	—
920	,, 17	c. 5 a.	,,	4	—
921	,, 17	c. 5.20 a.	,,	—	—
922	,, 17	5.32 a.	,,	8	c.98000;c.98000
923	,, 17	c. 5.40 or 5.45a.	,,	pr. 3	—
924	,, 17	c. 6.15 a.	,,	pr. 3	c. 870
925	1897 July 19	3.49 a.	,,	4	—
926	1898 Jan. 28	c. 10.5 p.	Oakham	3	130
927	Mar. 29	c. 10.25 p.	Helston	4	74
928	Apr. 1	9.55 p.	,,	4	76; 175
929	,, 2	c. 3 p.	,,	3	—
930	Aug. 22	c. 7.15 a.	Comrie	3	—

NO.	DATE	HOUR	CENTRE	INT.	DIST. AREA, ETC.
931	1899 Dec. 18	c. 6.50 a.	Invergarry	—	—
932	1900 Sept. 17	3.30 p.	Menstrie	—	—
933	„ 17	10.5 p.	„	—	—
934	„ 17	10.15 p.	„	4	117
935	„ 22	4.30 p.	„	4	60
936	1901 July 9	4.23 p.	Carlisle	5	2390; c. 3700
937	„ 9	c. 4.26 p.	„	4	—
938	„ 9	4.45 p.	„	4	c. 380; 1130
939	„ 11	c. 11.10 p.	„	—	—
940	Sept. 16	6.4 p.	Inverness	4	c. 108
941	„ 18	1.24 a.	„	8	c.33000; c.33000
942	„ 18	c. 1.35 a.	„	c. 4	c. 88
943	„ 18	c. 2 a.	„	—	—
944	„ 18	c. 2.30 a.	„	—	—
945	„ 18	c. 3 a.	„	—	—
946	„ 18	3.56 a.	„	⊀ 5	c. 2200
947	„ 18	9 a.	„	5	250
948	„ 23	c. 7.30 a.	„	3	—
949	„ 26	11.40 a.	„	4	—
950	„ 27	1.47 p.	„	5	—
951	„ 28	c. 4 a.	„	5	—
952	„ 29	9.6 p.	„	4	c. 57
953	„ 30	3.39 a.	„	7	—
954	Oct. 1	c. 4.35 a.	„	5	—
955	„ 6	4.24 a.	„	5	—
956	„ 13	4.24 p.	„	pr. 4	c. 35
957	„ 19	c. 7.25 p.	Framlingham	4	31; 31
958	„ 22	c. 9.15 a.	„	4	c. 31; c. 31
959	„ 22	c. 10.15 a.	Inverness	—	—
960	Nov. 15	c. noon	„	—	—
961	1902 Apr. 13	c. 11.50 a.	Doncaster	4	c. 600; c. 3000
962	Oct. 14	c. 5.15 p.	Strontian	5	1180; 1180
963	1903 Mar. 24	1.30 p.	Derby	7	c.12000; c.12000
964	„ 24	c. 1.45 p.	„	—	—
965	„ 24	c. 5 p.	„	—	—
966	May 3	9.22 p.	„	5	c. 179; c. 585
967	„ 15	6.15 p.	Menstrie	—	—
968	June 19	c. 4.25 a.	Carnarvon	—	—
969	„ 19	10.4 a.	„	7	c.15600; 25000
970	„ 19	10.9 a.	„	3	c. 219
971	„ 19	10.12 a.	„	3	—
972	„ 19	10.16 a.	„	3	—
973	„ 19	11.8 a.	„	3	219
974	„ 21	8.6 a.	„	3	c. 219
975	July 1	1.16 a.	Bala	4	c. 14
976	1904 Mar. 3	c. 1.5 p.	Penzance	5	c. 230; c. 230
977	June 21	c. 3.30 a.	Leicester	3	—
978	„ 21	5.28 a.	„	5	681; c. 1200
979	July 3	2.28 p.	Derby	3	—
980	„ 3	3.21 p.	„	7	10120; c.25000
981	„ 3	11.8 p.	„	4	125; 427
982	Sept. 18	4.7 a.	Dunoon	5	564; 564
983	Oct. 21	c. 6.5 a.	Beddgelert	4	—
984	1905 Jan. 20	1.50 a.	St Agnes	5	c. 530; c. 530
985	Apr. 23	0.15 a.	Menstrie	—	—
986	„ 23	c. 1.30 a.	Doncaster	—	—
987	„ 23	1.37 a.	„	7	c.10700; c.17000
988	July 23	0.15 a.	Menstrie	5	136; 136
989	„ 26	6.3 p.	„	4	—
990	Aug. 3	c. 6 p.	„	—	—

NO.	DATE		HOUR	CENTRE	INT.	DIST. AREA, ETC.
991	1905 Sept.	21	11.33 p.	Menstrie	6	700; c. 1000
992	,,	22	c. 1.30 a.	,,	—	—
993	,,	25	—	,,	—	—
994	,,	30	9.45 p.	,,	—	—
995	Oct.	29	10.53 a.	,,	—	—
996	Dec.	22	9.15 p.	,,	—	—
997	1906 May	7	8.20 p.	Fort William	—	—
998	June	27	9.45 a.	Swansea	8	37800; 66700
999	,,	29	3.2 a.	Carnarvon	4	c. 182
1000	July	3	2.15 p.	Menstrie	3	—
1001	,,	4	3.45 a.	,,	4	—
1002	,,	7	5.29 a.	,,	4	—
1003	Aug.	24	5.25 p.	,,	—	—
1004	,,	27	5.56 a.	Derby	5	c. 1360; c. 2100
1005	Sept.	28	0.25 p.	Menstrie	—	—
1006	Oct.	3	4.32 a.	,,	—	—
1007	,,	8	7.24 a.	,,	5	90; 90
1008	,,	8	8.16 a.	,,	4	45
1009	,,	12	7.20 a.	,,	—	—
1010	,,	20	7.15 a.	,,	—	—
1011	,,	24	7.11 p.	,,	—	—
1012	,,	26	7.15 p.	,,	—	—
1013	,,	30	0.15 p.	,,	—	—
1014	Dec.	28	4.12 p.	,,	6	c. 74
1015	,,	29	1.30 p.	,,	—	—
1016	,,	30	c. 1 a.	,,	—	—
1017	,,	30	2.10 p.	,,	3	—
1018	,,	30	4.15 p.	,,	6	c. 82
1019	,,	31	1 a.	,,	—	—
1020	1907 Jan.	17	1.54 p.	Oban	≮6	c. 3100
1021	Feb.	10	5.40 p.	Menstrie	—	—
1022	Mar.	19	7.33 p.	,,	5	—
1023	Apr.	7	11.11 p.	,,	—	—
1024	,,	7	11.19 p.	,,	—	—
1025	,,	8	6.45 a.	,,	—	—
1026	,,	11	5.30 a.	,,	4	—
1027	,,	11	5.40 a.	,,	4	—
1028	,,	11	6.5 a.	,,	—	—
1029	June	14	1.59 a.	,,	—	—
1030	,,	30	3.36 p.	,,	4	—
1031	July	3	3.40 a.	Swansea	4	c. 250
1032	,,	5	9.48 p.	Menstrie	—	—
1033	,,	21 or 28	—	,,	—	—
1034	Sept.	18	c. 5.30 p.	,,	—	—
1035	,,	27	8.12 a.	Malvern	5	206; c. 800
1036	1908 Jan.	19	1.27 a.	Menstrie	—	—
1037	May	1	6.54 p.	,,	4	—
1038	,,	2	7.5 a.	,,	4	—
1039	,,	10	0.48 a.	,,	4	—
1040	,,	10	0.58 a.	,,	4	—
1041	June	21	3 a.	,,	—	—
1042	,,	21	4.20 a.	,,	—	—
1043	July	3	6.15 a.	Dunoon	4	c. 400
1044	,,	17	5.27 p.	Menstrie	—	—
1045	Sept.	2	8.16 a.	,,	3	—
1046	,,	2	8.51 a.	,,	3	—
1047	Oct.	16	9.53 p.	,,	4	—
1048	,,	19	9.18 a.	,,	—	—
1049	,,	19	9.39 a.	,,	—	—
1050	,,	20	4.8 p.	,,	7	c. 1000

NO.	DATE			HOUR	CENTRE	INT.	DIST. AREA, ETC.
1051	1908	Oct.	20	4.13 p.	Menstrie	5	c. 123
1052		„	20	9.26 p.	„	—	—
1053		Nov.	6	4.45 p.	„	—	—
1054	1909	Jan.	19	5.28 a.	„	4	—
1055		„	19	5.29 a.	„	—	—
1056		„	23	0.15 p.	„	3	—
1057		„	24	0.15 p.	„	3	—
1058		„	24	2.28 p.	„	—	—
1059		„	24	3.35 p.	„	—	—
1060		„	27	1.40 p.	„	—	—
1061		Feb.	22	7.26 p.	„	—	—
1062		Mar.	19	9.35 a.	„	—	—
1063		May	22	3.23 p.	„	5	—
1064		„	22	5.1 p.	„	4	—
1065		„	22	5.24 p.	„	—	—
1066		„	22	8.23 p.	„	5	—
1067		Oct.	21	8.37 a.	„	4	—
1068		„	21	9.53 a.	„	—	—
1069		„	22	6.55 a.	„	—	—
1070		„	22	7.57 a.	„	4	—
1071		„	22	9.8 p.	„	—	—
1072	1910	Feb.	8	9.33 a.	„	4	—
1073		„	10	9.30 a.	„	—	—
1074		„	10	9.50 a.	„	4	—
1075		„	11	10.55 p.	„	4	—
1076		Apr.	17	1.45 a.	„	4	—
1077		May	19	9.45 a.	„	—	—
1078		„	20	2.40 p.	„	4	—
1079		„	20	2.50 p.	„	4	—
1080		„	20	3.5 p.	„	4	—
1081		„	20	3.15 p.	„	—	—
1082		„	27	0.30 a.	„	4	—
1083		July	10	3.7 p.	„	4	—
1084		„	11	0.1 a.	„	4	—
1085		„	22	11.59 p.	„	4	—
1086		Oct.	25	8.55 p.	„	—	—
1087		„	29	7.41 p.	„	—	—
1088		Nov.	20	1.57 p.	„	—	—
1089		Dec.	7	5.54 p.	„	—	—
1090		„	8	3.45 p.	„	—	—
1091		„	14	8.55 p.	Glasgow	5 or 6	250; c. 580
1092		„	14	8.57 p.	„	—	—
1093	1911	May	16	8.50 a.	Grasmere	6	495; 495
1094		July	27	8.10 a.	Menstrie	—	—
1095		„	27	5.15 p.	„	—	—
1096		„	28	7.46 p.	„	—	—
1097		Oct.	12	1.50 a.	„	4	—
1098		„	12	c. 2.20 a.	„	—	—
1099		„	12	3.45 a.	„	—	—
1100		„	17	6.35 a.	„	—	—
1101		„	21	1.25 p.	„	5	—
1102	1912	Jan.	20	c. 4 a.	„	—	—
1103		„	26	3.59 a.	„	5	—
1104		„	28	4 a.	„	4	c. 80
1105		„	28	4.40 p.	„	—	—
1106		„	28	5.25 p.	„	—	—
1107		„	30	5.25 p.	„	4	—
1108		„	31	5.29 p.	„	3	—
1109		Feb.	1	5.25 p.	„	4	—
1110		„	9	1.23 p.	„	4	—

NO.	DATE		HOUR	CENTRE	INT.	DIST. AREA, ETC.
1111	1912 Feb.	9	1.32 p.	Menstrie	5	nearly 80
1112	,,	9	1.39 p.	,,	—	—
1113	,,	11	3.45 a.	,,	3	—
1114	,,	24	3.15 a.	,,	—	—
1115	,,	24	4.53 a.	,,	—	—
1116	Mar.	3	2.24 p.	,,	—	—
1117	,,	20	7.50 a.	,,	—	—
1118	,,	21	10.20 p.	,,	—	—
1119	,,	21	10.25 p.	,,	—	—
1120	,,	21–22	midn.	,,	—	—
1121	,,	22	5.30 a.	,,	—	—
1122	,,	22	7.50 a.	,,	5	—
1123	,,	22	10.15 p.	,,	5	—
1124	,,	22	11.15 p.	,,	—	—
1125	,,	23	5.33 a.	,,	5	—
1126	,,	23	6.8 a.	,,	—	—
1127	Apr.	3	10.33 a.	,,	—	—
1128	,,	3	5.20 p.	,,	—	—
1129	,,	3	5.35 p.	,,	—	—
1130	,,	18	8.14 p.	,,	—	—
1131	,,	18	10.15 p.	,,	—	—
1132	,,	19	8.11 p.	,,	4	—
1133	,,	19	9.10 p.	,,	—	—
1134	,,	19	10.10 p.	,,	—	—
1135	,,	19	10.30 p.	,,	—	—
1136	,,	19–20	midn.	,,	—	—
1137	,,	20	2 a.	,,	—	—
1138	,,	20	2.5 a.	,,	—	—
1139	,,	20	2.8 a.	,,	4	—
1140	,,	20	2.10 a.	,,	—	—
1141	,,	20	2.12 a.	,,	—	—
1142	,,	20	2.14 a.	,,	—	—
1143	,,	20	2.18 a.	,,	—	—
1144	,,	20	4.15 a.	,,	—	—
1145	,,	21	1.5 a.	,,	—	—
1146	,,	21	2.47 p.	,,	—	—
1147	,,	22	7.17 a.	,,	—	—
1148	,,	23	6.10 a.	,,	—	—
1149	,,	26	c. 4 a.	,,	4	—
1150	May	3	6.20 a.	,,	—	—
1151	,,	3	c. 2.15 p.	,,	—	—
1152	,,	3	4.13 p.	,,	7	605 ; 605
1153	,,	3	11.45 p.	,,	—	—
1154	,,	4	0.10 a.	,,	—	—
1155	,,	4	2 a.	,,	3	—
1156	,,	9	11.28 p.	,,	—	—
1157	,,	10	9.20 p.	,,	—	—
1158	,,	11	c. 11.23 p.	,,	—	—
1159	,,	14	6.54 a.	,,	3	—
1160	,,	14	6.58 a.	,,	3	—
1161	,,	18	4 a.	,,	—	—
1162	,,	19	3.50 a.	,,	—	—
1163	,,	19	11.50 p.	,,	—	—
1164	July	5	1.50 a.	,,	—	—
1165	,,	6	1.52 a.	,,	—	—
1166	,,	6	c. 2.7 a.	,,	—	—
1167	,,	7	10.20 a.	,,	—	—
1168	,,	9	1.25 p.	,,	—	—
1169	,,	17	2.25 a.	,,	4	—
1170	,,	17	c. 3.20 a.	,,	—	—

D B E

3

NO.	DATE	HOUR	CENTRE	INT.	DIST. AREA, ETC.
1171	1912 July 18	3.15 a.	Menstrie	—	—
1172	,, 29	9.22 a.	,,	—	—
1173	Sept.	c. midn.	,,	—	—
1174	Oct. 18	c. 4.15 p.	,,	—	—
1175	Nov. 6	5.15 a.	,,	—	—
1176	1913 Aug. 27	4.30 a.	,,	—	—
1177	Oct. 28	4.30 a.	,,	4	—
1178	1914 Dec. 18	0.54 a.	,,	—	—
1179	1916 Jan. 14	7.29 p.	Stafford	7	12500; 50200
1180	,, 15	10.45 a.	,,	4	c. 640
1181	Oct. 21	2.55 p.	Menstrie	—	—
1182	,, 21	3.50 p.	,,	—	—
1183	,, 21	4.10 p.	,,	—	—
1184	,, 21	4.15 p.	,,	—	—
1185	,, 24	3.35 a.	,,	—	—
1186	1920 July 21	—	Comrie	3	—
1187	Sept. 9	c. 11.15 p.	Barnstaple	4	230
1188	,, 10	c. 0.15 a.	,,	—	—
1189	,, 14	—	Comrie	4	—
1190	1921 Apr. 30	10.35 a.	,,	4	—
1191	1924 Jan. 26	c. 6.10 a.	Hereford	4	c. 350

CHAPTER III

EARTHQUAKES OF INVERNESS AND OF OTHER CENTRES NEAR THE GREAT GLEN FAULT

IN this and the following chapters, the earthquakes of a district are grouped under the headings of certain centres. Those which are numbered consecutively under each centre are probably connected with that centre. In addition, there may be other records of two different kinds—of disturbances which I believe to be true earthquakes, but the connexion of which with the particular centre is doubtful, and disturbances of which the seismic origin is for one or more reasons uncertain. The former are distinguished by possessing a catalogue-number. Among the latter are included many shocks, usually very slight shocks, which may be quite genuine earthquakes, but which depend on the evidence of only one observer.

In naming an earthquake and the corresponding earthquake-centre, I have used whenever possible the name of the nearest well-known town, and I have sometimes followed this rule even if the shock were not felt in the town mentioned. When the origin of the shock is indefinite or unknown, I have used the name of the county, and in some, especially in the earlier, earthquakes the name of the country. In two earthquakes—and these as late as the years 1786 and 1852— even the country of origin is uncertain[1].

In the heading of each earthquake I have given, whenever possible, the date and hour of occurrence, the catalogue-number (see Chap. ii), the intensity of the shock according to the scale given on p. 9, the latitude and longitude of the centre of the innermost isoseismal or of the disturbed area, the number of records and of places of observation, and the authorities on which the account depends. From and after the year 1889, such references are usually to my own papers, and I have not thought it necessary then to repeat the author's name. In order to save room, references to newspapers are omitted, unless they should be the only or main authority, though the evidence which they provide is often of the greatest value. In the case of an earthquake-series, the references, if the same throughout, are given with the first shock only.

The earthquakes considered in this chapter belong to seven centres. The earthquakes of Inverness (by far the most important of the series),

---

[1] In some of my papers in which particular earthquakes are described, I used such terms as Exmoor, Loch Broom and Ochil earthquakes. For these, I have now, for the sake of uniformity, substituted Barnstaple, Ullapool and Menstrie earthquakes.

Fort William and Ballachulish are probably due to movements along the Great Glen fault. Those of the Oban centre may be connected with the same fault, or with one of a series of parallel faults on its south-east side. The Torosay, Phladda and Dhu Heartach centres lie near the line of continuation of the great fault.

INVERNESS EARTHQUAKES

1768 Oct. 24, an earthquake, accompanied by a loud noise, was felt at Inverness and Ruthven. The latter place is about 29 miles S. of Inverness (Cat. No. 143; Milne, vol. 31, p. 104).

1. **1769** NOV. 14 (*about*). Cat. No. 146; Intensity 8 (Milne, vol. 31, p. 105).

At Inverness, a shock which "threw down houses, and killed several persons."

About the year 1776, and again about the year 1800, a slight earth-quake "seems to have been felt" at Inverness (Lauder, pp. 366, 368).

2. **1816** AUG. 13, 10.45 *p.m.* Cat. No. 267; Intensity 8; Epicentre near Inverness; No. of records 52 from 45 places (Lauder, pp. 364–377; Milne, vol. 31, pp. 116–117; G. Inglis, *Phil. Mag.* vol. 48, 1816, pp. 431–433).

This is the strongest known earthquake felt, not only in the Inverness district, but in all Scotland. In Inverness, the injury to property was considerable. Chimney-tops were thrown down or damaged in every quarter of the town; many slates and tiles fell from roofs; and several walls were cracked. In the Mason Lodge, occupied as an hotel, one wall was rent, the north stalk of the chimney was partially thrown down, and one of the coping-stones, estimated to weigh 50 or 60 pounds, fell on the other side of the street, a distance of not less than 20 yards. The octagonal spire of the county jail was broken through about 5 or 6 feet from the top, and the upper part was twisted round so that the angles of the octagon were turned nearly to the middle of the flat sides below.

The isoseismal 5 lay outside the boundary of the disturbed area of the earthquake of 1888 (Fig. 1), for the shock was of intensity 5 at Aberdeen, Logie, Montrose and Perth. It must therefore have contained more than 20,000 square miles.

The disturbed area must have included all Scotland, except the south-western counties. The shock was felt at Edinburgh (114 miles from Inverness), Glasgow (115 miles) and, it is said, at Coldstream

(149 miles). Excluding the last place, the record from which is some-what doubtful, the total disturbed area must have contained about 50,000 square miles.

The sound-area probably extended nearly as far as the isoseismal 5, for the sound was heard at Perth (81 miles from Inverness), Aberdeen (83 miles) and Peterhead (92 miles). It probably included about 38,000 square miles.

3. **1816** AUG. 13, *about* 11.15 *p.m.* Cat. No. 268 (Lauder, p. 375; *Gent. Mag.* vol. 86, 1816, p. 269; etc.).

A slight shock, felt over a wide area, it is said even as far as Aberdeen (83 miles from Inverness).

4. **1816** AUG. 19. Cat. No. 269 (Milne, vol. 31, p. 117).

A shock felt at Inverness and in the neighbourhood.

5. **1816** AUG. 20. Cat. No. 270 (Milne, vol. 31, p. 117).

A shock felt at Inverness and in the neighbourhood.

1817 Apr. 26, 6.30 a.m., a smart shock was felt at Glasgow, Inverness and Greenock, and slightly at Leith (Milne, vol. 31, p. 118).

6. **1817** JUNE 10, 1.20 *p.m.* Cat. No. 273 (Milne, vol. 31, p. 118).

A smart shock at Urquhart, Dores and near Inverness.

7. **1817** JUNE 16, 5 *a.m.* Cat. No. 274 (Milne, vol. 31, p. 118).

Two smart shocks at Urquhart, Dores and near Inverness.

1817 June 23, 5 a.m. Two smart shocks are said to have been felt at Inverness, but more strongly at Urquhart, Dores and other places in the neighbourhood of Loch Ness. As the same record refers to another shock at the same places six days before, it would seem probable that the date (June 23) is incorrectly given for June 16.

8. **1817** JUNE 30. Cat. No. 275 (*Phil. Mag.* vol. 51, 1818, p. 193).

Two strong shocks at Inverness and in the neighbourhood.

1817 Aug. 7. Mr Milne (p. 118) records a shock at 8.20– at Urquhart, Dores and near Inverness. As he does not mention the next shock it is possible that the date may be given incorrectly.

9. **1817** AUG. 17, 8.20 *a.m.* Cat. No. 276 (*Edin. Adver.*).

A slight shock felt at Inverness, but more strongly at Urquhart, Aird and to the west of Inverness.

10. **1817** AUG. 31. Cat. No. 277 (Milne, vol. 31, p. 118).

A smart shock at Urquhart, Dores and near Inverness.

1817 Sept 2. Perrey (p. 151) records a shock on this day at Inverness, not mentioned by Milne, and adds that it is the fiftieth felt since the beginning of August.

11. 1817 SEPT. 7, *about* 3.30 *a.m.* Cat. No. 278 (*Edin. Adver.*).
A smart shock at Inverness.

12. 1818 FEB. 20, 1.20–. Cat. No. 283 (Milne, vol. 31, p. 118).
A shock felt at Inverness.

13. 1818 SEPT. 1. Cat. No. 288 (Milne, vol. 31, p. 118).
A shock at Inverness.

14. 1818 NOV. 9, 9 *p.m.* Cat. No. 289 (*Gent. Mag.* vol. 88, 1818, p. 557; *Phil. Mag.* vol. 52, 1818, pp. 394–395).
A slight shock at Inverness.

15. 1818 NOV. 9–10, *about midnight.* Cat. No. 290 (*Phil. Mag.* vol. 52, 1818, pp. 394–395).
A slight shock, accompanied by a faint sound, at Inverness.

16. 1818 NOV. 10, 0.20 *a.m.* Cat. No. 291; Intensity 6 (*Gent. Mag.* vol. 88, 1818, p. 557).
A smart shock, accompanied by a loud noise like thunder, at Inverness; felt with peculiar violence along the banks of Loch Ness.

17. 1818 NOV. 10, 4 *a.m.* Cat. No. 292 (*Gent. Mag.* vol. 88, 1818, p. 557; *Phil. Mag.* vol. 52, 1818, pp. 394–395).
A slight shock at Inverness.

18. 1853 FEB. 19, *about* 2.30 *p.m.* Cat. No. 711 (*Times*).
A slight shock, accompanied by a rumbling noise, felt at Inverness, Culloden, Strathglass and Forres.

1853 between Feb. 26 and Mar. 5, a slight shock at Inverness and in various parts of the Highlands (*Ill. London News*, 1853, Mar. 12).

19. 1888 FEB. 2, 5.1 *a.m.* Cat. No. 810; Intensity probably 7; No. of records 87 from 75 places; Fig. 1 (C. A. Stevenson, *Edin. Roy. Soc. Proc.* vol. 15, 1889, pp. 260–266; and other sources).

This earthquake differs from others of the district in its comparatively large disturbed area, and in the apparent absence of both fore-shocks and after-shocks. The position of its epicentre is unknown, but the map of the disturbed area, bounded by the broken-line in Fig. 1, seems to connect it with the Inverness centre.

The disturbed area is about 180 miles long from N.E. to S.W., about 140 miles wide, and contains about 20,000 square miles. There are also records of the shock from the following places to the south of the boundary—Lanrick (18 miles from the line), Dollar (28 miles) and Falkirk (37 miles). The sound was heard at most of the places which determine the boundary of the disturbed area, and also, it is said, at Comrie (9 miles), Crieff (11 miles) and Edinburgh (50 miles). At none of these six places, however, is the time of observation given.

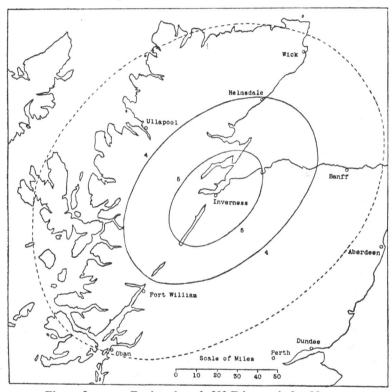

Fig. 1. Inverness Earthquakes of 1888 Feb. 2 and 1890 Nov. 15.

20. 1890 NOV. 15, 5.50 *p.m.* Cat. No. 838; Intensity 7; Centre of isoseismal 5 in lat. 57° 23·4′ N., long. 4° 12·2′ W.; No. of records 160 from 115 places; Fig. 1 (*Quart. Journ. Geol. Soc.* vol. 47, 1891, pp. 618–632).

At a few places, the shock was strong enough to produce slight damage to buildings. In the south-east part of Inverness, a chimney was thrown down, and, in the east part, a wall was cracked. Between

Dores and Inverness, another wall was cracked. At Clunes, some fissures in walls were re-opened; and at Beauly a chimney-pot was thrown down.

The isoseismals 5 and 4 are represented approximately by the continuous lines in Fig. 1. The former is 52 miles long, 33 miles wide, and includes an area of about 1350 square miles. Its centre lies about 1 mile S.E. of Inverness and its longer axis is directed about N. 43° E. The isoseismal 4 is about 105 miles long, 69 miles wide, and 5700 square miles in area. The disturbed area, which extends as far north as Braemore and Dunbeath Castle in Caithness, contains about 7500 square miles[1].

In the neighbourhood of the epicentre, the shock began with a tremor which rapidly increased in strength, culminating in a quick lurching motion, and ending with a tremor. At places near the boundary of the disturbed area, the shock was only perceptible as a weak tremulous motion. The mean duration of the shock throughout the disturbed area was 5·2 seconds.

The sound-area coincides approximately with the disturbed area, the sound being heard by 99 per cent. of all the observers. It was compared to passing waggons, etc., in 62 per cent. of the records, to thunder in 10, wind in 8, loads of stones falling in 3, the fall of a heavy body in 6, explosions in 9, and miscellaneous sounds in 2, per cent. The sound as a rule began faintly, grew louder to a maximum, and then gradually died away. The change of intensity was especially noticeable at places near the epicentre.

21. **1890** NOV. 15, 6.15 *p.m.* Cat. No. 839; Intensity 4; Centre of disturbed area in lat. 57° 26·3′ N., long. 4° 15·3′ W.; Records from 20 places; Fig. 2.

The disturbed area, the boundary of which is represented by the dotted line in Fig. 2, is 25½ miles long, 18½ miles wide, and contains about 370 square miles. Its centre is 3 miles S. 24° W. of Inverness, and the longer axis is directed N. 47° E. The boundary of the area corresponds to an intensity not much more than 3, the intensity being 4 only at Balnafettack, Inverness and Torbreck, all of which are close to the epicentre.

The shock at Inverness was similar to that of the principal earthquake, but of much less intensity. At Clunes, there was a swaying

---

[1] These figures differ slightly from those given in my original paper on this earthquake. The conclusions with regard to the origin of the earthquake are also revised.

motion followed by a tremor. The duration of the shock was 2 seconds at Inverness and 3 seconds at Clachnaharry.

The sound-area and disturbed area coincided approximately.

Nov. 15, 6.25 p.m. A shock similar to the preceding was felt by a single observer at Clunes station.

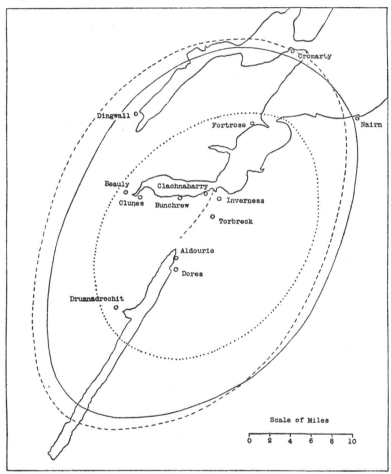

Fig. 2. Inverness Earthquakes of 1890: After-shocks.

22. 1890 NOV. 16, 9.15 *a.m.* Cat. No. 840; Intensity 4; Records from 12 places.

The centre of the disturbed area was probably close to a point 3½ miles S. 20° W. of Inverness. The duration of the shock was 3 seconds. The sound was heard at 4 places.

23. **1890** NOV. 16, 8.30 *p.m.* Cat. No. 841; Intensity 5; Centre of disturbed area in lat. 57° 26·2′ N., long. 4° 19·2′ W.; No. of records 37 from 21 places; Fig. 2.

The disturbed area, which is bounded by an isoseismal line of intensity 3, represented by the broken-line in Fig. 2, is 41 miles long, 27½ miles wide, and contains about 910 square miles. Its centre is about 5 miles S. 51° W. of Inverness, and the longer axis is directed N. 38° E. The intensity was 5 at Clunes, and 4 at Inchbreen and Inverness.

At Clunes, a swaying motion was felt, followed by a violent trembling. An undulating shock was felt at Beauly. In other parts only a slight tremor was observed. The duration of the shock was about 3 seconds. The sound was heard at 13 places.

Nov. 17, shortly after midnight, a shock was felt by one observer at Inverness.

Nov. 17, 4.30 a.m., two vibrations, each accompanied by a sharp report, were felt by one observer at Inverness.

24. **1890** NOV. 18, 2.20 *a.m.* Cat. No. 843.

A shock felt by several persons at Drumnadrochit, and similar to the preceding shock (No. 23).

25. **1890** NOV. 19, 1.40 *a.m.* Cat. No. 845.

A shock felt only at Inverness. The movement was rather vertical than horizontal. The epicentre must be near Inverness.

26. **1890** NOV. 19, 2.10 *a.m.* Cat. No. 846.

The shock was felt at Inverness and Torbreck, at the latter place being accompanied by a slight noise. The epicentre probably lies to the south-west of Inverness.

27. **1890** NOV. 19, 4.10 *a.m.* Cat. No. 847; Intensity 4.

The shock was felt at Balnafettack, Clachnaharry, Clunes, Inverness and Torbreck. The sound was heard at Balnafettack, Clachnaharry and Torbreck, but not at Clunes and Inverness. The epicentre is probably a short distance to the south-west of Inverness.

Nov. 19, 9.35 p.m., a slight tremor, felt by one observer at Inverness.

28. **1890** DEC. 1, 0.15 *a.m.* Cat. No. 848; Intensity 4.

The shock was felt at Aldourie, Balnafettack, Bunchrew, Clunes and Inverness. At Inverness, its intensity was 4, the movement being a gentle swaying, accompanied by sound. The epicentre was probably a short distance to the south-west of Inverness.

Dec. 1, about 3 a.m., a slight shock felt by one person at Bunchrew.

Dec. 3, 2.45 a.m., the same.

Dec. 6, 9.7 a.m., an earth-sound like the report of a distant cannon and lasting between 2 and 3 seconds, was heard by one observer at Clachnaharry.

29. **1890** DEC. 8, 4.10 *a.m.* Cat. No. 850; Intensity 5.

The shock was felt at Aldourie, Balnafettack, Bunchrew, Clunes, Dores and Inverness, the sound being heard at all the places except Inverness. The intensity was 5 at Balnafettack, 4 at Dores, and less than 4 at Clunes. The epicentre was probably near Balnafettack.

Dec. 12, 11.20 p.m., a slight shock felt by one observer at Bunchrew.

30. **1890** DEC. 14, 3.30 *a.m.* Cat. No. 851; Intensity 6; Centre of disturbed area in lat. 57° 26·0′ N., long. 4° 17′ W.; No. of records 20 from 12 places and negative records from 6 places; Fig. 2.

The disturbed area, the boundary of which is represented by the continuous line in Fig. 2, is 38 miles long, 27 miles wide, and contains about 820 square miles. Its centre is 4 miles S. 34° W. of Inverness, and the longer axis is directed about N. 42° E. The intensity was 6 at Balnafettack, 5 at Aldourie and Dores, and not less than 4 at Beauly, Clunes and Inchbreen.

The shock was distinguished from that of the principal earthquake by its sharpness and short duration. At Inverness, there was a sharp and sudden upheaval, followed by a rather strong tremulous motion.

The sound-area was probably co-extensive with the disturbed area.

31. **1901** SEPT. 16, 6.4 *p.m.* Cat. No. 940; Intensity 4; Centre of disturbed area in lat. 57° 24·9′ N., long. 4° 18·5′ W.; No. of records 9 from 8 places; Fig. 4 (*Quart. Journ. Geol. Soc.* vol. 58, 1902, pp. 377–397).

The disturbed area is roughly circular in form, about 12 miles in diameter, and containing about 108 square miles. Its centre lies about 1½ miles S. of Dochgarroch. The shock was extremely slight, the vibrations being hardly perceptible except at Dochgarroch. As a rule, the sound was also faint, though more prominent than the tremor.

Sept. 17, 11 p.m., a quivering, lasting for 2 seconds, at Inverness.

Sept. 18, 1.15 a.m., a tremor, accompanied by sound, at Dochgarroch.

32. 1901 SEPT. 18, 1.24 *a.m.* Cat. No. 941; Intensity 8; Centre of isoseismal 8 in lat. 57° 26·8′ N., long. 4° 15·8′ W.; No. of records 710 from 381 places and 77 negative records from 68 places; Fig. 3.

Isoseismal Lines and Disturbed Area. Five isoseismal lines are shown in Fig. 3, corresponding to intensities 8 to 4.

Fig. 3. Inverness Earthquake of 1901 Sept. 18.

The isoseismal 8 is 12 miles long, 7 miles wide, and contains 67 square miles. The centre of the curve is about 1½ miles E.N.E. of Dochgarroch. The longer axis of the curve is directed N. 33° E. In Inverness, the damage to buildings, though never serious, was by no means inconsiderable. One brick building used as a smithy collapsed, several chimneys or parts of them fell, and many chimney-pots were displaced or overthrown. At Dochgarroch and other places within the isoseismal 8, walls were cracked, chimneys thrown down, and lintels loosened.

The isoseismal 7 is 53½ miles long, 35 miles wide, and 1500 square miles in area. Its longer axis is directed N. 32° E. The distance between the isoseismals 8 and 7 is 9 miles on the north-west side and 14 miles on the south-east side. The isoseismal 6 includes a district 105 miles long, 87 miles wide, and 7300 square miles in area. The distance between the isoseismals 7 and 6 is 21½ miles on the north-west side and 31 miles on the south-east side. The isoseismal 5 is 157 miles long, 143 miles wide, and contains about 17,000 square miles. The distance between the isoseismals 6 and 5 is 21½ miles on the north-west side and 34½ miles on the south-east side. The isoseismal 4, which bounds the disturbed area, is 215 miles long from N.E. to S.W., 198 miles wide, and includes about 33,000 square miles. Its distance from the isoseismal 5 is 20 miles on the north-west side and 35 miles on the south-east side.

Nature of the Shock. There was but little variation in the nature of the shock throughout the disturbed area, and the following accounts may be regarded as typical for places respectively near to, and far from, the epicentre. At Inverness, there was first a gentle movement, followed by an extraordinary quivering which increased in force for 2 or 3 seconds and then decreased for 2 or 3 seconds; just as the quivering was about to cease, there was a distinct lurch or heave, after which the vibration was much more severe than before, and lasted several seconds longer than the first part of the shock. At Aberdeen, the shock consisted of two distinct parts, the first a tremble, followed, after a few seconds, by a swinging movement of longer duration than the tremble.

Throughout the disturbed area, the shock consisted of two distinct parts, the second being of greater duration and intensity than the first, and consisting of vibrations of longer period. At places near the epicentre (as at Inverness), there was no interval between the two parts; but, at a distance (as at Aberdeen), the intermediate tremors were imperceptible, and the parts were separated by an interval of rest and quiet lasting 2 or 3 seconds. The average duration of the shock, according to 59 observers who were awake when the earthquake began, was 4·7 seconds.

It is evident that the two series of vibrations were produced by two distinct impulses, the stronger impulse succeeding the other after an interval of a few seconds. It is possible that the corresponding foci were nearly or quite detached; but it is more probable that the focus of the earlier impulse was overlapped by, or included within, that of the second.

Sound-Phenomena. The boundary of the sound-area is not represented in Fig. 3. It lies, however, between the isoseismals 5 and 4, and must have included about 27,000 square miles. Throughout the whole disturbed area, 84 per cent. of the observers heard the sound, the percentages of audibility being 97 within the isoseismal 8, and 93, 91, 90 and 68 within the zones bounded by successive pairs of isoseismals.

The sound was compared to passing waggons, etc., in 39 per cent. of the records, to thunder in 25, wind in 14, loads of stones falling in 8, the fall of heavy bodies in 3, explosions in 4, and miscellaneous sounds in 7, per cent.

The isoseismal 7 approximately bounds the district in which the sound was very loud from that in which it was distinctly fainter. With one exception, the same curve includes all places at which explosive crashes were heard with the strongest vibrations.

Earth-Fissure. An effect of the earthquake unusual for this country was a long crack in the northern bank of the Caledonian Canal near Dochgarroch Locks. It was formed in the middle of the towing-path, and could be traced at intervals for a distance of 200 yards to the east of the locks and 400 yards to the west, being often a mere thread, and in no place more than half an inch wide. Shortly after the earthquake, the fissure was obliterated by heavy showers of rain.

In the following pages are given records of 46 after-shocks and 10 earth-sounds. Of these, 16 after-shocks and 1 earth-sound were noticed by several or many observers; the remainder depend on the evidence of only one person. The list, however, is far from complete. Thus, on 1901 Sept. 18, between the two prominent after-shocks at 3.56 and 9 a.m., there is only one record given below, though, according to an observer at Dochgarroch, 18 slight shocks were felt there in the same interval. Several shocks were also felt at Dores, and many earth-sounds were heard at Bunchrew, besides those contained in the register; while one observer at Lochend (Aldourie) estimates the total number of shocks up to Oct. 23 at about 70.

33. 1901 SEPT. 18, *about* 1.35 *a.m.* Cat. No. 942; Intensity about 4; Centre of disturbed area in lat. 57° 24·9′ N., long. 4° 16·8′ W.; No. of records 7 from 7 places; Fig. 4.

The boundary of the disturbed area is nearly circular in form, about 10½ or 11 miles in diameter, and contains about 88 square miles. Its centre is about $1\frac{3}{4}$ miles S.E. of Dochgarroch. The shock was very slight, and the sound a low rumble like distant thunder.

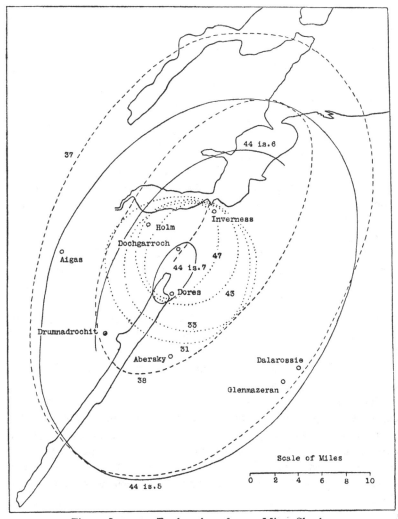

Fig. 4. Inverness Earthquakes of 1901: Minor Shocks.

34. 1901 SEPT. 18, *about 2 a.m.* Cat. No. 943; No. of records 2 from 1 place.

A slight shock, accompanied by a noise like distant thunder, was felt at Glenmazeran, near Dalarossie, in the valley of the Findhorn.

35. 1901 SEPT. 18, *about* 2.30 *a.m.* Cat. No. 944; No. of records 2 from 2 places.

A slight tremor was felt at Inverness and Abersky, and at the latter place it was accompanied by a noise like that of a passing vehicle.

36. 1901 SEPT. 18, *about* 3 *a.m.* Cat. No. 945; No. of records 2 from 2 places.

A tremor was observed by the engineer at the Inverness District Asylum (2 miles W.S.W. of Inverness), and a slight rumbling at Aigas (near Beauly).

37. 1901 SEPT. 18, 3.56 *a.m.* Cat. No. 946; Intensity not less than 5; Centre of isoseismal 5 in lat. 57° 25·3′ N., long. 4° 15·9′ W.; No. of records 90 from 43 places; Fig. 4.

The only isoseismal line which can be drawn is that of intensity 5, which is 38 miles long, 25 miles wide, and contains about 750 square miles. Its centre lies 2 miles E.S.E. of Dochgarroch, the direction of its longer axis being N. 35° E.

Outside this isoseismal, the shock was felt at Little Scatwell, Relugas, Dunphail and Grantown, which are respectively 18, 24, 25 and 26 miles from the centre of the isoseismal 5. The disturbed area therefore contains about 2200 square miles. There are also records of its occurrence at Banff, Comrie, Crieff, Deskford and Ordiquhill, which lie between 57 and 76 miles from the centre. The absence of observations at intermediate places lessens, however, the value of these records.

The shock consisted of two or three distinct oscillations, the mean duration being 2·5 seconds.

The sound was heard by 62 per cent. of the observers, and was compared to passing waggons, etc., in 14 per cent. of the records, to thunder in 43, wind in 14, the fall of a heavy body in 7, explosions in 14, and miscellaneous sounds in 7, per cent. The sound-area was probably excentric with regard to the isoseismal lines, overlapping the isoseismal 5 by several miles on the north-west side.

Sept. 18, about 6.25 a.m., a noise heard at Inverness.

38. 1901 SEPT. 18, 9 *a.m.* Cat. No. 947; Intensity 5; Centre of disturbed area in lat. 57° 27·0′ N., long. 4° 15·1′ W.; No. of records 26 from 18 places; Fig. 4.

The disturbed area of this shock is elliptical in form, and is 26 miles long, 12½ miles wide, and 250 square miles in area. Its centre is situated 2 miles E.N.E. of Dochgarroch. The longer axis is directed N. 36° E.

The shock consisted of a single series of vibrations, and was accompanied by a sound compared to a passing vehicle, thunder, or the discharge of cannon.

Sept. 18, noon, a tremor at Inverness District Asylum, strong enough to knock down some loose plaster which was lying on the joists of buildings.

Sept. 18, about 11.25 p.m., a low, but distinct, rumble at Inverness.

Sept. 20, 4 p.m., a slight shock at Dores.

Sept. 21, 10.45 a.m., a shock, accompanied by a noise like the discharge of cannon at Holm (near Lentran).

Sept. 21, 11.20 a.m. and 3.15 p.m., the same.

39. **1901** SEPT. 23, *about* 7.30 *a.m.* Cat. No. 948; Intensity 3; No. of records 4 from 4 places.

The shock was felt at Dores, Holm, the Inverness District Asylum, and Kirkhill. The sound resembled the passing of a coach or the discharge of cannon. The epicentre probably lies in the neighbourhood of Dochgarroch.

Sept. 23, 9 a.m., a shock and rumbling noise at Teanassie (near Beauly).

Sept. 24, 5.15 a.m., a shock, accompanied by a sound like the discharge of cannon, at Holm.

Sept. 24, 7.10 a.m. and 4.20 p.m.; Sept. 26, 8.8 a.m. and 9.25 a.m., the same.

40. **1901** SEPT. 26, 11.40 *a.m.* Cat. No. 949; Intensity 4; No. of records 3 from 3 places.

The shock was felt at Dores and Holm. At Drumälan (near Drumnadrochit) there was no tremor, and the sound was like that of a train gradually approaching, passing, and receding from, the house.

Sept. 26, 9.39 p.m., a shock (intensity less than 4), accompanied by a sound like that of distant cannon, at Holm.

41. **1901** SEPT. 27, 1.47 *p.m.* Cat. No. 950; Intensity 5; No. of records 2 from 2 places.

The intensity of this shock was 5 at Holm and probably 4 at Aigas (6 miles S.W. of Beauly). The sound, which was heard at both places, resembled that of distant gun-firing.

42. 1901 SEPT. 28, *about* 4 *a.m.* Cat. No. 951; Intensity 5; No. of records 2 from 1 place.

A shock felt at Inverness.

Sept. 28, 11.50 a.m., a slight shock (intensity less than 4) and sound in Glen Urquhart. The exact position is uncertain, but it is probably near Loch Ness.

Sept. 28, 1.40 p.m., a slight tremor felt in a boat on Loch Ness.

Sept. 29, about 4.30 a.m., a slight shock, accompanied by a faint sound like distant thunder, at Inverness.

43. 1901 SEPT. 29, 9.6 *p.m.* Cat. No. 952; Intensity 4; Centre of disturbed area in lat. 57° 26·1′ N., long. 4° 17·0′ W.; No. of records 6 from 5 places; Fig. 4.

The places at which this earthquake was observed lie within an area which is probably circular in form, about 8½ miles in diameter, and including about 57 square miles. The centre is nearly 1 mile E. of Dochgarroch. The shock was slight, and the sound faint and of brief duration.

Sept. 29, 11 p.m., a rumbling sound heard at Aigas.

Sept. 29–30, two slight shocks felt at Drumälan (Drumnadrochit) before 3.39 a.m., one of which may be identical with the earth-sound at Aigas.

44. 1901 SEPT. 30, 3.39 *a.m.* Cat. No. 953; Intensity 7; Centre of isoseismal 7 in lat. 57° 24·5′ N., long. 4° 19·3′ W.; No. of records 54 from 33 places; Fig. 4.

On the map are shown the isoseismal 7 (approximately), part of the isoseismal 6, and the isoseismal 5. The isoseismal 7 is about 5 miles long, 2½ miles wide, and contains about 10 square miles. Its centre is 2 miles S. 11° W. of Dochgarroch. The isoseismal 5 is 33 miles long, 23 miles wide and 595 square miles in area. Its longer axis is directed N. 34° E. The distance between the isoseismals 7 and 5 is 8 miles on the north-west side and 12½ miles on the south-east side. Outside the isoseismal 5, the records of the shock are few in number. The intensity of the shock was 4 at Lochluichart, which is 13 miles on the north-west side from the isoseismal 5. If the isoseismal 4 passes through this place and is coaxial with the isoseismal 5, its area would be about 2100 square miles.

The shock consisted as a rule of horizontal vibrations of longer period than usual. Its mean duration was 3 seconds. The sound was heard by at least 80 per cent. of the observers. It was compared to passing waggons, etc., in 25 per cent. of the records, to thunder in 50, wind in 4, loads of stones falling in 4, explosions in 8, and miscellaneous sounds in 8, per cent.

Sept. 30, about 4.10 a.m., a slight shock at Aldourie.

45. 1901 OCT. 1, *about* 4.35 *a.m.* Cat. No. 954; Intensity 5; No. of records 2 from 2 places.

A shock, preceded, accompanied, and followed by a sound like thunder, was felt at Dalarossie and Coignafuinternach.

Oct. 1, about 3 p.m., a faint earth-sound, like distant thunder, at Dalarossie.

Oct. 1, 5.6 p.m. and Oct. 2, 2.7 p.m., a shock (intensity less than 4), accompanied by a sound, as of distant thunder, at Holm.

46. 1901 OCT. 6, 4.24 *a.m.* Cat. No. 955; Intensity 5; No. of records 2 from 2 places.

At Dochgarroch, two vibrations (intensity 5) were felt, without any sound. The shock was also felt at Holm. It is probable that the epicentre was close to Dochgarroch, and that the depth of the focus was small.

Oct. 9, 7.40 p.m., a shock (intensity probably 5), preceded, accompanied and followed by a sound like that of a light carriage passing, at Dalarossie.

Oct. 12, 8.40 a.m., a single vibration (intensity 5), accompanied by a rumbling sound, at Dochgarroch.

Oct. 12, 0.56 p.m., a slight shock at Holm.

Oct. 12, about 4 p.m., a shock at Inverness, as if a heavy body fell against the house and was then dragged along the side of it.

Oct. 13, 0.30 p.m., a slight earth-sound at Dalarossie.

47. 1901 OCT. 13, 4.24 *p.m.* Cat. No. 956; Intensity probably 4; Centre of disturbed area in lat. 57° 26·1′ N., long. 4° 18·0′ W.; No. of records 7 from 5 places; Fig. 4.

The records are too few to determine the boundary of the disturbed area with any approach to accuracy; but it is of elongated form, with its longer axis approximately parallel to those of the isoseismal lines of previous shocks. The length of the curve represented in Fig. 4 is 8 miles, the width 5½ miles, and the area contained by it about 35 square miles. Its centre is about ¼ mile S.E. of Dochgarroch.

The shock consisted of two or three vibrations and was accompanied by a loud report like a shot from a gun.

Oct. 13, about 5.30 p.m., a very distinct shock at Holm.

Oct. 14, 1 a.m., a shock, preceded by a noise like the soughing of the wind through trees, at Inverness.

Oct. 14, 5 p.m., a shock (intensity less than 4), accompanied by a sound like that of a heavily loaded lorry running on a country road, at Dochgarroch Locks.

Oct. 22, 5.30 a.m., a distinct tremor at Drumnadrochit.

48. 1901 Oct. 22, *about* 10.15 *a.m.* Cat. No. 959; No. of records 2 from 2 places.

At Aldourie, eight or nine sounds were heard between 9.45 a.m. and noon. The sound at about 10.15 a.m. was very distinct, and resembled the roar of a furnace when the door is opened or of an underground train; it lasted 2 or 3 seconds. No distinct shock was felt, but there was evidently a weak tremor. At Drumnadrochit, only a sound was heard; it resembled distant thunder, growing louder and then dying away.

Oct. 22, 0.55 p.m., an earth-sound, like distant thunder, at Drumnadrochit.

Oct. 22, 8.20 p.m., a slight tremor, accompanied by a rumbling sound, at Bunchrew.

Oct. 22, 8.25 p.m., a rumbling sound at Bunchrew.

Nov. 5, 0.12 a.m., an earth-sound, like distant thunder and lasting 4 or 5 seconds, at Dalarossie.

49. 1901 NOV. 15, *about noon.* Cat. No. 960.

An underground rumbling sound was heard by several persons at Dochgarroch.

Nov. 15, an earth-sound heard during the night, near the end of Loch Ness.

Nov. 21, a slight vibration and rumbling sound during the night at Dochgarroch.

SOUND-PHENOMENA OF THE AFTER-SHOCKS OF 1901

The percentage of audibility of all the after-shocks is 77. That of the shock of Sept. 18 (no. 37) is 62; and that of the shock of Sept. 30 (no. 44), 80. Omitting these, which are the most important after-shocks, the percentage for the remainder is 85, showing that the sound, though much fainter than that which accompanied the principal earthquake, was nevertheless a comparatively important feature.

Taking all the after-shocks together, the comparisons to passing waggons, etc., occur in 26 per cent. of the records, to thunder in 43, wind in 7, loads of stones falling in 2, the fall of a heavy body in 2, explosions in 17, and miscellaneous sounds in 3, per cent. Omitting the two strong after-shocks on Sept. 18 and 30 (nos. 37 and 44), the corresponding figures are: passing waggons, etc., 33, thunder 33, wind 5, and explosions 28, per cent. These are approximately the proportions met with in slight earthquakes with short foci, and indicate that, as a rule, the foci of the after-shocks were of short linear dimensions.

ORIGIN OF THE INVERNESS EARTHQUAKES

The evidence with regard to the position of the fault in which the Inverness earthquakes originated is unusually complete. The direction of the longer isoseismal axes in eight earthquakes from 1890 Nov. 15 to 1901 Sept. 30 (nos. 20, 21, 23, 30, 32, 37, 38 and 44) varies between N. 33° E. and N. 47° E. The average of all eight estimates is N. 38° E. The hade of the fault is given by two earthquakes—those of 1901 Sept. 18 and 30 (nos. 32 and 44)—and both agree in assigning a hade to the S.E. Thus, the fault-line must be directed about N. 38° E. and must lie a short distance on the north-west side of the epicentre.

The chief structural feature of the district is the great boundary fault which runs from Tarbat Ness along the eastern coast of Ross-shire, and follows the line of the Great Glen. The mean direction of this fault between Inverness and Loch Ness is about N. 35° E., and its hade is to the S.E. Its course

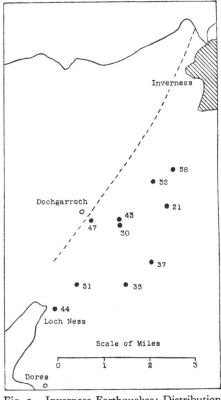

Fig. 5.   Inverness Earthquakes: Distribution of Epicentres.

within the epicentral district is represented by the broken-line in Fig. 5, and it will be noticed that all the known epicentres, ten in number, lie on the south-east side of the fault-line.

With regard to the earthquakes before 1890, the evidence is inconclusive. It is probable, however, that the earthquakes of 1816 and 1888 were due to slips along this fault. There can be little doubt that all the more important shocks from 1890 onwards and the majority of the slighter shocks were so caused. The exceptional earthquakes were those of 1901 Sept. 18 and Oct. 1 (nos. 34 and 45), which seem to be due to local, but sympathetic, movements in the valley of the Findhorn.

Earthquakes of 1816. This important series consisted of the principal shock on Aug. 13, followed by 16 noticeable after-shocks, of which 3 occurred in 1816, 6 in 1817, and 6 in 1818. The epicentre of the first after-shock (no. 3) is unknown, but it was probably near Inverness. Assuming this to be the case, 10 after-shocks originated in the central region near Inverness, 5 near Dores, Urquhart, etc., and 1 beneath Loch Ness. Denoting the central region by C and the other by W, the after-shocks occurred in the following order:

C, C, C, W, W, C, W, W, C, C, C, C, C, W, C;

showing a certain westerly migration of the focus, ending with a general return to the central region.

The evidence of the earthquake of 1888 is insufficient to throw any light on the growth of the fault.

Earthquakes of 1890. The series consisted of the principal shock on Nov. 15 (no. 20), the epicentre of which was near Inverness, followed by 10 after-shocks, of which 5 occurred near the central region and 5 farther to the south-west. Using the same notation as before, the order of the after-shocks was as follows:

W, W, W, W, C, C, C, C, C, W;

showing again a westerly migration of the focus, followed by a general return to the central region. In three of the earthquakes—the principal earthquake of Nov. 15 and the after-shocks of Nov. 15 and Dec. 14 (nos. 20, 21 and 30)—the position of the epicentre is known with some approach to accuracy. Its distance from the fault-line was about 2 miles in the principal earthquake, 1·2 miles in the first after-shock, and about ·5 mile in the last, indicating, though the evidence is but slight, a gradual approach of the focus towards the surface

Earthquakes of 1901. In this series, there was at least one fore-shock (no. 31) on Sept. 16, the principal earthquake (no. 32) on Sept. 18, 15 after-shocks (nos. 33, 35–44, 46–49) from Sept. 18 to Nov. 15, and 2 sympathetic shocks (nos. 34 and 45) on Sept. 18 and Oct. 1.

The first slip (no. 31) occurred in a small region of the fault about $\frac{1}{2}$ mile N.E. of the end of Loch Ness. Two other slight fore-slips, if we may rely on solitary records, took place farther to the north-east, one near Inverness, the other near Dochgarroch. As the principal focus extended nearly from Loch Ness to Inverness, it follows that the first fore-slip occurred in its western margin, and the later doubtful fore-slips in its eastern margin and central region, respectively.

By these fore-slips, small obstructions were removed and the effective stress was equalised along several miles of the fault; so that the next displacement, which resulted in the principal shock, took place over a region not less than 5 miles in length. There were two distinct slips in rapid succession, with continuous slight motion between them, the second slip being greater in amount and extending over a region which probably overlapped, if it did not entirely include, that within which the first slip took place.

The sequence of subsequent events will be rendered clearer if, for the present, we disregard the very small slips (some of which may have been due to merely local variations of stress) and confine our attention to the six chief after-slips (nos. 33, 37, 38, 43, 44 and 47), three of which (nos. 37, 38 and 44) were of much greater importance than the rest and affected several miles of the fault-surface.

An interval of only 10 minutes separated the principal earthquake and the first after-shock. This was caused by a small slip (no. 33) near the south-west margin of the principal focus. After $2\frac{1}{4}$ hours, the chief after-slip (no. 37) occurred; its centre migrated about $\frac{1}{2}$ mile to the north-east; but, as its focus was several miles in length, its south-west margin extended some distance beyond that of the principal focus. The seat of action was then transferred to the other side of that focus, a long after-slip (no. 38) taking place after the lapse of about 5 hours; its centre was approximately $\frac{1}{2}$ mile N.E. of the principal centre, and its focus probably extended a short distance beyond the north-east margin of the principal focus. During the next $11\frac{1}{2}$ days, there were no movements of any consequence; but, at the end of that time, a small slip (no. 43) occurred about 1 mile S.W. of the principal centre and close to Dochgarroch. This was followed, in $6\frac{1}{2}$ hours, by the third long after-slip (no. 44), the centre of which lay to the south-west of the principal focus, and the slip itself must have extended 2 or 3 miles beneath Loch Ness. Again, after a further lapse of $13\frac{1}{2}$ days, there was a slip (no. 47) about 2 miles long, in the immediate neighbourhood of Dochgarroch.

Thus, of the six chief after-slips, one originated in the region of the fault on the north-east side of the principal centre, and five in that on the south-west side.

There remain nine after-shocks recorded by more than one observer (nos. 35, 36, 39–42, 46, 48 and 49). Of four of these, the epicentre is undetermined; of the others, the focus of one lay to the north-east, and the foci of four to the south-west, of the principal centre.

Further light is thrown on the nature of the fault-movements by the numerous tremors and earth-sounds recorded by single observers; for, in such slight disturbances (assuming them to be of seismic origin), the epicentres must have been close to the places of observation. The numbers recorded in different districts near the boundary-fault are as follows: Inverness and Bunchrew 8, Dochgarroch and Holm 16, Dores, Aldourie, etc., 3, and Drumnadrochit, etc., 6; or 8 on the north-east side of the principal centre and 25 on the south-west side.

Thus, while the great slip reached nearly from Loch Ness to Inverness and was greatest at a point about halfway between, the three chief after-slips extended the area of principal displacement in both directions along the fault-surface probably less than $\frac{1}{2}$ mile to the north-east and 6 miles or more to the south-west. The minor slips (some of them mere creeps) were most numerous in three regions— one about a mile south-west of the principal centre, the others near Inverness and Drumnadrochit, places about 14 miles apart, which lie near the extremities of the displaced area of the fault.

In addition to the migration of the focus in the direction of the fault, there was also, in the six principal after-shocks, a continuous decrease in the depth of the focus, for the distances of the epicentres of these shocks from the fault-line are respectively 1·5, 1·7[1], 1·0, 0·6, 0·5 and 0·1 miles.

FORT WILLIAM EARTHQUAKES

1869 Mar. 9, about 8 p.m., a shock with sound at Fort William, Monessie, Strone and in Lochaber, which may be connected with the Fort William centre (Cat. No. 748; *Times*, Mar. 16). Five days later, on Mar. 14, Perrey records a tremor in the north of Scotland

---

[1] In this case (No. 37), the distance refers to the centre of the isoseismal 5, which would be somewhat farther to the south-east than the true epicentre.

which may refer to the same earthquake (*Mém. Cour.* vol. 24, 1875, p. 13).

1. **1870** END *of* NOV. Cat. No. 754 (*Brit. Ass. Rep.* 1871, p. 197).

A feeble shock felt in Lochaber, especially in Glen Spean and the lower part of the Great Glen.

2. **1870** DEC. 3, *about 2 a.m.* Cat. No. 755 (Perrey, *Mém. Cour.* vol. 24, 1875, p. 144).

A strong shock at Fort William.

3. **1889** MAY 22, 1.58 *p.m.* Cat. No. 823; Intensity about 4; Epicentre near Ben Nevis (*Geol. Mag.* vol. 8, 1891, pp. 364–365).

A slight shock felt only in the Ben Nevis observatory.

4. **1894** JAN. 12, *about* 11.50 *p.m.* Cat. No. 901; Intensity 4; Centre of disturbed area in lat. 56° 46′ N., long. 5° 7′ W.; No. of records 17 from 14 places and negative records from 5 places; Fig. 6 (*Geol. Mag.* vol. 7, 1900, pp. 112–114).

The boundary of the disturbed area is an oval curve 17 miles long, 10½ miles wide, and contains 140 square miles. Its centre is about 4 miles S. of Fort William and the direction of the longer axis N. 32° E. As a rule, the shock resembled the tremulous motion felt in a house when a heavy cart is passing. At Nether Lochaber manse, the movement was more distinct, being like that felt in a heavy carriage in rapid motion on a rough road, and culminating in a strong vibration. The sound

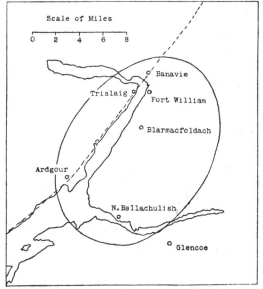

Fig. 6. Fort William Earthquake of 1894 Jan. 12.

was compared to the rumbling of a waggon passing or to distant thunder. At the Low-Level observatory at Fort William, the shock was registered by the seismometer, the direction being from S.W. to N.E.

At about the same time, a slight shock was felt at three places
—Arisaig, Glenmoidart and Roshven—in the extreme west of Inverness-shire. As these places are separated from the disturbed area
of the Fort William earthquake by a band (including Kinlochiel,
Duisky and Kingairloch) in which no movement was observed, it is
probable that there were two distinct shocks at about the same time.
At Glenmoidart, it is said that another, though weaker, shock was
felt in the daytime some days afterwards, and a third at night.

5. 1895 JAN. 9, *between 5.45 and 5.50 a.m.* Cat. No. 908 (*Geol. Mag.*
vol. 7, 1900, pp. 114–115).

An earth-sound, resembling distant thunder, was heard at Fort
William, and also at Blarmacfoldach and Lundavra, which are 3 and
5 miles, respectively, to the south of Fort William.

6. 1896 JUNE 5, *early morning.* Cat. No. 912 (*Geol. Mag.* vol. 7,
1900, p. 168).

A slight shock was felt in Glen Nevis, being strongest at Achreach
(about 4 miles S.E. of Fort William). One of the seismographs in the
Low-Level observatory at Fort William gave a somewhat doubtful
indication of the shock.

7. 1906 MAY 7, 8.20 *p.m.* Cat. No. 997 (*Geol. Mag.* vol. 5, 1908,
p. 300).

A slight shock at Fort William, lasting about 2 seconds, and preceded by a rumbling noise.

BALLACHULISH EARTHQUAKE

1885 JUNE 18, *about* 1 *a.m.* Cat. No. 798 (*Nature,* vol. 32, 1885,
p. 176).

A shock was distinctly felt at Ballachulish, in many of the houses
in the slate quarry district of Glencoe, and at Clachaig in Glencoe
(3 miles E. of Ballachulish).

OBAN EARTHQUAKES

1. 1841 APR. 19, 5.30 *a.m.* Cat. No. 506 (Milne, vol. 36, p. 76).
A shock felt at and near Oban.

2. 1841 APR. 19, 11 *a.m.* Cat. No. 507 (Milne, vol. 36, p. 76).
A shock felt at and near Oban.

3. 1841 APR. 19, 2.30 *p.m.* Cat. No. 508 (Milne, vol. 36, p. 76).
A shock felt at and near Oban.

4. **1841 APR. 21, 1.35 *a.m.*** Cat. No. 509 (Milne, vol. 36, p. 76).

A shock, of intensity probably less than 4, was felt at Oban, Lismore lighthouse (6 miles N.W. of Oban) and Connel (4 miles N.E.). At Lismore, a loud noise was heard like that of a cannon discharged at a short distance.

5. **1877 MAR. 11, 11.30 *a.m.*** Cat. No. 777; Intensity 4 (D. Stevenson, *Edin. Roy. Soc. Proc.* vol. 9, 1878, pp. 403–405).

A smart shock felt at Hynish lighthouse (island of Tyree) and Sound of Mull lighthouse, and preceded, accompanied and followed by a noise like distant thunder, the shock being felt at the time when the sound was loudest.

1877 Mar. 20, a shock was felt in the island of Mull (Fuchs, p. 138). In *Nature* for Apr. 26 (vol. 15, 1877, p. 561), it is said that, a short time before, a shock was felt at Tobermory (island of Mull).

6. **1877, *between* APR. 15 *and* 21.** Cat. No. 778 (*Nature*, vol. 15, 1877, p. 561).

A slight shock felt in the island of Kerrera.

7. **1877 APR. 23, *about* 3.40 *a.m.*** Cat. No. 779; Intensity 4 (D. Stevenson, *Edin. Roy. Soc. Proc.* vol. 9, 1878, pp. 403–405; *Nature*, vol. 15, 1877, p. 561).

A sharp shock felt at Lismore lighthouse, Oban and Phladda lighthouse, the two lighthouses being about 20 miles apart; a rumbling noise accompanied the shock.

8. **1880 NOV. 28, *about* 5.52 *p.m.*** Cat. No. 785; Intensity 5 or 6; Centre of disturbed area in lat. 56° 29′ N., long. 5° 37′ W.; No. of records 77 from 72 places and negative records from 20 places; Fig. 7 (C. A. Stevenson, *Edin. Roy. Soc. Proc.* vol. 11, 1881, pp. 176–187; etc.).

This earthquake is remarkable for its extensive disturbed area (the boundary of which is represented by the continuous line in Fig. 7), which includes the larger part of Scotland and the north-eastern counties of Ireland. The area is about 315 miles long from N.E. to S.W., 215 miles wide, and contains about 50,000 square miles. The boundary is, however, an isoseismal of low intensity, as many of the determining stations are lighthouses. The centre is about 5 miles N.W. of Oban, and about 1 mile W. of Lismore lighthouse.

In most places, the shock was apparently simple in character. At MacArthur's Head lighthouse, the shock was double, each part lasting a few seconds, the latter part being the stronger.

A 11 p.m., on the same day, a second and much slighter shock was felt in Inverary castle.

9. **1907** JAN. 17, 1.54 *p.m.* Cat. No. 1020; Intensity not less than 6; Centre of disturbed area in lat. 56° 26′ N., long. 5° 21′ W.; No. of records 57 from 36 places; Fig. 7 (*Geol. Mag.* vol. 5, 1908, pp. 303–304).

Fig. 7. Oban Earthquakes of 1880 Nov. 28 and 1907 Jan. 17.

The boundary of the disturbed area, which is represented by the broken-line in Fig. 7, is about 69 miles long, 58 miles wide, and contains about 3100 square miles. Its centre is about 5 miles E. of Oban, and the direction of its longer axis is about N. 35° E. The intensity of the shock was probably 6, but the shock was strong enough to cause some slight damage to buildings at Kilninver and Cuilfail Hotel (Kilmelford).

The shock consisted of two distinct parts, of which the first was regarded as the stronger at Aros and Kilmore (near Oban), and the second at Delavich and Inverinan (near Kilchrenan) and Port Appin. The mean duration of the shock was 5·5 seconds, and that of the interval between the two parts 2·3 seconds.

The sound was heard by 98 per cent. of the observers, and was compared to passing waggons, etc., in 44 per cent. of the records, thunder in 23, wind in 2, loads of stones falling in 10, the fall of a heavy body in 4, and explosions in 17, per cent.

The Great Glen fault skirts the northern shore of Loch Linnhe. To the south of the loch, there are three short faults parallel to the Great Glen fault, one of which passes through Oban and a third follows the line of Loch Etive. The positions of the epicentres of the earthquakes of 1880 and 1907 are not known with sufficient accuracy to determine which of these faults may be connected with the Oban earthquakes. Both earthquakes seem to have originated at a considerable depth.

The remaining centres (Torosay, Phladda and Dhu Heartach) lie near the line of continuation of the Great Glen fault.

TOROSAY (MULL) EARTHQUAKE

1843 NOV. 1, 11.15 *p.m.* Cat. No. 618 (Milne, reprint, p. 244).

A shock felt in the east of Mull at the manse of Torosay (Loch-Don-Head) and Lochbuy, accompanied by a deep rumbling sound.

PHLADDA EARTHQUAKE

1882 APR. 8, 7.37 *p.m.* Cat. No. 788 (C. A. Stevenson, *Edin. Roy. Soc. Proc.* vol. 15, 1889, pp. 259–260).

A shock was felt at Phladda Island lighthouse, dying away with a noise like distant thunder which lasted about 3 seconds. It was felt also in the neighbouring islands.

DHU HEARTACH EARTHQUAKE

1874 JAN. 6, 8.13 *p.m.* Cat. No. 769 (*Nature*, vol. 9, 1874, p. 242).

A tremulous motion, lasting about 2 seconds and accompanied by a rumbling noise like the booming of a cannon, at the Dhu Heartach lighthouse (about 15 miles S.W. of Iona).

CHAPTER IV

EARTHQUAKES OF COMRIE AND OF OTHER CENTRES NEAR THE HIGHLAND BORDER FAULT

THE most important earthquakes described in this chapter are those of the Comrie centre, that are evidently caused by movements of the great Highland Border fault, which traverses Scotland in a south-westerly direction from Stonehaven. The Dunkeld earthquakes may be connected with the same fault; the Dunoon and Rothesay earthquakes with a fault which appears to be associated with the Highland Border fault. It is doubtful whether the Kintyre earthquakes are connected with this fault or with a parallel fault on its north-west side.

DUNKELD EARTHQUAKES

1. **1816** AUG. 6, 10.45 *p.m.* Cat. No. 266 (Milne, vol. 31, p. 116).

A shock felt at Perth and Dunkeld and in the Carse of Gowrie and Strathearn. Its epicentre probably lies a short distance to the south-west of Dunkeld, close to the Highland Border fault.

2. **1822** APR. 22, 9.30 *a.m.* Cat. No. 312 (Milne, vol. 31, p. 119).

A shock at Dunkeld.

COMRIE EARTHQUAKES

The earthquakes of Comrie have been fortunate in their chroniclers, and it is only during the early years of their activity that our information is incomplete. In 1789, the Rev. Ralph Taylor lived at Ochtertyre, about 4 miles east of Comrie. He was at once interested in the earthquakes, which had begun in the previous autumn, and wrote a valuable account of them that was communicated to the Royal Society of Edinburgh[1]. Unfortunately, the shocks, at that time as afterwards, were almost confined to Comrie, and only a few of several hundred felt there were sensible as far as Ochtertyre. Mr Taylor left the district in February 1791, and his record closes with the slight shock of 1791 Sept. 2.

He was succeeded, as "Secretary to the Earthquakes," by the Rev. S. Gilfillan (1762–1826), Secession minister, who lived at Comrie for about thirty years. During this period he kept a close watch on the Comrie earthquakes and noted in his journal their times of occurrence and more striking phenomena. He referred to them occasionally in

[1] *Edin. Roy. Soc. Trans.* vol. 3, 1794, pp. 240–246.

papers of a religious character, but his valuable records would probably have been lost to science had not one of his sons sent extracts from his journal to Mr David Milne. The period covered by the journal is not quite certain, but it seems to have begun with the earthquake of 1792 Oct. 10, and closed with that of 1822 Apr. 13, during which time at least 68 shocks were felt.

From 1826 to 1839, the register may be again incomplete, but the period was one almost barren in earthquakes. When the activity of the Comrie centre was renewed in October 1839, the earthquakes at once came under the notice of two interested and careful observers— Mr Peter Macfarlane, postmaster of Comrie, and Mr James Drummond, a shoemaker in that village—the former attending chiefly to the shocks and the latter to the earth-sounds. In addition to keeping his register, Mr Macfarlane constructed pendulums for detecting the occurrence of earthquakes and tracing the direction of their movements. He soon realised the need of some means of representing the relative intensity of the shocks, and we are indebted to him for one of the earliest seismic scales ever devised. This scale contains ten degrees, the highest corresponding to the intensity of the principal earthquake of 1839 Oct. 23, and the lowest to that of a just sensible shock[1]. Mr Macfarlane was encouraged in his study of the earthquakes by Mr David Milne, and for several years his records were published in papers and reports written by Mr Milne. Mr Drummond's list of Comrie earthquakes is valuable as an independent record, and the portion we possess (from Oct. 1839 to Dec. 1841) makes one regret that he spent his time afterwards in speculating on the origin of the earthquakes rather than in continuing his very useful work as an observer[2].

In 1839, Mr David Milne (afterwards Milne Home) became interested in the earthquakes of Great Britain and especially in those of the Comrie district. Mr Milne belonged to that small body of

---

[1] In this chapter, estimates of the intensity given in brackets are referred to the Comrie scale; those given in the heading of the earthquake to the scale on p. 9 as usual. The scales may be roughly correlated as follows:

Comrie scale	1	2	3	4	5	6	7	8	9	10
Davison scale			3		4	5	6	7		8	

A year or two later, a similar scale to denote the intensity of the sound was added.

[2] J. Drummond, "A table of shocks of earthquakes, from September 1839 to the end of 1841, observed at Comrie, near Crieff," *Phil. Mag.* vol. 20, 1842, pp. 240–247. (The earthquake register begins in October 1839.) In 1875, Mr Drummond published a small volume on *The Comrie Earthquakes*. It does not, however, add to our knowledge of the earthquakes, and its main effect is to revive and preserve the memory of a painful controversy.

workers to whom British science is so greatly indebted—those who are relieved by ample fortune from the cares of this world. One of his first acts was to obtain the appointment of a Committee of the British Association "to register the shocks of earthquakes in Scotland and Ireland." The Committee at first consisted of six members, four (including Mr Milne as secretary and Prof. J. D. Forbes) representing Scotland, and two (including Mr James Bryce) representing Ireland. In the following year, English earthquakes were brought within the scope of the Committee, three new members (including Prof. W. Buckland) being added for the purpose. Grants were awarded (though only partly spent) of £20 in 1840 and £100 in each of the years 1841 and 1842. Four reports were presented, the first in 1841 and the last in 1844[1]; but, after the latter year, earthquakes became infrequent at Comrie, and the Committee was not re-appointed.

At the present time, the chief value of these reports lies in their inclusion of Mr Macfarlane's registers. To the Committee, however, the main work seemed to be the construction of instruments that would record the time and direction of the shocks. A few points of historical interest may be referred to. The first is Mr Milne's invention of the term *seismometer*, our oldest seismological term, and the parent no doubt of *seismograph* and possibly of *seismology*; the second Prof. J. D. Forbes' design of a seismometer[2], of which an inverted pendulum was the essential part; and the third the realisation that the position of the epicentre could be determined by the intersection of two or more lines of direction.

Towards the close of 1840, three seismometers were erected in the Comrie district, namely, an inverted pendulum 10 ft. 8 ins. long placed in the steeple of Comrie parish church; a second inverted pendulum 39 ins. long at Comrie House (about ¼ mile north of Comrie); and a common pendulum 39 ins. long at Garrichrew, close to Cluan Hill (2 miles W.N.W. of Comrie). During the following year, seven new instruments were provided, namely, four inverted pendulums erected at Crieff (6 miles E. of Comrie), St Fillans (4½ miles W.), Invergeldie (4 miles N.W.), and Kinlochmoidart near Strontian in Argyllshire; an instrument consisting of four horizontal glass tubes slightly turned up at each end and filled with mercury, placed in Mr Macfarlane's house at Comrie; and, lastly, two instruments for indicating vertical

---

[1] *Brit. Ass. Rep.* 1841, pp. 46–50; 1842, pp. 92–98; 1843, pp. 120–127; 1844, pp. 85–90.
[2] *Edin. Roy. Soc. Trans.* vol. 15, 1844, pp. 219–228.

motion. The latter consisted of a horizontal bar fixed to a solid wall by means of a strong flat watch-spring and loaded at the free end. One of them was put in Mr Macfarlane's house, the other at Kinlochmoidart. Instruments were also ordered for two other places—Ardvoirlich and Tyndrum, about 9 and 28 miles respectively to the west of Comrie—but it is uncertain whether they reached their destinations. The records of these seismometers will be noted in the following catalogue of Comrie earthquakes. The instruments, however, were not sensitive enough for the purpose. For instance, during the first half of 1841, they were displaced only twice. In the same months, 27 shocks were distinctly felt at Comrie.

While the British Association reports were appearing, Mr Milne contributed a valuable series of papers to the *Edinburgh New Philosophical Journal* under the title "Notices of earthquake-shocks felt in Great Britain, and especially in Scotland[1]." In the first of these papers, Mr Milne included the extracts from the Rev. S. Gilfillan's journal already referred to. The third and succeeding articles contained a full account of the principal Comrie earthquake of 1839 Oct. 23—the fullest yet published of any British earthquake—and of the after-shocks of the next three years. On this admirable work, it will be seen, the following catalogue of Comrie earthquakes mainly depends.

In 1844, 32 earthquakes were felt in the Comrie district. After this year, they fell off rapidly in frequency and strength, the numbers being 14 in 1845, 9 in 1846, 11 in 1847, 7 in 1848, and 7 from 1849 to 1874. During this period, Mr Macfarlane kept up his register, the results of his observations being published in M. Perrey's annual catalogues. With his death in the latter year, the Comrie earthquakes lost their last and most persevering "Secretary," who in 36 years must have felt and recorded more than 300 earthquakes, the largest number experienced by any student of British earthquakes.

A few years before Mr Macfarlane's death there were, in 1869, slight signs of a renewal of activity in the Comrie area, and this seems to have led to the appointment of a new Committee of the British Association "for the purpose of continuing the researches on earthquakes in Scotland." Dr James Bryce (1806–1877), who served on the old Committee, was secretary, and Mr Milne Home one of the members. One at least of the seismometers erected at Comrie, apparently that designed by Prof. Forbes, was still in good order,

[1] See List of Authorities: Catalogues of Earthquakes.

but it was not sufficiently sensitive for the registration of slight shocks. Failing to find a suitable instrument, the Committee adopted an apparatus suggested by Mr Mallet. In a small building at Dunearn (about half a mile from the parish church), series of upright cylinders were set up in two lines at right angles to one another. The cylinders were of gradually diminishing diameter, and it was hoped that the size of the largest overthrown would measure roughly the intensity of the shock and that the direction of the movement would be given by that in which the cylinders fell. Unfortunately, the renewal of activity in 1869 proved to be both slight and temporary, and the Mallet apparatus was never put to any practical test at Comrie. The Committee presented seven brief reports from 1870 to 1876[1], and lapsed on the death of the secretary in the latter year. The "earthquake house" at Dunearn is, however, still in existence, though much in need of repair[2].

1. **1788** NOV. 11. Cat. No. 164 (Milne, vol. 31, p. 108).

A shock felt at Comrie, Crieff, etc.

During May 1789, unusual rumbling noises were heard by some in the neighbourhood of Killin; and towards the end of August, two or three shocks are said to have been felt at Dundurn, Dunira Lodge and Comrie (Taylor, p. 241).

2. **1789** SEPT. 2, *about* 11 *p.m.* Cat. No. 166 (Taylor, p. 241).

A smart shock, felt at Comrie. At Ochtertyre, a rumbling noise was heard like that of a large table being dragged along the floor upstairs.

During the months of September and October, many feeble sounds and several smart shocks were observed in Glen Lednoch and in the neighbourhood of Comrie. At Lawers House, rumbling noises, like that of distant thunder, were heard at intervals (Taylor, p. 241; Lauder, p. 367).

3. **1789** NOV. 5, 5.57 *or* 5.58 *p.m.* Cat. No. 167; Intensity 5; Centre of isoseismal 5 in lat. 56° 22·8′ N., long. 3° 56·4′ W.; No. of records 12, from 9 places; Fig. 8 (Lauder, p. 367; Milne, vol. 31, p. 108; Taylor, p. 242).

The isoseismal 5, represented by the broken-line in Fig. 8, is $8\frac{1}{2}$ miles long, 7 miles wide, and contains about 47 square miles. Its

---

[1] *Brit. Ass. Rep.* 1870, pp. 48–49; 1871, pp. 197–198; 1872, pp. 240–241; 1873, pp. 194–197; 1874, p. 241; 1875, pp. 64–65; 1876, p. 74.
[2] *Nature*, vol. 87, 1911, p. 415.

centre lies about ¼ mile south-east of Lawers House, or about 1 mile north-west of the fault-line. The shock was, however, felt for several miles beyond this isoseismal; for instance, at Ardoch about 5 miles to the south. The disturbed area may thus have contained as much as 250 square miles.

At Lawers, the shock was as if the foundations of the house had been violently struck with a great mallet. At Ochtertyre, only a

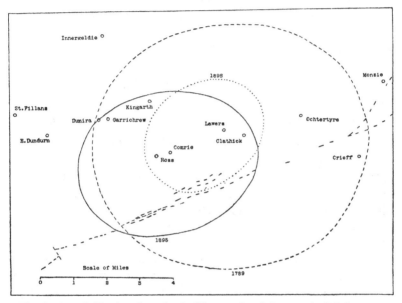

Fig. 8. Comrie Earthquakes of 1789 Nov. 5, 1895 July 12 and 1898 Aug. 22.

tremulous motion was felt. The sound was compared to the rapid passing of a heavily loaded waggon on a hard road, to the discharge of distant artillery, etc. On the Loch of Monivaird, near Ochtertyre, the wild fowl were observed to scream and flutter.

This shock was succeeded by a number of slight earth-sounds, not less than 30 being counted within the first two hours. They were not, however, heard to the east, north and west of Comrie (Taylor, p. 243).

4. 1789 NOV. 10, *about 3 p.m.* Cat. No. 168; Intensity 5 (Milne, vol. 31, p. 108; Taylor, p. 243).

The shock is described by Taylor as of approximately the same direction, intensity and extent of disturbed area as that of Nov. 5. It was market-day at Comrie. "The hardware exposed for sale in the shops and booths shook and clattered, and the horses crowded together with signs of unusual terror."

5. **1789 NOV. 11, *morning.*** Cat. No. 169 (Milne, vol. 31, pp. 108–109; Lauder, p. 367).

The shock was much stronger than that of Nov. 5. It was felt towards the west at Killin (14 miles W.N.W. of Comrie) and Ardvoirlich (9 miles W. of Comrie); and towards the east at Glenalmond (11 miles E.N.E. of Comrie). The disturbed area must therefore have been at least 25 miles in length from east to west. The shock was strong enough to shiver the ice on a sheet of water near Lawers House.

6. **1789 DEC. 29, 1 *p.m.*** Cat. No. 170 (Milne, vol. 31, p. 109; Taylor, p. 243).

A rather strong shock, felt at Comrie, Ochtertyre and Crieff.

7. **1791 SEPT. 2, 5.5 *p.m.*** Cat. No. 172 (Taylor, p. 246).

A slight shock, felt at Ochtertyre.

Mr Taylor left Ochtertyre in February 1791, and from that time until the end of January 1793, when his paper was finished, he only visited the district occasionally. He notes, however, that, after the slight earthquake of 1791 Sept. 2, "shocks have been observed at different times till within these few weeks past" (Jan. 1793). We have records of three shocks—on Oct. 10, between Oct. 12 and Nov. 18, and Nov. 18, 1792. They were evidently of greater strength than usual, if we may judge from a letter written from Comrie on 1792 Nov. 30. "We have of late," the writer says, "been greatly alarmed with several very severe shocks of an earthquake. They were more sensible and alarming than any felt formerly, and the noise attending them was uncommonly loud and tremulous....The houses were greatly shaken." "It must be no small force," he adds, "that can shake a country to the extent of between twenty and thirty miles" (quoted by W. Creech in *Edinburgh Fugitive Pieces*, 1815, pp. 126–127). Sir T. Dick Lauder also notes that in 1792, the neighbourhood of Comrie was at different times disturbed by several smart shocks, much more distinctly sensible and alarming than any previously experienced in that district (*Ann. of Phil.* vol. 8, 1816, p. 367).

8. **1792 OCT. 10, *morning.*** Cat. No. 175 (Milne, vol. 31, p. 109).

A shock, felt at Comrie and Crieff, the houses being much shaken and many people awakened; a very loud noise heard.

9. **1792**, *between* OCT. 12 *and* NOV. 18, 7.30 *p.m.* Cat. No. 176 (Milne, vol. 31, p. 109).

A smart shock at Comrie.

10. **1792** NOV. 18, 11 *a.m.* Cat. No. 178 (Milne, vol. 31, p. 109).

An alarming shock at Comrie.

11. **1793** FEB. 3, *evening.* Cat. No. 179 (Milne, vol. 31, p. 109).

Two violent shocks at Comrie.

12. **1793** FEB. 25, 10.30 *p.m.* Cat. No. 180 (Milne, vol. 31, p. 109).

A great noise, accompanied by a slight shock, was heard at Comrie.

13. **1793** MAY —. Cat. No. 181 (Milne, vol. 31, p. 109).

A shock felt at Comrie, the movement being horizontal from north to south. Dykes [stone walls without mortar] were thrown down.

14. **1794** MAY 2, 4 *p.m.* Cat. No. 183 (Milne, vol. 31, p. 110).

A very severe shock at Comrie, with loud noise. Dogs barked, and some cattle ran as they do during thunder.

15. **1794** SEPT. 28, 3 *p.m.* Cat. No. 184 (Milne, vol. 31, p. 110).

Earthquake at Comrie.

16. **1794** OCT. 1, 5.30 *p.m.* Cat. No. 185 (Milne, vol. 31, p. 110).

A loud noise heard at Comrie, like that which usually accompanies shocks.

17. **1794** OCT. 2, 11 *p.m.* Cat. No. 186 (Milne, vol. 31, p. 110).

A shock felt at Comrie.

18. **1794** OCT. 18, 1 *a.m.* Cat. No. 187 (Milne, vol. 31, p. 110).

A violent shock, with very loud noise, at Comrie.

19. **1794** DEC. 3, 5.30 *p.m.* Cat. No. 188 (Milne, vol. 31, p. 110).

A severe shock, with loud rumbling noise, at Comrie.

20. **1794** DEC. 4, 10 *p.m.* Cat. No. 189 (Milne, vol. 31, p. 110).

A shock felt at Comrie.

21. **1794** DEC. 25, 1.15 *p.m.* Cat. No. 190 (Milne, vol. 31, p. 110).

A severe shock and great noise at Comrie.

22. **1794** DEC. 30, 8 *p.m.* Cat. No. 191 (Milne, vol. 31, p. 110).

A shock felt at Comrie.

23. **1795** JAN. 2, 1.50 *a.m.* Cat. No. 192 (Milne, vol. 31, p. 110; *Gent. Mag.* vol. 65, 1795, p. 74; *Edin. Mag.*).

A smart shock and very loud noise at Comrie. The movement in this earthquake was vertical, though apparently horizontal in previous shocks. The trembling noise, which did not last so long as usual, seemed to begin directly below the village and ended suddenly. The shock was felt to a distance of 20 miles around Comrie.

24. **1795** JAN. 22, 2.40 *p.m.* Cat. No. 193 (Milne, vol. 31, p. 110).

A shock, with long continuing noise, at Comrie.

25. **1795** MAR. 12, 11 *p.m.* Cat. No. 194 (Milne, vol. 31, p. 110; *Edin. Mag.*).

Two very smart shocks at Comrie, separated by an interval of 3 seconds. The first shock was the strongest so far known at Comrie, and was felt to the west as far as Tyndrum (28 miles). The noise that preceded, accompanied and followed the shocks was unusually loud. Cattle started in their stalls and dogs ran about alarmed. The disturbed area probably contained more than 600 square miles.

During the two hours after this earthquake, many rumbling noises were heard at Comrie.

26. **1795** MAR. 13. Cat. No. 195 (Milne, vol. 31, p. 110).

Rumbling noises heard at Comrie.

27. **1795** MAR. 16. Cat. No. 196 (Milne, vol. 31, p. 110).

Rumbling noises heard at Comrie.

28. **1795** MAR. 21. Cat. No. 197 (Milne, vol. 31, p. 110).

Rumbling noises heard at Comrie during the morning and preceding night.

29. **1795** MAR. 23. Cat. No. 198 (Milne, vol. 31, p. 110).

Rumbling noises heard at Comrie in the evening.

30. **1795** MAR. 27, 8 *p.m.* Cat. No. 199 (Milne, vol. 31, p. 110).

Rumbling noises heard at Comrie.

31. **1795** APR. 8, 3 *p.m.* Cat. No. 200 (Milne, vol. 31, p. 111).

A shock and rumbling noise at Comrie.

32. **1795** APR. 25, *about* 6 *p.m.* Cat. No. 201 (Milne, vol. 31, p. 111; *Edin. Mag.*).

A smart shock at Comrie, from north-west to south-east, accompanied by a very loud noise, closely resembling thunder but of longer duration than an ordinary peal.

33. **1795** JUNE 19. Cat. No. 202 (Milne, vol. 31, p. 111).
A rumbling noise at Comrie.

34. **1795** JULY 14, *evening*. Cat. No. 203 (Milne, vol. 31, p. 111).
Two or three rumbling noises at Comrie.

35. **1795** JULY 15, *evening*. Cat. No. 204 (Milne, vol. 31, p. 111).
A rumbling noise at Comrie.

36. **1795** JULY 25, 6.30 *p.m.* Cat. No. 205 (Milne, vol. 31, p. 111).
A smart shock and loud noise at Comrie.

37. **1795** SEPT. 1, *at night*. Cat. No. 206 (Milne, vol. 31, p. 111).
Earth-sounds heard at Comrie.

38. **1795** SEPT. 4, *between 2 and 3 p.m.* Cat. No. 207 (Milne, vol. 31, p. 111).
A shock, with some accompanying sounds, at Comrie.

39. **1795** OCT. 4. Cat. No. 208 (Milne, vol. 31, p. 111).
A shock felt at Comrie.

40. **1796** JAN., *last week of*. Cat. No. 210 (Milne, vol. 31, p. 112).
Some earth-sounds heard at Comrie.

41. **1796** MAR. 16, *morning*. Cat. No. 211 (Milne, vol. 31, p. 112).
A shock felt at Comrie.

42. **1797** FEB. 8, *about 7 p.m.* Cat. No. 212 (Milne, vol. 31, p. 112).
A slight shock and loud noise at Comrie.

43. **1797** FEB. 10, 0.20 *a.m.* Cat. No. 213 (Milne, vol. 31, p. 112; *Edin. Mag.*).
A very smart shock at Comrie, the movement being horizontal, accompanied by a loud noise which lasted about half a minute.

44. **1797** FEB. 10, 6 *a.m.* Cat. No. 214 (Milne, vol. 31, p. 112; *Edin. Mag.*).
A very smart shock at Comrie, not quite so strong as the preceding, and accompanied by a loud noise.

45. **1797** FEB. 17, *afternoon*. Cat. No. 215 (Milne, vol. 31, p. 112).
A slight shock, with noise, at Comrie.

46. **1797** MAY 12, *night*. Cat. No. 216 (Milne, vol. 31, p. 112).
Two shocks at Comrie.

47. **1797** AUG. 24, *night*. Cat. No. 217 (Milne, vol. 31, p. 112).
A shock at Comrie. Mr Milne adds that this shock was felt in Argyllshire, but, if so, the latter shock was probably of different origin.

48. **1797** NOV. 19, 11 *a.m.* Cat. No. 218 (Milne, vol. 31, p. 113).
A shock, with long and loud noise, at Comrie.

49. **1797** DEC. 19, 5 *a.m.* Cat. No. 219 (Milne, vol. 31, p. 113).
A slight shock, with loud noise, at Comrie.

50. **1798** APR. 19, *morning*. Cat. No. 220 (Milne, vol. 31, p. 113).
Three shocks at Comrie during the morning, one very smart.

51. **1798** MAY 6, 10 *a.m.* Cat. No. 221 (Milne, vol. 31, p. 113).
A very smart shock and uncommon noise at Comrie.

52. **1799** JAN. 17. Cat. No. 222 (Lauder, p. 367; Milne, vol. 31, p. 113).
A shock, with motion from west to east, and lasting about 2 seconds, at Comrie. The noise was of longer duration. This shock and the next are described as more violent and alarming than any felt so far.

53. **1799** FEB. 24, 1.50 *p.m.* Cat. No. 223 (Lauder, p. 367; Milne, vol. 31, p. 113; etc.).
A very smart shock at Comrie, with motion from west to east, and lasting about 2 seconds. It was accompanied and followed by a loud noise.

54. **1799**, *between* FEB. 24 *and* MAR. 3. Cat. No. 224 (Milne, vol. 31, p. 113).
Two loud rumbling noises at Comrie during this week.

55. **1800** DEC. 8, 9 *a.m.* Cat. No. 225 (Milne, vol. 31, p. 113).
A violent shock at Comrie, the noise being very loud and long continued.

56. **1801** JAN. 11, 7 *a.m.* Cat. No. 226 (Milne, vol. 31, p. 113).
A violent shock, with loud noise, at Comrie.

57. **1801** JAN. 11, *after* 11 *a.m.* Cat. No. 227 (Milne, vol. 31, p. 113).

A violent shock, with loud noise, at Comrie. The shock was felt at Lochearnhead (11 miles W. of Comrie), Killin (14 miles W.N.W.), Tyndrum (28 miles W.), Glenfinlas (17 miles S.W.), Perth (21 miles E.) and Edinburgh (44 miles S.E.). The disturbed area thus probably contained about 1500 square miles.

58. **1801** SEPT. 6, 1.15 *p.m.* Cat. No. 229 (Milne, vol. 31, p. 113; *Edin. Mag.*).

A strong shock, with very loud noise, at Comrie. It was followed by a slight shock and several hollow sounds.

59. **1801** SEPT. 7, 4 *a.m.* Cat. No. 230 (Milne, vol. 31, p. 114; *Phil. Mag.* vol. 10, 1801, pp. 368–370; *Edin. Mag.*).

A shock felt at Comrie, and Grangemouth (27 miles S.E. of Comrie).

60. **1801** SEPT. 7, 6 *a.m.* Cat. No. 231; Intensity 7; Epicentre close to Comrie; No. of records 69 from 36 places (Lauder, p. 368; Milne, vol. 31, pp. 113–114; *Phil. Mag.* vol. 10, 1801, pp. 368–370; *Edin. Adver.* Sept. 8, 1801; *Edin. Mag.*).

This was the strongest of all the Comrie earthquakes with the exception of those on 1839 Oct. 23 and 1841 July 30. It was felt at Tyndrum (28 miles W. of Comrie); between S.S.W. and S. at Glasgow (38 miles), Renfrew (39 miles), Paisley (41 miles), Hamilton (42 miles) and Ayr (68 miles); towards the S.E. at Edinburgh and Leith (44 miles) and Portobello (46 miles). From the district to the north, observations are almost entirely wanting, the most distant place being Dunkeld (20 miles N.E.). If we assume that the boundary of the disturbed area was a circle with its centre at Comrie and passing through Ayr, the area included must have been about 3600 square miles. At Comrie slates fell from some houses and portions of stone dykes fell—an amount of damage not entitling the intensity to be reckoned above 7. The gable of an old barn near Edinburgh fell down and two reapers were killed and a third buried; and a large tenement in Paterson's Court at Edinburgh sank so much that it was abandoned by its inhabitants and was ordered by the magistrates to be pulled down. It is difficult to believe, however, that the earthquake was in any way responsible for this damage.

At a few places, the shock is described as double. At Crieff, there were two strong shocks separated by a very brief interval. Two slight shocks were felt at Dunfermline and even as far as Ayr. The sound, which is described as resembling thunder or the fall of a load of stones, was heard at Ayr, though to a few observers (as at Dunfermline, Leith and Edinburgh) it was inaudible.

During the next ten or fifteen minutes, the shock was followed by about 20 earth-sounds.

61. **1801** SEPT. 18, *night*. Cat. No. 232 (Milne, vol. 31, p. 114).

A shock felt at Comrie. Mr Milne adds that several rumbles were heard during the preceding part of the week, that is, from Sept. 13 to 18.

62. **1801** SEPT. 25, 10 *a.m.* Cat. No. 233 (Milne, vol. 31, p. 114).

A slight shock at Comrie.

63. **1802** JUNE 10, 11 *p.m.* Cat. No. 234 (Milne, vol. 31, p. 114).

A slight shock and loud noise at Comrie.

64. **1802** OCT. 3–8. Cat. No. 237 (Milne, vol. 31, p. 114).

Some slight shocks at Comrie.

65. **1804** MAR. 4, *about* 4 *p.m.* Cat. No. 239 (*Edin. Adver.*).

A slight shock and several earth-sounds at Comrie.

66. **1804** MAR. 11, *about* 11 *a.m.* Cat. No. 240 (*Edin. Adver.*).

A more violent shock at Comrie, attended by an interrupted sound, unlike the usual sound.

67. **1804** MAR. 14, *about* 2 *a.m.* Cat. No. 241 (*Edin. Adver.*).

A rather smart shock at Comrie.

68. **1806** MAY 29. Cat. No. 243 (Milne, vol. 31, p. 114).

Two smart shocks at Comrie, with very loud noise.

69. **1809** JAN. 9, *about* 5.30 *a.m.* Cat. No. 244 (Milne, vol. 31, p. 114; etc.).

A violent shock at Comrie, with very loud and prolonged noise; the strongest shock since 1801 Sept. 7 (no. 60).

70. **1810** NOV. 14–15. Cat. No. 254 (Milne, vol. 31, p. 115).

A smart shock at Comrie, with loud noise. The month during which this earthquake occurred is doubtful, as the entry (as given above) in Mr Milne's catalogue is immediately before that of a shock dated Oct. 12.

71. **1812** SEPT. 10, *night*. Cat. No. 260 (Milne, vol. 31, p. 115).

A shock at Comrie.

72. **1814** NOV. 20–25. Cat. No. 263 (Milne, vol. 31, p. 115).

Some slight shocks were felt at Comrie during this interval.

73. **1818** JUNE 19. Cat. No. 287 (Milne, vol. 31, p. 118).

Two shocks, at an interval of a minute, at Comrie.

74. **1819** NOV. 28, 1.30 *a.m.* Cat. No. 295; Intensity 4 (Milne, vol. 31, p. 118; *Ann. de Chim. et de Phys.* vol. 12, 1819, p. 428; *Journ. of Sci.* vol. 9, 1820, p. 205; *Phil. Mag.* vol. 54, 1819, pp. 467–468).

A stronger shock than usual at Comrie, lasting for nearly 10 seconds, and accompanied by a sound like distant thunder. It was felt for several miles around the village.

Earthquakes were felt in the neighbourhood of Comrie on 1820 Dec. 25 and 1821 Oct. 22 and 23, but it is not certain that they had any connexion with the Comrie series.

75. **1822** APR. 13, *about* 9.30 *a.m.* Cat. No. 311 (Milne, vol. 31, p. 119; *Ann. de Chim. et de Phys.* vol. 33, 1826, p. 406; *Phil. Mag.* vol. 59, 1822, p. 398).

A strong shock at Comrie, the strongest felt for twenty years, accompanied by two loud reports; a number of pots and pans were thrown down by the shock; the noise is said to have lasted half a minute and to have been louder than any thunder.

76. **1822** MAY 18, *between* 9 *and* 10 *a.m.* Cat. No. 313; Intensity 6 (*Phil. Mag.* vol. 59, 1822, p. 398).

A strong shock felt at Crieff and in the surrounding country.

77. **1824** AUG. 8, *morning.* Cat. No. 315; Intensity 4 (*Phil. Mag.* vol. 64, 1824, p. 233).

A smart shock at Comrie, accompanied by a noise like that of a heavy waggon being driven rapidly over a paved causeway.

78. **1828** DEC. 9. Cat. No. 325 (Milne, vol. 31, p. 119).

A shock at Comrie. According to Mr Milne, the third within the last three months.

79. **1834** AUG. 25. Cat. No. 335 (*New Stat. Acc. of Scotland*, vol. 10 Perth, 1845, p. 266).

A shock at Comrie.

80. **1838** NOV. —. Cat. No. 349 (Milne, vol. 31, p. 121).

A shock at Crieff.

81. **1839** MAY 24, 2 *a.m.* Cat. No. 351 (Milne, vol. 31, p. 122).

Two shocks at Crieff, each of which lasted 2 seconds, accompanied by a noise of much longer duration.

1839 Oct. 2, Perrey (p. 156) records an earthquake in Scotland, with its epicentre probably at Comrie.

82. **1839** OCT. 3, 3.30 *p.m.* Cat. No. 355 (Milne, vol. 32, p. 109).

A feeble shock (int. 2) at Comrie. At Blairmore (10½ miles E. of Comrie), an observer, walking, felt a movement of the ground at about 3 p.m. as if a heavy carriage were passing at a short distance; it was preceded by a sound like distant thunder.

1839 Oct. 4, an earth-sound heard by Mr Drummond at Comrie.

83. **1839** OCT. 7, 4.30 *a.m.* Cat. No. 356 (Milne, vol. 32, p. 108).

A very slight shock at Comrie.

84. **1839** OCT. 7, 11.30 *a.m.* Cat. No. 357 (Milne, vol. 32, p. 108).

A very slight shock at Comrie. At Cultoquehey (2 miles N.E. of Comrie), a rumbling noise without accompanying shock.

85. **1839** OCT. 7, 3 *p.m.* Cat. No. 358 (Milne, vol. 32, p. 110).

A very slight shock at Comrie; at Cultoquehey, a decided shock, accompanied by a noise like thunder.

86. **1839** OCT. 7, 4 *p.m.* Cat. No. 359 (Milne, vol. 32, p. 110).

A very slight shock at Comrie; at Cultoquehey, a slight noise without attendant shock.

87. **1839** OCT. 7, 5.30 *p.m.* Cat. No. 360 (Milne, vol. 32, p. 107).

A slight shock (int. 2) at Comrie.

On this day (Oct. 7), one earth-sound was heard at Comrie by Mr Drummond and another on Oct. 8.

88. **1839** OCT. 9, 2.15 *a.m.* Cat. No. 361 (Milne, vol. 32, p. 110).

A very feeble shock felt at Comrie. At Blairmore (10½ miles E. of Comrie), a tremulous movement was felt; the sound was louder and harsher than usual, like a heavy gust of wind.

89. **1839** OCT. 10, 4.15 *a.m.* Cat. No. 362 (Milne, vol. 32, pp. 110–111).

A shock (int. 5) at Comrie; felt also at Dunira (2 miles W.N.W.) and at Strathallan (10 miles S.E.). At Ardvoirlich (9 miles W.) and Monzie (7 miles E.N.E.), a sound was heard but no shock was felt.

The disturbed area contains about 80 square miles. In addition to this earthquake, Mr Drummond records one earth-sound on this day.

90. **1839** OCT. 11, 10 *p.m.* Cat. No. 363 (Milne, vol. 32, p. 107).

A very feeble shock (int. 1) at Comrie. Mr Drummond records one earth-sound.

On Oct. 12, Mr Macfarlane records 10 shocks at Comrie, of which the strongest occurred at 1 p.m., 3 p.m. and 4 p.m. According to Mr Drummond, there were on this day 5 earth-sounds and 5 shocks at Comrie, the times of the latter being 1 p.m., 2.30 p.m., 3 p.m., 3.30 p.m. and 4.30 p.m.

91. **1839** OCT. 12, 1 *p.m.* Cat. No. 364; Intensity 5 (Milne, vol. 32, pp. 111–114).

The shock was felt at Dunira, Comrie (int. 6), Crieff and Kincardine ($\frac{1}{2}$ mile E. of Crieff). The noise at Comrie was said to be louder than the loudest thunder. The shock was felt to a distance of 14 miles from Comrie.

92. **1839** OCT. 12, 2.30 *p.m.* Cat. No. 365 (Drummond, p. 242).

A shock felt at Comrie and to a distance of 8 miles.

93. **1839** OCT. 12, 3 *p.m.* Cat. No. 366; Intensity 7 (Milne, vol. 32, pp. 111–116).

The shock was felt at Dunira, Comrie (where some slates fell from houses and some loose stones from walls), between Comrie and Lawers, at Crieff, Monzie, Glenshee (about 30 miles N.E. of Comrie) and Kenmore (14 miles N. of Comrie). The disturbed area contains about 700 square miles. The shock was preceded and accompanied by two loud noises of long duration. The trees on the road between Comrie and Lawers were at the time wet with rain, which was thrown off by the shock. A pond in Monzie Park was ruffled as if by the wind, though it was perfectly calm at the time.

94. **1839** OCT. 12, 4 *p.m.* Cat. No. 367 (Milne, vol. 32, p. 107).

A shock (int. 6) at Comrie. This is probably the shock recorded by Mr Drummond at 3.30 p.m., which, he states, was felt to a distance of 12 miles.

95. **1839** OCT. 12, *about* 4.30 *p.m.* Cat. No. 368 (Drummond, p. 242).

A tremor felt at Comrie and to a distance of 6 miles.

96. **1839** OCT. 13, 9 *a.m.* Cat. No. 369 (Milne, vol. 32, p. 107).

A shock (int. 3) at Comrie.

97. **1839** OCT. 13, 11 *a.m.* Cat. No. 370 (Milne, vol. 32, p. 107).

A shock (int. 2) at Comrie.

98. **1839** OCT. 13, 11.30 *a.m.* Cat. No. 371 (Milne, vol. 32, p. 107).

A shock (int. 3) at Comrie. Mr Drummond records only 2 earth-sounds on this day.

99. **1839** OCT. 14, 2.45 *p.m.* Cat. No. 372; Intensity 5 (Milne, vol. 32, pp. 116–118).

The shock was felt at Dunira (2 miles W.N.W. of Comrie), Tully-banocher ($\frac{1}{2}$ mile W.), Monzie (2 miles N. of Crieff), Glenalmond (6 miles N.E. of Crieff), and Kenmore (14 miles N. of Comrie). The disturbed area must thus contain about 150 square miles. At Comrie, the movement seems to have been nearly vertical. At Dunira, soot and lime fell down several chimneys. The sound was evidently very loud. Mr Macfarlane records 4 shocks on this day, of which the strongest occurred at 2.45 p.m.; Mr Drummond heard 3 earth-sounds and felt one vertical shock.

100. **1839** OCT. 15, 3 *p.m.* Cat. No. 373 (Milne, vol. 32, p. 107).

A feeble shock at Comrie (int. 2). Mr Drummond heard 2 earth-sounds on this day.

101. **1839** OCT. 16, 2.30 *a.m.* Cat. No. 374; Intensity 5 (Milne, vol. 32, pp. 118–119).

The shock was felt at Ardvoirlich (9 miles W. of Comrie), Dunira, Comrie, Kenmore (14 miles N.) and probably at Blairhill (about 21 miles S.E.). The area disturbed probably exceeded 150 square miles. At Dunira, the sound resembled thunder and was as loud as a severe clap.

102. **1839** OCT. 16, 6 *a.m.* Cat. No. 375 (Milne, vol. 32, p. 107).

A comparatively slight shock at Dunira and Comrie (int. 2).

103. **1839** OCT. 16, 5.45 *p.m.* Cat. No. 376 (Milne, vol. 32, p. 107).

A shock (int. 5) at Comrie. On this day, Mr Drummond records 2 earth-sounds, and shocks at 2.30 a.m. and 5.45 p.m.

104. **1839** OCT. 17. Cat. No. 377 (Milne, vol. 32, p. 107).

Four very slight shocks at Comrie (the times not given), and 4 earth-sounds heard by Mr Drummond.

105. **1839** OCT. 18. Cat. No. 378 (Milne, vol. 32, p. 107).

Three very feeble shocks (int. 1) at Comrie. According to Mr Drummond, 3 earth-sounds were heard on this day at Comrie.

106. **1839** OCT. 19–22. Cat. No. 379 (Milne, vol. 32, p. 107).

Six very feeble shocks, none of higher intensity than 1 (Comrie scale), felt at Comrie on these days. According to Mr Drummond, 2 earthquakes were heard at Comrie on Oct. 19, 2 on Oct. 20, and 1 on Oct. 21.

107. **1839** OCT. 23, 10.15 *p.m.* Cat. No. 380; Intensity 8; Epicentre near lat. 56° 2′ N., long. 3° 59′ W.; No. of records 112 from 91 places; Fig. 9 (Milne, vol. 32, pp. 119–127, 362–378; vol. 33, pp. 372–388; vol. 34, pp. 85–106; vol. 35, pp. 137–159; *New Stat. Account of Scotland*, vol. 10, 1847, pp. 707, 902 n.; vol. 12, 1845, p. 559; etc.).

This earthquake, which is the strongest known in the Comrie district, disturbed the greater part of Scotland. The three northern counties, the Western Isles, and portions of the southern counties were the only regions unaffected. Several chimneys were shattered and chimney-pots thrown down, some walls were cracked, or dry stone-dykes were overthrown at Dundurn (4 miles W. of Comrie), Comrie, Lawers (1½ miles E.), Clathick (2 miles E.), Monzievaird (3 miles E.) and Crieff (6 miles E.). These six places lie in the valley of the Earn, which here runs nearly east and west. The total length of the isoseismal 8 cannot thus be less than 10 miles. The mid-point between Dundurn and Crieff is 1 mile east of Comrie, but it is probable that the epicentre lies a short distance to the north of this point. Mr Milne places it about 1 mile north-west of Comrie. It was clearly in the immediate neighbourhood of the village.

On the map (Fig. 9) are shown two isoseismal lines, those of intensities 5 and 4. The isoseismal 5, which is somewhat roughly drawn, is 150 miles long from north-east to south-west, 105 miles wide, and about 12,000 square miles in area. The isoseismal 4 forms the boundary of the disturbed area, and is 210 miles long from north-east to south-west, 160 miles wide, and contains about 26,500 square miles.

At most places, the shock consisted of two or three strong vibrations followed by a tremulous motion. At Comrie, there were three such vibrations in a nearly vertical direction, while, at greater distances (as, for instance, near Cupar and at Edinburgh), two or three undulations were felt. A tremulous motion at some places preceded the principal vibrations, and, in a few cases, was separated from them by a brief interval, so that the shock appeared double, the second part being the stronger. The average duration of the shock was about 3 seconds.

The sound-area coincided with the disturbed area approximately, the sound being heard at most of the places which determine the boundary of the latter. The sound was compared to passing waggons etc., in 41 per cent. of the records, thunder in 22, wind in 14, loads of stones falling in 5, the fall of a heavy body in 4, explosions in 9, and miscellaneous sounds in 5, per cent.

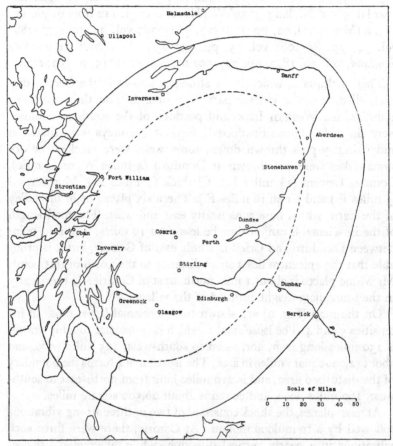

Fig. 9. Comrie Earthquake of 1839 Oct. 23.

In the neighbourhood of Amulree (about 10 miles N. of Comrie), several fissures were formed in the ground, running generally in the N.N.E. direction. One of these was about 200 yards and another about 50 yards long. On the side of the Tay opposite to Perth (21 miles E. of Comrie), a crack was formed during the night of Oct. 23, on the side of the road above a steep bank. Two days later, a slice of the road, about 25 yards in length, along the line of the crack slipped down.

The principal earthquake was followed by an unusually large number of after-shocks and earth-sounds. The alarm caused by the earthquake was so great that many persons sat up all night. Mr Macfarlane, who attended to shocks only, recorded three on Oct. 23 —at 10.30 p.m., 11 p.m. and midnight. Mr Drummond notes the occurrence of 8 earth-sounds and 3 shocks on that day—the latter at 10.15 p.m., 11 p.m. and midnight. Mr D. Graham, schoolmaster of Dunira and Comrie, sat up all night with friends and counted, I am informed, 53 shocks on Oct. 23. At Leichdin ($\frac{1}{4}$ mile N.W. of Comrie), earth-sounds were heard 12 times before midnight, being only once accompanied by a tremor. At Tullybanocher ($\frac{1}{2}$ mile W. of Comrie), the principal shock was followed in rapid succession by 10 other shocks, 7 within the first twenty minutes, the last soon after midnight. They all began with a sound like an explosion, which was followed immediately by the shaking, the noise lasting twice as long as the shock. At Lawers, one observer counted 12 or 15 earth-sounds during the night, none of which was accompanied by any tremor.

108. **1839** OCT. 23, *about* 10.45 *p.m.* Cat. No. 381; No. of records 14 from 12 places (Milne, vol. 32, pp. 107, 364, 365, 372, 375, 376, 387; vol. 34, p. 92; reprint, p. 156).

The shock was felt at Comrie (int. 6), Leichdin, Lawers, Clathick and Crieff; also at Glenshee (30 miles N.E. of Comrie), Dunning (16 miles E.S.E.) and Dollar (20 miles S.E.). The area disturbed was probably about 700 square miles. The sound was also heard at Dull (17 miles N. of Comrie), Kingussie (about 50 miles N.) and Preston-pans (about 48 miles S.E.). The shock was as a rule a mere tremor, much weaker than the principal shock at 10.15.

109. **1839** OCT. 23–24, *about midnight.* Cat. No. 382 (Milne, vol. 32, p. 107; etc.).

A slight shock felt at Comrie.

110. **1839** OCT. 24, 5 *a.m.* Cat. No. 383 (Drummond, p. 242).

A tremor felt at Comrie and to a distance of 6 miles. Mr Macfarlane records the occurrence (without stating the times) of 12 very slight shocks, only one of which attained the intensity 2 of the Comrie scale. Mr Drummond observed 7 earth-sounds and the tremor at 5 a.m.

111. **1839** OCT. 25, 7 *p.m.* Cat. No. 384 (Drummond, p. 242).

A tremor felt at Comrie and to a distance of 8 miles. Mr Macfarlane states that there were 2 very slight shocks on this day (each of int. 1). Mr Drummond records 2 earth-sounds and the tremor at 7 p.m.

112. **1839** OCT. 26, 7 *p.m.* Cat. No. 385 (Milne, vol. 32, p. 107).

A shock (int. 4) at Comrie, and felt to a distance of 8 miles.

113. **1839** OCT. 26, *between 7 and* 11.30 *p.m.* Cat. No. 386 (Milne, vol. 32, p. 107).

A very slight shock (int. 1) at Comrie.

114. **1839** OCT. 26, 11.30 *p.m.* Cat. No. 387 (Milne, vol. 32, p. 107).

A slight shock (int. 2) at Comrie. On this day, Mr Drummond observed 3 earth-sounds and the tremor at 7 p.m.

115. **1839** OCT. 27. Cat. No. 388 (Milne, vol. 32, p. 107).

Two very slight shocks (int. 1) at Comrie, the times of which are not given. Mr Drummond heard 2 earth-sounds.

116. **1839** OCT. 28, 11.30 *p.m.* Cat. No. 389 (Milne, vol. 32, p. 107).

A shock (int. 4) at Comrie. Mr Macfarlane felt another shock (int. 1) earlier in the day. Mr Drummond heard 3 earth-sounds.

117. **1839** OCT. 29, 2.45 *a.m.* Cat. No. 390 (Milne, vol. 32, p. 107).

A shock (int. 4) at Comrie. Mr Drummond records 2 earth-sounds.

118. **1839** OCT. 30. Cat. No. 391 (Milne, vol. 32, p. 107).

A very slight shock (int. 1) at Comrie. Mr Drummond heard 2 earth-sounds.

119. **1839** OCT. 31. Cat. No. 392 (Milne, vol. 32, p. 107).

Two very slight shocks (int. 1) at Comrie. Mr Drummond heard 2 earth-sounds.

120. **1839** NOV. 1, 2.30 *p.m.* Cat. No. 393 (Milne, vol. 32, p. 107).

A very slight shock (int. 1) at Comrie. Mr Drummond records only 1 earth-sound.

During the next week (Nov. 2–8), slight tremors and earth-sounds continued to be observed at Comrie. Mr Macfarlane (without giving times) records the occurrence of 9 very slight shocks, the three stronger being of intensity 1 (Comrie scale). Mr Drummond, in addition to the tremor on Nov. 7, heard 10 earth-sounds—2 on Nov. 2, 1 on Nov. 3, 1 on Nov. 4, 2 on Nov. 5, 1 on Nov. 6, 1 on Nov. 7, and 2 on Nov. 8.

121. **1839** NOV. 7, 4 *a.m.* Cat. No. 394 (Drummond, p. 243).

A slight tremor at Comrie, felt to a distance of 8 miles.

On Nov. 8, Mr Drummond heard 2 earth-sounds at Comrie.

122. **1839** NOV. 9. Cat. No. 395 (Milne, vol. 32, p. 107).

Mr Macfarlane felt 6 very slight shocks at Comrie (the three stronger of int. 1). Mr Drummond heard 1 earth-sound.

During the nine days, Nov. 10–18, no shock is recorded by Mr Macfarlane; Mr Drummond heard 6 earth-sounds—1 on Nov. 10, 1 on Nov. 12, 2 on Nov. 15, 1 on Nov. 17, and 1 on Nov. 18.

123. **1839**, *between* NOV. 19 *and* 28, 4 *a.m.* Cat. No. 396 (Milne, vol. 32, p. 107).

A very slight shock (int. 1) at Comrie, the particular day not being mentioned. During these ten days, Mr Drummond heard 4 earth-sounds, one on each of the days Nov. 23, 25, 26 and 28.

124. **1839** NOV. 29. Cat. No. 397 (Milne, vol. 32, p. 107).

A shock (int. 1) at Comrie; 2 earth-sounds heard by Mr Drummond.

125. **1839** NOV. 30. Cat. No. 398 (Milne, vol. 32, p. 107).

Three shocks (ints. 2, 1 and 1) at Comrie, no times being given; 3 earth-sounds heard by Mr Drummond.

On Dec. 1 Mr Drummond heard 1 earth-sound at Comrie.

126. **1839** DEC. 2. Cat. No. 399 (Milne, vol. 32, p. 107).

A very slight shock (int. 1) at Comrie; 2 earth-sounds heard by Mr Drummond.

127. **1839** DEC. 3. Cat. No. 400 (Milne, vol. 32, p. 107).

A very slight shock (int. 1) at Comrie; 1 earth-sound heard by Mr Drummond.

128. **1839** DEC. 4. Cat. No. 401 (Milne, vol. 32, p. 107).

A very slight shock (int. 1) at Comrie; 1 earth-sound heard by Mr Drummond.

129. **1839** DEC. 5. Cat. No. 402 (Milne, vol. 32, p. 107).

A very slight shock (int. 1) at Comrie; 2 earth-sounds heard by Mr Drummond.

130. **1839** DEC. 6, 2 *a.m.* Cat. No. 403 (Milne, vol. 32, p. 107).

A shock (int. 4) at Comrie; according to Mr Drummond, there were 1 earth-sound and 1 tremor at 3 a.m. (possibly the same as the above), felt to a distance of 8 miles.

131. **1839** DEC. 7. Cat. No. 404 (Milne, vol. 32, p. 107).

A very slight shock (int. 1) at Comrie; 1 earth-sound heard by Mr Drummond.

132. **1839** DEC. 8. Cat. No. 405 (Milne, vol. 32, p. 107).

Two very slight shocks (int. 1) at Comrie; 2 earth-sounds heard by Mr Drummond.

On Dec. 9 an earth-sound was heard at Comrie by Mr Drummond.

133. **1839** DEC. 11, 9.30 *p.m.* Cat. No. 406 (Milne, vol. 32, p. 107).

A shock (int. 3) at Comrie; Mr Drummond observed 1 earth-sound, and also a tremor at 9.15 p.m. (probably the same as the above), felt to a distance of 6 miles.

134. **1839** DEC. 12, 3 *a.m.* Cat. No. 407 (Milne, vol. 32, p. 107).

A shock (int. 2) at Comrie; Mr Drummond observed 1 earth-sound and a tremor at 3 a.m., felt to a distance of 8 miles.

135. **1839** DEC. 13–18. Cat. No. 408 (Milne, vol. 32, p. 107).

Four shocks (the two stronger of int. 1) felt at Comrie. During this interval, Mr Drummond heard 6 earth-sounds—2 on Dec. 13, 1 on Dec. 15, 1 on Dec. 17, and 2 on Dec. 18.

On Dec. 19 Mr Drummond heard 1 earth-sound at Comrie.

136. **1839** DEC. 20, 3.15 *a.m.* Cat. No. 409 (Milne, vol. 32, p. 107).

A shock (int. 3) at Comrie; 1 earth-sound heard by Mr Drummond.

On Dec. 21 and 23, 1 earth-sound was heard on each day at Comrie by Mr Drummond.

137. **1839** DEC. 24. Cat. No. 411 (Milne, vol. 32, p. 107).

A very slight shock (int. 1) at Comrie; 2 earth-sounds heard by Mr Drummond.

On Dec. 26 and 27, 1 earth-sound was heard on each day at Comrie by Mr Drummond.

138. **1839** DEC. 28, 3 *a.m.* Cat. No. 412 (Milne, vol. 32, p. 107).

A very slight shock, hardly sensible, at Comrie.

139. **1839** DEC. 28, *noon.* Cat. No. 413 (Milne, vol. 32, p. 107).

A very slight shock, hardly sensible, at Comrie. On this day Mr Drummond heard 2 earth-sounds.

On Dec. 30, 1 earth-sound was heard at Comrie by Mr Drummond.

140. **1839** DEC. 31, 2 *a.m.* Cat. No. 414 (Milne, vol. 31, p. 107).

A very slight shock (int. 1) at Comrie; 1 earth-sound heard by Mr Drummond.

On 1840 Jan. 1 an earth-sound was heard at Comrie by Mr Drummond.

141. **1840** JAN. 2, 1.30 *a.m.* Cat. No. 415 (Milne, vol. 32, p. 107).

A shock (int. 1) at Comrie; 1 earth-sound heard by Mr Drummond.

142. **1840** JAN. 4, 11.45 *p.m.* Cat. No. 416 (Milne, vol. 32, p. 107).

A shock (int. 2) at Comrie; 1 earth-sound heard by Mr Drummond.

On Jan. 6 and 7 an earth-sound was heard on each day at Comrie by Mr Drummond.

143. **1840** JAN. 8, 0.30 *a.m.* Cat. No. 417 (Milne, vol. 32, p. 107).

A shock (int. 2) felt at Comrie.

144. **1840** JAN. 8, 1 *a.m.* Cat. No. 418 (Milne, vol. 32, p. 107).

A shock (int. 3) at Comrie; 2 earth-sounds heard by Mr Drummond on this day.

On Jan. 9 an earth-sound was heard by Mr Drummond at Comrie.

145. **1840** JAN. 11, 2 *a.m.* Cat. No. 419 (Milne, vol. 32, p. 107).

A very slight shock (int. 1) at Comrie.

146. **1840** JAN. 11, *noon.* Cat. No. 420 (Milne, vol. 32, p. 107).

A very slight shock (int. 1) at Comrie; 1 earth-sound heard by Mr Drummond on this day.

On each of the days Jan. 12, 14, 16 and 17, 1 earth-sound was heard by Mr Drummond at Comrie.

147. **1840** JAN. 18, 9.45 *p.m.* Cat. No. 421; Intensity 4 (Milne, vol. 32, p. 107; vol. 36, pp. 73–74).

A shock felt at Comrie (int. 4) and at Clathick. According to Mr Drummond, a lateral shock (probably the same as this) was felt at Comrie at 9.15 p.m. and to a distance of 18 miles.

148. **1840** JAN. 18, 10 *p.m.* Cat. No. 422 (Milne, vol. 32, p. 107; vol. 36, p. 74).

A shock (int. 2) felt at Comrie; at Clathick, a single report was heard, accompanied by a very slight movement; Mr Drummond also heard 1 earth-sound on this day.

149. **1840** JAN. 19, 5 *a.m.* Cat. No. 423 (Milne, vol. 32, p. 107).

A very slight shock (int. 1) at Comrie.

150. 1840 JAN. 19, 9 *a.m.* Cat. No. 424 (Milne, vol. 32, p. 107).

A very slight shock (int. 1) at Comrie; 2 earth-sounds were heard by Mr Drummond on this day.

151. 1840 JAN. 20, 9 *a.m.* Cat. No. 425 (Milne, vol. 32, p. 107).

A very slight shock (int. 1) at Comrie; 1 earth-sound heard by Mr Drummond.

On each of the days Jan. 23, 25 and 26, 1 earth-sound was heard by Mr Drummond at Comrie.

152. 1840 JAN. 27, 6 *a.m.* Cat. No. 426 (Milne, vol. 32, p. 107).

A shock (int. 3) at Comrie, and, according to Mr Drummond, felt to a distance of 5 miles; in addition, Mr Drummond heard 2 earth-sounds.

Two earth-sounds were heard by Mr Drummond at Comrie on Jan. 28, 1 on Jan. 31, and 1 on Feb. 4.

153. 1840 FEB. 6, 8.30 *a.m.* Cat. No. 427 (Milne, vol. 32, p. 107).

A very slight shock (int. 1) at Comrie; 1 earth-sound heard by Mr Drummond on this day.

154. 1840 FEB. 9, 9 *a.m.* Cat. No. 428 (Milne, vol. 32, p. 107).

A very slight shock (int. 1) at Comrie; 1 earth-sound heard by Mr Drummond on this day.

155. 1840 FEB. 10, 4 *a.m.* Cat. No. 429 (Milne, vol. 32, p. 107).

A shock (int. 4) felt at Comrie, and to a distance of 6 miles; 1 earth-sound also heard by Mr Drummond.

On Feb. 12 an earth-sound was heard by Mr Drummond at Comrie.

156. 1840 FEB. 14, 1 *a.m.* Cat. No. 430 (Milne, vol. 32, p. 107).

A very slight shock (int. 1) at Comrie; 1 earth-sound heard by Mr Drummond.

On each of the days Feb. 16, 18, 20, 22 and 24, 1 earth-sound was heard by Mr Drummond at Comrie.

157. 1840 FEB. 25, 2 *p.m.* Cat. No. 431 (Milne, vol. 32, p. 107).

A shock (int. 3) at Comrie, felt, according to Mr Drummond, to a distance of 6 miles; 1 earth-sound also heard by him.

158. 1840 FEB. 26, 1 *a.m.* Cat. No. 432 (Milne, vol. 32, p. 108).

A very slight shock (int. 1) at Comrie; 1 earth-sound heard by Mr Drummond.

On each of the days Feb. 28 and Mar. 7 an earth-sound was heard by Mr Drummond at Comrie.

159. **1840** MAR. 8, 6 *a.m.* Cat. No. 433 (Milne, vol. 32, p. 108).

A very slight shock (int. 1) at Comrie.

160. **1840** MAR. 8, 4 *p.m.* Cat. No. 434 (Milne, vol. 32, p. 108).

A very slight shock (int. 1) at Comrie; 2 earth-sounds heard by Mr Drummond on this day.

161. **1840** MAR. 9, 5.30 *a.m.* Cat. No. 435 (Milne, vol. 32, p. 108).

A shock (int. 3) at Comrie; Mr Drummond records on this day 1 earth-sound and 1 tremor at 5.30 p.m., felt to a distance of 8 miles.

On Mar. 10 an earth-sound was heard by Mr Drummond at Comrie.

162. **1840** MAR. 11, 5 *a.m.* Cat. No. 436 (Milne, vol. 32, p. 108).

A shock (int. 1) at Comrie.

163. **1840** MAR. 11, 6 *a.m.* Cat. No. 437 (Milne, vol. 32, p. 108).

A shock (int. 1) at Comrie.

164. **1840** MAR. 11, *probably after* 6 *a.m.* Cat. No. 438 (Milne, vol. 32, p. 108).

A shock (int. 1) at Comrie; 3 earth-sounds heard by Mr Drummond on this day.

On Mar. 12 an earth-sound was heard by Mr Drummond at Comrie.

165. **1840** MAR. 13, 8 *p.m.* Cat. No. 439 (Milne, vol. 32, p. 108).

A shock (int. 3) at Comrie and felt to a distance of 6 miles; 1 earth-sound also heard by Mr Drummond.

166. **1840** MAR. 14, 9.30 *p.m.* Cat. No. 440 (Milne, vol. 32, p. 108).

A shock (int. 3) felt at Comrie and to a distance of 6 miles; 1 earth-sound also heard by Mr Drummond.

167. **1840** MAR. 21, 0.30 *a.m.* Cat. No. 441 (Milne, vol. 32, p. 108).

A shock (int. 1) at Comrie; 1 earth-sound heard by Mr Drummond.
On Mar. 23 an earth-sound was heard by Mr Drummond at Comrie.

168. **1840** MAR. 24, 8.30 *a.m.* Cat. No. 442 (Milne, vol. 32, p. 108).

A shock (int. 2) at Comrie.

169. **1840** MAR 24, 11 *a.m.* Cat. No. 443 (Milne, vol. 32, p. 108).

A shock (int. 1) at Comrie; 2 earth-sounds heard by Mr Drummond on this day.

170. **1840** MAR. 25, 1 *p.m.* Cat. No. 444 (Milne, vol. 32, p. 108).

A shock (int. 1) at Comrie; 1 earth-sound heard by Mr Drummond.
On Mar. 26 an earth-sound was heard by Mr Drummond at Comrie.

171. 1840 MAR. 27, 1.30 *p.m.* Cat. No. 445 (Milne, vol. 32, p. 108).
A shock (int. 1) at Comrie.

172. 1840 MAR. 27, 2.30 *p.m.* Cat. No. 446 (Milne, vol. 32, p. 108).
A shock (int. 1) at Comrie; 1 earth-sound heard by Mr Drummond
on this day.

173. 1840 APR. 1, 1 *p.m.* Cat. No. 447 (Milne, vol. 32, p. 108).
A shock (int. 1) at Comrie.

174. 1840 APR. 1, 1.30 *p.m.* Cat. No. 448 (Milne, vol. 32, p. 108).
A shock (int. 1) at Comrie; 2 earth-sounds heard by Mr Drummond
on this day.
On Apr. 3 and 5, 1 earth-sound heard each day by Mr Drummond
at Comrie.

175. 1840 APR. 7, 4 *p.m.* Cat. No. 449 (Milne, vol. 32, p. 108;
vol. 36, p. 75).
A shock (int. 6) felt at Comrie, also strongly at Crieff, and to a
distance of 12 miles from Comrie; at Crieff it was accompanied by a
noise like distant thunder.

176. 1840 APR. 7, 4.15 *p.m.* Cat. No. 450 (Milne, vol. 32, p. 108).
A shock (int. 2) at Comrie.

177. 1840 APR. 7, 5 *p.m.* Cat. No. 451 (Milne, vol. 32, p. 108).
A shock (int. hardly 2) at Comrie; 2 earth-sounds also heard on
this day by Mr Drummond.
On Apr. 10 an earth-sound was heard by Mr Drummond at Comrie.

178. 1840 APR. 11, 2.45 *a.m.* Cat. No. 452 (Milne, vol. 32, p. 108).
A shock (int. 1) at Comrie.

179. 1840 APR. 12, 1 *p.m.* Cat. No. 454 (Milne, vol. 32, p. 108).
A shock (int. 1) at Comrie; 1 earth-sound heard by Mr Drummond.

180. 1840 APR. 13, 1 *p.m.* Cat. No. 455 (Milne, vol. 32, p. 108).
A shock (int. 1) at Comrie; 1 earth-sound heard by Mr Drummond.

181. 1840 MAY 19, 1.45 *p.m.* Cat. No. 456 (Milne, vol. 32, p. 108).
A shock (int. 3) felt at Comrie and to a distance of 6 miles.

182. 1840 MAY 19, 2 *p.m.* Cat. No. 457 (Milne, vol. 32, p. 108).
A shock (int. 1) at Comrie.

183. **1840** MAY 19, 2.15 *p.m.* Cat. No. 458 (Milne, vol. 32, p. 108).

A shock (int. 1) at Comrie; 2 earth-sounds observed by Mr Drummond on this day in addition to the tremor at 1.45 p.m.

184. **1840** MAY 22, 1.30 *a.m.* Cat. No. 459 (Milne, vol. 32, p. 108).

A shock (int. 2) at Comrie.

185. **1840** MAY 22, 5.30 *a.m.* Cat. No. 460 (Milne, vol. 32, p. 108).

A shock (int. 2) at Comrie; 1 earth-sound heard by Mr Drummond on this day.

On May 29 and (probably) June 16, 1 earth-sound was heard each day by Mr Drummond; on July 2, 2 earth-sounds.

186. **1840** JULY 3, 11 *a.m.* Cat. No. 461 (Milne, vol. 32, p. 108).

A shock (int. 1) at Comrie.

187. **1840** JULY 11, 11 *a.m.* Cat. No. 462 (Milne, vol. 32, p. 108).

A shock (int. 1) at Comrie.

On July 15, 2 earth-sounds were heard by Mr Drummond at Comrie.

188. **1840** JULY 16, 3.30 *a.m.* Cat. No. 463 (Milne, vol. 32, p. 108).

Two shocks (int. 1) at Comrie.

189. **1840** JULY 17, 8.30 *a.m.* Cat. No. 464 (Milne, vol. 32, p. 108).

A shock (int. 1) at Comrie.

190. **1840** JULY 23, 1 *a.m.* Cat. No. 465 (Milne, vol. 32, p. 108).

A shock (int. 1) at Comrie.

On July 25, 1 earth-sound was heard by Mr Drummond at Comrie.

191. **1840** AUG. 5, 6 *p.m.* Cat. No. 466 (Milne, vol. 32, p. 108).

A shock (int. 1) at Comrie.

192. **1840** AUG. 6, 2 *a.m.* Cat. No. 467 (Milne, vol. 32, p. 108).

A shock (int. 1) at Comrie; 1 earth-sound heard by Mr Drummond.
On Sept. 1, 1 earth-sound was heard by Mr Drummond at Comrie.

193. **1840** SEPT. 19, 3.30 *a.m.* Cat. No. 468 (Milne, vol. 32, p. 108).

A shock (int. 3) at Comrie; 2 earth-sounds heard by Mr Drummond.

194. **1840** SEPT. 21, 3 *p.m.* Cat. No. 469 (Milne, vol. 32, p. 108).

A shock (int. 1) at Comrie.

195. **1840** SEPT. 26, 3 *a.m.* Cat. No. 470 (Milne, vol. 32, p. 108).

A shock (int. 2) at Comrie.

On Sept. 27, 1 earth-sound was heard by Mr Drummond at Comrie.

196. **1840** OCT. 4, 9 *p.m.* Cat. No. 471 (Milne, vol. 32, p. 108).

A shock (int. 2) at Comrie.

197. **1840** OCT. 20, 11 *a.m.* Cat. No. 472 (Milne, vol. 32, p. 108).

A shock (int. 2) at Comrie.

On Oct. 21 (probably), 1 earth-sound was heard by Mr Drummond at Comrie.

198. **1840** OCT. 26, 6.45 *p.m.* Cat. No. 473 (Milne, vol. 32, p. 108; vol. 36, p. 75).

A shock (int. 4) felt at Comrie and to a distance of 12 miles, the motion, according to Mr Drummond, being vertical. The instruments erected at Comrie were affected by this shock, two of them indicating upward movements of $\frac{3}{4}$ and $\frac{1}{2}$ inch, respectively; in a third, designed to measure the horizontal movement, the lower end made a furrow in the flour half an inch from the centre in the direction W. by N. Mr Drummond also heard 1 earth-sound on this day.

On Oct. 27 (probably), 1 earth-sound was heard by Mr Drummond at Comrie.

199. **1840** NOV. 12, 2.30 *p.m.* Cat. No. 474 (Milne, vol. 32, p. 108).

A shock (int. 1) at Comrie; 2 earth-sounds heard by Mr Drummond on this day.

200. **1840** NOV. 13, 2.30 *a.m.* Cat. No. 475 (Milne, vol. 32, p. 108).

A shock (int. 1) at Comrie.

201. **1840** NOV. 13, 4 *a.m.* Cat. No. 476 (Milne, vol. 32, p. 108).

A shock (int. 1) at Comrie.

202. **1840** NOV. 13, 7 *a.m.* Cat. No. 477 (Milne, vol. 32, p. 108).

A shock (int. 1) at Comrie; 3 earth-sounds heard by Mr Drummond on this day.

On Nov. 14, 1 earth-sound was heard by Mr Drummond at Comrie.

203. **1840** NOV. 16, 3 *a.m.* Cat. No. 478 (Milne, vol. 32, p. 108).

A shock (int. 1) at Comrie.

204. **1840** NOV. 16, 6 *a.m.* Cat. No. 479 (Milne, vol. 32, p. 108).

A shock (int. 1) at Comrie; 1 earth-sound heard by Mr Drummond on this day.

On Nov. 17, 2 earth-sounds were heard by Mr Drummond at Comrie.

205. 1840 NOV. 24, 4 *a.m.* Cat. No. 480 (Milne, vol. 32, p. 108).
A shock (int. 1) at Comrie.

206. 1840 DEC. 6, 7 *p.m.* Cat. No. 481 (Milne, vol. 32, p. 108).
A shock (int. 1) at Comrie.

207. 1840 DEC. 7. Cat. No. 482 (Milne, vol. 32, p. 108).
A shock (int. 2) at Comrie.

208. 1840 DEC. 8. Cat. No. 483 (Milne, vol. 32, p. 108).
A shock (int. 2) at Comrie.

209. 1840 DEC. 10, 2 *a.m.* Cat. No. 484 (Milne, vol. 32, p. 108).
A shock (int. 2) at Comrie.
On Dec. 13, 2 earth-sounds were heard at Comrie by Mr Drummond; on Dec. 15, 1 earth-sound.

210. 1840 DEC. 18, 2 *a.m.* Cat. No. 485 (Milne, vol. 32, p. 108).
A shock (int. 1) at Comrie.
On Dec. 24, 2 earth-sounds were heard by Mr Drummond at Comrie; on Dec. 25, 1 earth-sound.

211. 1841 JAN. 6. Cat. No. 486 (Perrey, p. 159).
A very feeble shock at Comrie.
On Jan. 10, 2 earth-sounds were heard by Mr Drummond at Comrie; on Jan. 15, 1; and on Jan. 17, 3.

212. 1841 JAN. 18. Cat. No. 487 (Milne, vol. 32, p. 108).
A shock (int. 1) at Comrie.

213. 1841 JAN. 31, 2 *a.m.* Cat. No. 489 (Milne, vol. 32, p. 108).
A shock (int. 1) at Comrie.

214. 1841 JAN. 31, 9 *p.m.* Cat. No. 490 (Milne, vol. 32, p. 108).
A shock (int. 1) at Comrie.

215. 1841 JAN. 31, 11.30 *p.m.* Cat. No. 491 (Milne, vol. 32, p. 108).
A shock (int. 1) at Comrie.

216. 1841 FEB. 1, 7.30 *p.m.* Cat. No. 492 (Milne, vol. 32, p. 108).
A shock (int. 1) at Comrie; 1 earth-sound heard by Mr Drummond.

217. 1841 FEB. 14, 9.45 *a.m.* Cat. No. 493 (Milne, vol. 32, p. 108).
A shock (int. 1) at Comrie; 1 earth-sound heard by Mr Drummond.

218. 1841 FEB. 16. Cat. No. 494 (Milne, vol. 32, p. 108).
A shock (int. 1) at Comrie.

219. **1841** MAR. 6, 9 *a.m.* Cat. No. 495 (Milne, vol. 32, p. 108).

A shock (int. 1) at Comrie; 1 earth-sound heard by Mr Drummond.

220. **1841** MAR. 10, 4.45 *p.m.* Cat. No. 496 (Milne, vol. 32, p. 108; vol. 36, p. 76).

A rather strong lateral shock (int. 3) felt at Comrie and to a distance of 15 miles. The two inverted pendulums (39 ins. and 128 ins. long) at Comrie had their points thrown half an inch to the west.

221. **1841** MAR. 11, 1 *a.m.* Cat. No. 497 (Milne, vol. 32, p. 108).

A very slight shock (int. 1) at Comrie.

222. **1841** MAR. 22, 1 *a.m.* Cat. No. 498 (Milne, vol. 32, p. 108).

A very slight shock (int. 1) at Comrie.

223. **1841** MAR. 22, 6.15 *a.m.* Cat. No. 499 (Milne, vol. 32, p. 108; vol. 36, p. 76).

A slight shock (int. 2) felt at Comrie and to a distance of 6 miles. The two inverted pendulums (39 ins. and 128 ins. long) at Comrie were affected by both shocks on this day, as on Mar. 10, though to a less extent.

224. **1841** MAR. 23, 1 *a.m.* Cat. No. 500 (Milne, vol. 32, p. 108).

A very slight shock (int. 1) at Comrie.

225. **1841** APR. 3, 8 *a.m.* Cat. No. 501 (Milne, vol. 32, p. 108).

A very slight shock (int. 1) at Comrie; 1 earth-sound heard by Mr Drummond.

226. **1841** APR. 9, 7.45 *a.m.* Cat. No. 502 (Milne, vol. 32, p. 108).

A vertical shock (int. 2) felt at Comrie and to a distance of 10 miles.

227. **1841** APR. 12, 5 *a.m.* Cat. No. 503 (Milne, vol. 32, p. 108).

A tremor (int. 1) felt at Comrie and to a distance of 5 miles.

228. **1841** APR. 14, 5 *a.m.* Cat. No. 504 (Milne, vol. 32, p. 108).

A very slight shock (int. 1) at Comrie.

229. **1841** APR. 17. Cat. No. 505 (Milne, vol. 32, p. 108).

A very slight shock (int. 1) at Comrie.

230. **1841** APR. 24, 1.45 *p.m.* Cat. No. 510 (Drummond, p. 246).

A vertical shock felt at Comrie and to a distance of 10 miles.

231. **1841** APR. 24, 9.45 *p.m.* Cat. No. 511 (Milne, vol. 32, p. 108).

A very slight shock (int. 1) at Comrie.

232. 1841 APR. 24, *after* 9.45 *p.m.* Cat. No. 512 (Milne, vol. 32, p. 108).

A very slight shock (int. 1) at Comrie.

233. 1841 APR. 25, 1 *a.m.* Cat. No. 513 (Milne, vol. 32, p. 108).

A very slight shock (int. 1) at Comrie.

234. 1841 MAY 5, 9 *a.m.* Cat. No. 514 (Milne, vol. 32, p. 108).

A very slight shock (int. 1) at Comrie; 1 earth-sound heard on this day by Mr Drummond.

235. 1841 MAY 8, *noon.* Cat. No. 515 (Milne, vol. 32, p. 108).

A very slight shock (int. 1) at Comrie.

236. 1841 MAY 22, *noon.* Cat. No. 516 (Milne, vol. 32, p. 108).

A hardly sensible shock (int. 1) felt at Comrie and to a distance of 6 miles.

237. 1841 MAY 26. Cat. No. 517 (Milne, vol. 32, p. 108).

A hardly sensible shock (int. 1) at Comrie.

238. 1841 MAY 27, 8 *a.m.* Cat. No. 518 (Milne, vol. 32, p. 108).

A very slight shock (int. 1) at Comrie; 1 earth-sound heard by Mr Drummond.

239. 1841 MAY 28, 7 *a.m.* Cat. No. 519 (Milne, vol. 32, p. 108).

A very slight shock (int. 1) at Comrie; 1 earth-sound heard by Mr Drummond.

240. 1841 MAY 30, 6.45 *a.m.* Cat. No. 520 (Milne, vol. 32, p. 108).

A tremor (int. 1) felt at Comrie and to a distance of 5 miles.

241. 1841 JUNE 29. Cat. No. 521 (Milne, vol. 32, p. 108).

A very slight shock (int. 1) at Comrie.

242. 1841 JULY 2, 5.45 *a.m.* Cat. No. 522 (Milne, vol. 32, p. 108).

A very slight shock (int. 1) at Comrie; 1 earth-sound heard by Mr Drummond.

243. 1841 JULY 23, 1 *a.m.* Cat. No. 524 (Milne, vol. 36, p. 77; *Brit. Ass. Rep.* 1842, p. 93).

A slight shock (int. 2) at Comrie.

244. 1841 JULY 23, 4.15 *p.m.* Cat. No. 525 (Drummond, p. 246).

A lateral shock felt at Comrie and to a distance of 10 miles; also 1 earth-sound heard.

245. 1841 JULY 25, 4.45 *p.m.* Cat. No. 526 (Milne, vol. 36, p. 77; *Brit. Ass. Rep.* 1841, p. 49; 1842, p. 93).

A shock (int. 3) at Comrie.

246. 1841 JULY 26, 3 *a.m.* Cat. No. 527 (Milne, vol. 36, p. 77; *Brit. Ass. Rep.* 1842, pp. 93, 95).

A shock, accompanied by a rather loud noise, at Comrie. The instruments at Comrie recorded this shock and the preceding. The inverted pendulum in the steeple of Comrie parish church was thrown about half an inch to the west; an upward movement of half an inch was indicated by two other instruments. An earth-sound was heard on this day by Mr Drummond.

On July 27, 1 earth-sound was heard by Mr Drummond at Comrie.

On July 30, there was a marked renewal of activity. Twelve shocks were felt at Comrie, the strongest being at 8 a.m., 2.30 p.m. and 3.30 p.m; 12 earth-sounds were heard by Mr Drummond.

247. 1841 JULY 30, 8 *a.m.* Cat. No. 528 (Milne, vol. 32, p. 108; *Brit. Ass. Rep.* 1842, p. 93).

A rather smart vertical shock (int. 2) felt at Comrie and to a distance of 10 miles. This shock was recorded by the instruments at Comrie.

248. 1841 JULY 30, 2.30 *p.m.* Cat. No. 529; Intensity 8 (Milne, vol. 36, pp. 77–82; *Brit. Ass. Rep.* 1842, pp. 93, 95–96).

Since 1839 Oct. 23, this is the strongest shock felt in the Comrie district. Several chimneys were damaged or walls rent within a small area, about 2 miles in diameter, lying W.N.W. of Comrie, and including Dunira, Garrichrew and Ross; and dykes were thrown down in many places. The epicentre was thus probably about 1 mile N.W. of Comrie. The boundary of the disturbed area lies at least 10 miles within the isoseismal 5 of the earthquake of 1839 (Fig. 9), the shock being felt at Aberfeldy (17 miles N. of Comrie), Dalmally (37 miles W.), Alloa (21 miles S.), Newburgh (29 miles E.) and Dunkeld (20 miles N.E.). It thus contains less than 1100 square miles.

The shock was distinctly double, the second part being the stronger at Comrie. Two inverted pendulums in Mr Macfarlane's house vibrated to the extent of half an inch in the direction N.–S.; at Tomperran (about ½ mile E. of Comrie), an instrument on the principle of the common pendulum vibrated E.–W. The instruments for showing vertical movement were but slightly affected.

Immediately after this earthquake, two or three slight shocks were felt at Comrie by Mr Macfarlane.

249. **1841** JULY 30, 3.30 *p.m.* Cat. No. 530 (Milne, vol. 32, p. 108; *Brit. Ass. Rep.* 1842, p. 93).

A shock (int. 2), accompanied by a loud noise, at Comrie. This shock was recorded by the instruments at Comrie.

250. **1841** JULY 31, 8 *a.m.* Cat. No. 531 (Milne, vol. 32, p. 108; *Brit. Ass. Rep.* 1842, p. 93).

A vertical shock (int. 2) felt at Comrie and to a distance of 8 miles. Two other very slight shocks (int. 1) were felt at Comrie. One earth-sound was heard by Mr Drummond.

251. **1841** AUG. 1. Cat. No. 532 (Milne, vol. 32, p. 108; *Brit. Ass. Rep.* 1842, p. 93).

Two very slight shocks (both of int. 1) at Comrie; 1 earth-sound heard by Mr Drummond.

On each of the days Aug. 2, 3 and 7, 1 earth-sound was heard by Mr Drummond at Comrie.

252. **1841** AUG. 10. Cat. No. 533 (Milne, vol. 32, p. 108; *Brit. Ass. Rep.* 1842, p. 93).

Two very slight shocks (both of int. 1) at Comrie; 1 earth-sound heard by Mr Drummond.

253. **1841** AUG. 12, 10 *a.m.* Cat. No. 534 (Milne, vol. 32, p. 108; *Brit. Ass. Rep.* 1842, p. 93).

A very slight vertical shock felt at Comrie and to a distance of 10 miles. On Aug. 22 an earth-sound was heard by Mr Drummond at Comrie.

254. **1841** AUG. 30. Cat. No. 535 (Milne, vol. 32, p. 108; *Brit. Ass. Rep.* 1842, p. 93).

A very slight shock (int. 1) at Comrie.

On Sept. 1 an earth-sound was heard by Mr Drummond at Comrie.

255. **1841** SEPT. 8, 3 *a.m.* Cat. No. 536 (Milne, vol. 32, p. 108; *Brit. Ass. Rep.* 1842, p. 93).

Two shocks (int. 1) at Comrie, one at 3 a.m., the time of the other not given.

256. **1841** SEPT. 9, 11.45 *p.m.* Cat. No. 537 (Milne, vol. 32, p. 108; vol. 36, pp. 82–83; *Brit. Ass. Rep.* 1842, p. 93).

A shock (int. 5) at Comrie, recorded by two instruments; one in the steeple of Comrie parish church showed a trace $\frac{3}{4}$ inch long to the south; the other in Comrie House ($\frac{1}{2}$ mile to the north of the church) one $\frac{1}{2}$ inch long to the north.

257. 1841 SEPT. 10, 2.30 *a.m.* Cat. No. 538 (Milne, vol. 32, p. 108; *Brit. Ass. Rep.* 1842, p. 93).

A shock (int. 3) at Comrie.

258. 1841 SEPT. 10, 4.30 *a.m.* Cat. No. 539 (Milne, vol. 32, p. 109; *Brit. Ass. Rep.* 1842, p. 93).

A shock (int. 1) at Comrie.

259. 1841 SEPT. 10, 11.15 *a.m.* Cat. No. 540 (Milne, vol. 32, p. 108; *Brit. Ass. Rep.* 1842, p. 93).

A slight vertical shock felt at Comrie and to a distance of 20 miles.

260. 1841 SEPT. 11, 2.15 *a.m.* Cat. No. 541 (Drummond, p. 247).

A vertical shock felt at Comrie and to a distance of 15 miles.

261. 1841 SEPT. 16, *morning.* Cat. No. 542 (Milne, vol. 32, p. 108; *Brit. Ass. Rep.* 1842, p. 93).

A shock (int. 1) at Comrie.

262. 1841 SEPT. 16, 9.30 *p.m.* Cat. No. 543 (Milne, vol. 32, p. 108; *Brit. Ass. Rep.* 1842, p. 93).

A shock (int. 1) at Comrie.

263. 1841 SEPT. 17, 1 *a.m.* Cat. No. 544 (Milne, vol. 32, p. 108; *Brit. Ass. Rep.* 1842, p. 93).

A shock (int. 1) at Comrie.

264. 1841 SEPT. 17, 4.30 *a.m.* Cat. No. 545 (Milne, vol. 32, p. 108; *Brit. Ass. Rep.* 1842, p. 93).

A shock (int. 1) at Comrie.

265. 1841 SEPT. 22, 11.30 *p.m.* Cat. No. 546 (Milne, vol. 32, p. 108; *Brit. Ass. Rep.* 1842, p. 93).

A shock (int. 1) at Comrie; 1 earth-sound heard by Mr Drummond.

266. 1841 SEPT. 23, 1.15 *a.m.* Cat. No. 547 (Drummond, p. 247).

A vertical shock felt at Comrie and to a distance of 10 miles; also 1 earth-sound at Comrie.

267. 1841 SEPT. 23, 2.30 *a.m.* Cat. No. 548 (Milne, vol. 32, p. 108; *Brit. Ass. Rep.* 1842, p. 93).

A shock (int. 1) at Comrie.

268. 1841 SEPT. 25, 9 *p.m.* Cat. No. 549 (Drummond, p. 247).

A lateral shock felt at Comrie and to a distance of 8 miles; also 1 earth-sound at Comrie.

269. **1841 SEPT. 29,** *morning.* Cat. No. 550 (Milne, vol. 32, p. 108; *Brit. Ass. Rep.* 1842, p. 93).

A shock (int. 1) at Comrie.

270. **1841 SEPT. 29, 9.15** *a.m.* Cat. No. 551 (Milne, vol. 32, p. 108; *Brit. Ass. Rep.* 1842, p. 93).

A shock (int. 1) at Comrie.

271. **1841 OCT. 5,** *morning.* Cat. No. 552 (Milne, vol. 32, p. 108; *Brit. Ass. Rep.* 1842, p. 93).

A shock (int. 1) at Comrie.

272. **1841 OCT. 13,** *midnight.* Cat. No. 553 (Drummond, p. 247).

A tremor felt at Comrie and to a distance of 6 miles.

273. **1841 OCT. 23,** *midnight.* Cat. No. 554 (Milne, vol. 32, p. 108; *Brit. Ass. Rep.* 1842, p. 93).

A shock (int. 2) at Comrie.

Mr Drummond records a lateral shock at 1 a.m. on this day, felt at Comrie and to a distance of 8 miles, which may be the same shock as the preceding.

274. **1841 NOV. 3,** *midnight.* Cat. No. 555 (Milne, vol. 32, p. 108; *Brit. Ass. Rep.* 1842, p. 93).

A shock (int. 1) at Comrie; 3 earth-sounds heard by Mr Drummond on this day at Comrie.

275. **1841 NOV. 5, 1** *a.m.* Cat. No. 556 (Milne, vol. 32, p. 108; *Brit. Ass. Rep.* 1842, p. 93).

A slight tremor felt at Comrie and to a distance of 5 miles.

276. **1841 NOV. 6, 8** *a.m.* Cat. No. 557 (Milne, vol. 32, p. 108; *Brit. Ass. Rep.* 1842, p. 93).

A slight tremor felt at Comrie and to a distance of 6 miles.

277. **1841 NOV. 7.** Cat. No. 558 (Milne, vol. 32, p. 108; *Brit. Ass. Rep.* 1842, p. 93).

A shock (int. 1) at Comrie.

278. **1841 NOV. 8, 8** *a.m.* Cat. No. 559 (Milne, vol. 32, p. 108; *Brit. Ass. Rep.* 1842, p. 93).

A slight tremor felt at Comrie and to a distance of 5 miles; also 1 earth-sound heard by Mr Drummond at Comrie.

279. **1841** NOV. 18, 8 *a.m.* Cat. No. 560 (Milne, vol. 32, p. 108; *Brit. Ass. Rep.* 1842, p. 93).

A slight tremor felt at Comrie and to a distance of 5 miles; also 1 earth-sound heard by Mr Drummond.

280. **1841** NOV. 26, *about* 1 *p.m.* Cat. No. 561 (Milne, vol. 36, p. 84).

A slight shock, accompanied by sound, at Comrie and Dunira, strong enough at Dunira to make the trees shake.

281. **1841** DEC. 3, 8.30 *a.m.* Cat. No. 562 (Milne, vol. 32, p. 108; *Brit. Ass. Rep.* 1842, p. 93).

A slight tremor at Comrie; 1 earth-sound heard by Mr Drummond on this day.

282. **1841** DEC. 6, 3 *a.m.* Cat. No. 563 (Milne, vol. 32, p. 108; *Brit. Ass. Rep.* 1842, p. 93).

A tremor (int. 1) felt at Comrie and to a distance of 6 miles.

283. **1841** DEC. 7, 3 *a.m.* Cat. No. 564 (Milne, vol. 32, p. 108; *Brit. Ass. Rep.* 1842, p. 93).

A shock (int. 1) at Comrie; 1 earth-sound heard by Mr Drummond.

284. **1841** DEC. 19, 1.30 *a.m.* Cat. No. 565 (Drummond, p. 247).

A vertical shock felt at Comrie and to a distance of 5 miles.

285. **1842** JAN. 2, *morning.* Cat. No. 567 (*Brit. Ass. Rep.* 1842, p. 93).

A shock (int. 1) at Comrie.

286. **1842** JAN. 7, *morning.* Cat. No. 568 (*Brit. Ass. Rep.* 1842, p. 93).

Two shocks (int. 1) at Comrie.

287. **1842** MAR. 10, 1 *p.m.* Cat. No. 570 (*Brit. Ass. Rep.* 1842, p. 93).

A shock (int. 1) at Comrie.

288. **1842** APR. 21, 3 *p.m.* Cat. No. 571 (*Brit. Ass. Rep.* 1842, p. 93).

A shock (int. 1) at Comrie.

289. **1842** APR. 22, 10.30 *p.m.* Cat. No. 572 (*Brit. Ass. Rep.* 1842, p. 93).

A shock (int. 1) at Comrie.

290. **1842** APR. 22, 11 *p.m.* Cat. No. 573 (*Brit. Ass. Rep.* 1842, p. 93).

A shock (int. 2) at Comrie.

291. **1842** JUNE 1. Cat. No. 575 (*Brit. Ass. Rep.* 1842, p. 93).

A shock (int. 1) at Comrie.

292. **1842** JUNE 2, 1 *a.m.* Cat. No. 576 (*Brit. Ass. Rep.* 1842, p. 93).
A shock (int. 2) at Comrie.

293. **1842** JUNE 8, 1.30 *a.m.* Cat. No. 577 (*Brit. Ass. Rep.* 1842, p. 93).
A shock at Comrie.

294. **1842** JUNE 8, 1.45 *a.m.* Cat. No. 578 (*Brit. Ass. Rep.* 1842, p. 93).
A shock felt at Comrie. A pendulum in Mr Macfarlane's house was disturbed either by this shock or by the preceding or by both; it indicated a vertical movement upward of fully a quarter of an inch.

295. **1842** JULY 1, 2.30 *p.m.* Cat. No. 580 (*Brit. Ass. Rep.* 1843, pp. 126–127).
A shock (int. 1) at Comrie.

296. **1842** JULY 1, 4.30 *p.m.* Cat. No. 581 (*Brit. Ass. Rep.* 1843, pp. 126–127).
A shock (int. 1) at Comrie.

297. **1842** *between* JULY 1 *and* 10, 9 *a.m.* Cat. No. 582 (*Brit. Ass. Rep.* 1843, pp. 126–127).
A shock (int. 1) at Comrie.
On July 10, at 3.30 a.m., Perrey (p. 162) records a slight shock at Comrie and Dunblane.

298. **1842** AUG. 27, 11.45 *p.m.* Cat. No. 585 (*Brit. Ass. Rep.* 1843, pp. 126–127).
A shock (int. 1) at Comrie.

299. **1842** SEPT. 2, 3 *a.m.* Cat. No. 586 (*Brit. Ass. Rep.* 1843, pp. 126–127).
A shock (int. 1) at Comrie.

300. **1842** SEPT. 24, 5.53 *a.m.* Cat. No. 587 (Milne, reprint, pp. 224–225; *Brit. Ass. Rep.* 1843, pp. 126–127; etc.).
A double concussion (int. 3) at Comrie, consisting of a crash, slight tremor, and stronger tremor; the sound (int. 4) and the shock both lasted 3 seconds. The inverted pendulum in Comrie parish church was moved $\frac{1}{8}$ inch to the N.W., and the horizontal pendulum indicated a vertical movement of $\frac{1}{16}$ inch; the spring pendulum moved $\frac{1}{8}$ inch W.

301. **1842** SEPT. 24, 6.59 *p.m.* Cat. No. 588 (Milne, reprint, pp. 224–225; *Brit. Ass. Rep.* 1843, pp. 126–127; etc.).

A single concussion (int. 2) and sound (int. 3) at Comrie; both shock and sound lasted 3 seconds. The instruments at Comrie were not affected by this shock.

302. **1842** SEPT. 25, *morning.* Cat. No. 589 (*Brit. Ass. Rep.* 1843, pp. 126–127).

Two shocks (int. 1) at Comrie.

303. **1842** NOV. 18, *midnight.* Cat. No. 590 (*Brit. Ass. Rep.* 1843, pp. 126–127).

A shock (int. 1) at Comrie.

Perrey records a slight shock, with noise, at Comrie on Nov. 25 at 2 a.m., on the authority of Mr Macfarlane. It is not, however, given in the British Association list.

304. **1842** NOV. 29, 11 *p.m.* Cat. No. 591 (*Brit. Ass. Rep.* 1843, pp. 126–127).

A tremor (int. 1) at Comrie, lasting 1 second, accompanied by sound (int. 1) lasting 2 seconds.

305. **1842** DEC. 4, 2 *a.m.* Cat. No. 592 (*Brit. Ass. Rep.* 1843, pp. 126–127).

An earth-sound (int. 1), lasting 1 second, at Comrie.

306. **1842** DEC. 17, 8 *a.m.* Cat. No. 593 (*Brit. Ass. Rep.* 1843, pp. 126–127).

A tremor (int. 1) lasting 1 second at Comrie, accompanied by sound (int. 1) and lasting 1 second.

307. **1842** DEC. 17, *morning.* Cat. No. 594 (*Brit. Ass. Rep.* 1843, pp. 126–127).

Two tremors and sounds at Comrie, similar to the preceding.

308. **1843** MAR. 23, 8.30 *p.m.* Cat. No. 600 (*Brit. Ass. Rep.* 1843, pp. 126–127).

A shock (int. 2) at Comrie, accompanied by sound (int. 2); duration of shock 2 seconds, of sound 3 seconds.

309. **1843** MAR. 23, 11 *p.m.* Cat. No. 601 (*Brit. Ass. Rep.* 1843, pp. 126–127).

A slight tremor (int. 1) at Comrie, accompanied by sound (int. 1).

310. **1843** MAY 14, 1 *a.m.* Cat. No. 602 (*Brit. Ass. Rep.* 1843, pp. 126–127).

A slight tremor (int. 1) at Comrie, accompanied by sound (int. 1).

311. **1843** MAY 14, 2 *a.m.* Cat. No. 603 (*Brit. Ass. Rep.* 1843, pp. 126–127).

A slight tremor (int. 1) at Comrie, accompanied by sound (int. 1).

312. **1843** MAY 28, 0.20 *p.m.* Cat. No. 604 (*Brit. Ass. Rep.* 1843, pp. 126–127).

A shock (int. 1) and sound (int. 1) at Comrie.

313. **1843** JUNE 4, 0.15 *p.m.* Cat. No. 605 (Milne, vol. 36, p. 375; *Brit. Ass. Rep.* 1843, pp. 126–127).

A shock (int. 2) felt with almost equal strength at Comrie, St Fillans, Clathick and Invergeldie. At Comrie, the shock lasted 2 seconds; the sound (int. 2) lasted 3 seconds. The vertical pendulums at Comrie moved $\frac{1}{8}$ inch E.–W., the horizontal pendulum moved $\frac{1}{16}$ inch downwards.

314. **1843** JUNE 4, 3 *p.m.* Cat. No. 606 (*Brit. Ass. Rep.* 1843, pp. 126–127).

A shock felt only at Clathick.

315. **1843** JUNE 10, 2.30 *a.m.* Cat. No. 607 (*Brit. Ass. Rep.* 1843, pp. 126–127).

A shock (int. 2) and sound (int. 2) at Comrie; both shock and sound lasted 2 seconds.

316. **1843** JUNE 10, 3 *a.m.* Cat. No. 608 (*Brit. Ass. Rep.* 1843, pp. 126–127).

A shock (int. 1) and sound (int. 1) at Comrie; both shock and sound lasted 1 second.

317. **1843** JUNE 15. Cat. No. 609 (*Brit. Ass. Rep.* 1843, pp. 126–127).

A shock (int. 1) and sound (int. 1) at Comrie; both shock and sound lasted 1 second.

318. **1843** JUNE 17, 4 *a.m.* Cat. No. 610 (*Brit. Ass. Rep.* 1843, pp. 126–127).

A shock (int. 2) and sound (int. 2) at Comrie; both shock and sound lasted 2 seconds.

319. 1843 AUG. 25, 10.40 *a.m.* Cat. No. 611 (Milne, vol. 36, p. 376; *Brit. Ass. Rep.* 1844, pp. 86–87).

A shock (int. 5) felt at Comrie and, according to Mr Milne, over an area of more than 100 square miles; the shock being a double concussion, of which the first part was the stronger, and lasting 3 seconds. The duration of the sound (int. 5) was 4 seconds.

The shock was recorded by several instruments. At Comrie, the spiral spring in the steeple of the parish church indicated a horizontal movement of $\frac{3}{4}$ inch in the direction W. by N., and a vertical movement of $\frac{1}{4}$ inch; at Kingarth ($1\frac{1}{2}$ miles N.N.W. of Comrie), the point of the inverted pendulum was thrown $\frac{3}{4}$ inch to N.W.; at Invergeldie (4 miles N.W.) $\frac{3}{4}$ inch to S.W.; at Clathick (2 miles E.), the spiral pendulum and sand-glass were both affected, the sand in the tube sinking 2 inches; and at Crieff (6 miles E.), the point of the inverted pendulum was displaced $\frac{3}{4}$ inch to the W.

320. 1843 AUG. 25, 11 *a.m.* Cat. No. 612 (*Brit. Ass. Rep.* 1844, pp. 86–87).

A shock (int. 1) with sound (int. 1) at Comrie, both shock and sound lasting 1 second.

321. 1843 AUG. 25, 5 *p.m.* Cat. No. 613 (*Brit. Ass. Rep.* 1844, pp. 86–87).

A shock (int. 1) with sound (int. 1) at Comrie, both shock and sound lasting 1 second.

322. 1843 SEPT. 1, 5.15 *a.m.* Cat. No. 614 (*Brit. Ass. Rep.* 1844, pp. 86–87).

A slight concussion and tremor (int. 2) with sound (int. 2) at Comrie, both shock and sound lasting 2 seconds.

323. 1843 SEPT. 1–2, *midnight.* Cat. No. 615 (*Brit. Ass. Rep.* 1844, pp. 86–87).

A slight concussion and tremor (int. 2) lasting 3 seconds, with sound (int. 2) lasting 2 seconds, at Comrie.

324. 1843 SEPT. 2, 6 *a.m.* Cat. No. 616 (Perrey, p. 165).

A shock with sound, both lasting 1 second, at Comrie. During the night Sept. 1–2, a third slight shock, of which no details are given, was felt at Comrie.

325. 1843 SEPT. 16, 11 *a.m.* Cat. No. 617 (*Brit. Ass. Rep.* 1844, pp. 86–87).

A shock (int. 1) with sound (int. 1), both lasting 1 second, at Comrie.

326. **1843** NOV. 25, 2 *a.m.* Cat. No. 619 (*Brit. Ass. Rep.* 1844, pp. 86–87).

A shock (int. 1) with sound (int. 1), both lasting 1 second, at Comrie.

327. **1843** NOV. 26, 4 *a.m.* Cat. No. 620 (*Brit. Ass. Rep.* 1844, pp. 86–87).

A slight concussion and tremor (int. 2) with sound (int. 2), both lasting 2 seconds, at Comrie.

328. **1843** NOV. 30, *morning.* Cat. No. 621 (*Brit. Ass. Rep.* 1844, pp. 86–87).

Two slight tremors (both int. 1) with sound (both int. 1), at Comrie, the duration of shock and sound in each being 1 second.

329. **1844** JAN. 1, 2 *a.m.* Cat. No. 622 (*Brit. Ass. Rep.* 1844, pp. 86–87).

A slight concussion and tremor (int. 1) with sound (int. 1), both lasting 1 second, at Comrie.

330. **1844** JAN. 3, 3 *a.m.* Cat. No. 623 (*Brit. Ass. Rep.* 1844, pp. 86–87).

A slight concussion and tremor (int. 1) with sound (int. 1), both lasting 1 second, at Comrie.

331. **1844** JAN. 9, 5 *a.m.* Cat. No. 624 (*Brit. Ass. Rep.* 1844, pp. 86–87).

A slight concussion and tremor (int. 1) with sound (int. 1), both lasting 1 second, at Comrie.

On Jan. 14, there was a marked increase of activity. In addition to the eight shocks recorded below, Mr Milne states that 3 slight earth-sounds were heard before the first shock, 2 between the shocks, 1 about a quarter of an hour after the last shock, and 2 during the following night.

332. **1844** JAN. 14, 1 *p.m.* Cat. No. 625 (Milne, reprint, pp. 244–245; *Brit. Ass. Rep.* 1844, pp. 86–90).

A shock (int. 3) with considerable tremor, and sound (int. 2), at Comrie, the shock lasting 4 seconds and the sound 3 seconds. The shock was felt at Aberfeldy (17 miles N. of Comrie), Tyndrum (28 miles W.), Balquhidder (15 miles W.), Callendar (17 miles S.W.), Perth (21 miles E.), and Strathardle (40 miles E.N.E.), and, it is reported, at Kinlochmoidart (73 miles W.N.W.). As the pointer of the seismometer at Kinlochmoidart made a trace $\frac{5}{8}$ inch long, and as the instruments at Dunira and Ardvoirlich were not affected, it is probable that the disturbance at the former place had no connexion with the Comrie earthquake. The disturbed area, however, must have contained not less than 525 square miles. The sound-area was apparently co-extensive with the disturbed area. The spiral instrument in Comrie church steeple indicated a vertical motion of ⅜ inch, but no lateral movement.

333. 1844 JAN. 14, 1.10 *p.m.* Cat. No. 626 (*Brit. Ass. Rep.* 1844, pp. 86–87).

A slight concussion and tremor (int. 1), with sound (int. 1), both lasting 1 second, at Comrie.

334. 1844 JAN. 14, 1.20 *p.m.* Cat. No. 627 (*Brit. Ass. Rep.* 1844, pp. 86–87).

A slight concussion and tremor (int. 1), with sound (int. 1), both lasting 1 second, at Comrie.

335. 1844 JAN. 14, 1.37 *p.m.* Cat. No. 628 (*Brit. Ass. Rep.* 1844, pp. 86–87).

A considerable tremor (int. 2½), lasting 4 seconds, with sound (int. 2), lasting 3 seconds, at Comrie. The shock was also felt at Balquhidder, and probably at Aberfeldy and Strathardle.

336. 1844 JAN. 14, 2 *p.m.* Cat. No. 629 (*Brit. Ass. Rep.* 1844, pp. 86–87).

A slight concussion and tremor (int. 1), with sound (int. 1), both lasting 1 second, at Comrie.

337. 1844 JAN. 14, 9 *p.m.* Cat. No. 630 (*Brit. Ass. Rep.* 1844, pp. 86–87).

A slight tremor (int. 1), with sound (int. 1), both lasting 1 second, at Comrie.

338. 1844 JAN. 14, 11 *p.m.* Cat. No. 631 (*Brit. Ass. Rep.* 1844, pp. 86–87).

The same as the preceding, at Comrie.

339. 1844 JAN. 14, 11.45 *p.m.* Cat. No. 632 (*Brit. Ass. Rep.* 1844, pp. 86–87).

The same as the preceding, at Comrie.

340. 1844 JAN. 23, 2 *a.m.* Cat. No. 633 (*Brit. Ass. Rep.* 1844, pp. 86–87).

The same as the preceding, at Comrie.

341. 1844 FEB. 6, 4 *a.m.* Cat. No. 634 (*Brit. Ass. Rep.* 1844, pp. 86–87).

The same as the preceding, at Comrie.

342. 1844 FEB. 16, *morning.* Cat. No. 635 (*Brit. Ass. Rep.* 1844, pp. 86–87).

The same as the preceding, at Comrie.

343. 1844 FEB. 20, 6.30 *a.m.* Cat. No. 636 (*Brit. Ass. Rep.* 1844, pp. 86–87).

The same as the preceding, at Comrie.

344. 1844 MAR. 3, 7 *a.m.* Cat. No. 637 (*Brit. Ass. Rep.* 1844, pp. 86–87).

The same as the preceding, at Comrie.

345. 1844 MAR. 19, 1 *p.m.* Cat. No. 639 (*Brit. Ass. Rep.* 1844, pp. 86–87).

The same as the preceding, at Comrie.

346. 1844 APR. 1, 10.15 *a.m.* Cat. No. 640 (*Brit. Ass. Rep.* 1844, pp. 86–87).

The same as the preceding, at Comrie.

347. 1844 MAY 11, 10 *p.m.* Cat. No. 641 (*Brit. Ass. Rep.* 1844, pp. 86–87).

The same as the preceding, at Comrie.

348. 1844 MAY 14, 4 *a.m.* Cat. No. 642 (*Brit. Ass. Rep.* 1844, pp. 86–87).

The same as the preceding, at Comrie.

349. 1844 MAY 22, 7 *a.m.* Cat. No. 643 (*Brit. Ass. Rep.* 1844, pp. 86–87).

A slight tremor (int. 2), with sound (int. 1), both lasting 2 seconds, at Comrie.

350. 1844 MAY 25, 5.30 *a.m.* Cat. No. 644 (*Brit. Ass. Rep.* 1844, pp. 86–87).

A slight tremor (int. 1), with sound (int. 1), both lasting 1 second, at Comrie.

351. 1844 JULY 9, 3 *a.m.* Cat. No. 646 (*Brit. Ass. Rep.* 1844, pp. 86–87).

The same as the preceding, at Comrie.

352. **1844** AUG. 25, 3 *a.m.* Cat. No. 647 (Perrey, *Dijon Acad.* 1845–46, p. 397).

A slight shock and sound, both lasting 1 second, at Comrie. On the same day, in the morning, a similar tremor (int. 1), with sound (int. 1), both lasting 1 second, at Tomperran only.

353. **1844** SEPT. 1, 11.15 *p.m.* Cat. No. 648 (*Brit. Ass. Rep.* 1844, pp. 86–87).

A shock (int. 2), with sound (int. 1), both lasting 1 second, at Comrie.

354. **1844** SEPT. 4, 6.15 *p.m.* Cat. No. 649 (*Brit. Ass. Rep.* 1844, pp. 86–87).

A slight tremor (int. 1), with sound (int. 1), both lasting 1 second, at Lawers only.

355. **1844** SEPT. 13, 5.30 *a.m.* Cat. No. 650 (Perrey, *Dijon Acad.* 1845–46, p. 399).

A shock (int. 1) at Comrie.

356. **1844** SEPT. 22, 8.50 *p.m.* Cat. No. 651 (*Brit. Ass. Rep.* 1844, p. 90).

An earth-sound, like that produced by a cart passing over pavement, heard at Lawers; the sound lasted 4 seconds and was loudest in the middle. It was heard also at a spot about 2 miles S. of Comrie, and at Ardvoirlich.

357. **1844** OCT. 8, 5.45 *a.m.* Cat. No. 652 (Perrey, *Dijon Acad.* 1845–46, p. 399).

A feeble shock at Comrie.

358. **1844** OCT. 22, 9.15 *a.m.* Cat. No. 653 (Perrey, *Dijon Acad.* 1845–46, p. 399).

A rather strong shock (int. 3) at Comrie.

359. **1844** OCT. 24, 8.15 *p.m.* Cat. No. 654 (Perrey, *Dijon Acad.* 1845–46, p. 399).

A very feeble shock at Comrie.

360. **1844** OCT. 27, 8.15 *p.m.* Cat. No. 655 (Perrey, *Dijon Acad.* 1845–46, p. 399).

A very feeble shock at Comrie.

361. **1845** JAN. 31, 1 *a.m.* Cat. No. 656 (*Athenaeum*, 1845, p. 158).

A slight shock felt at Comrie. The writer of the notice adds that it is the only shock observed at Comrie since Dec. 12.

362. **1845** MAY 12, 1.30 *a.m.* Cat. No. 659 (Perrey, *Dijon Acad.* 1845–46, p. 168).

A feeble shock at Comrie.

363. **1845** AUG. 7, 1.15 *p.m.* Cat. No. 660 (Perrey, *Dijon Acad.* 1845–46, p. 407).

A shock (int. 4) at Comrie.

364. **1845** SEPT. 7, 7 *p.m.* Cat. No. 661 (Perrey, *Dijon Acad.* 1845–46, p. 408).

A shock (int. 1) at Comrie.

365. **1845** SEPT. 22, 2 *p.m.* Cat. No. 662 (Perrey, *Dijon Acad.* 1845–46, p. 409).

A shock (int. 1) at Comrie.

366. **1845** SEPT. 22, 2.30 *p.m.* Cat. No. 663 (Perrey, *Dijon Acad.* 1845–46, p. 409).

A shock (int. 1) at Comrie.

367. **1845** SEPT. 22, 4 *p.m.* Cat. No. 664 (Perrey, *Dijon Acad.* 1845–46, p. 409).

A shock (int. 1) at Comrie.

368. **1845** OCT. 5, 7.15 *p.m.* Cat. No. 665 (Perrey, *Dijon Acad.* 1845–46, p. 410).

A shock (int. 1) at Comrie.

369. **1845** OCT. 13, 2 *a.m.* Cat. No. 666 (Perrey, *Dijon Acad.* 1845–46, p. 412).

A moderately strong shock at Comrie.

370. **1845** OCT. 13, 9.15 *a.m.* Cat. No. 667 (Perrey, *Dijon Acad.* 1845–46, p. 412).

A moderately strong shock at Comrie.

371. **1845** OCT. 18, 9.45 *p.m.* Cat. No. 668 (Perrey, *Dijon Acad.* 1845–46, p. 412).

Three shocks (int. 2) at Comrie.

372. **1845** OCT. 19, *between 1 and 2 a.m.* Cat. No. 669 (Perrey, *Dijon Acad.* 1845–46, p. 412).

Three shocks, slighter than the preceding, at Comrie.

373. **1845** NOV. 12, 8.15 *p.m.* Cat. No. 670 (Perrey, *Dijon Acad.* 1845–46, p. 413).

A rather strong shock (int. 3) at Comrie.

374. **1845** DEC. 21, 10 *a.m.* Cat. No. 671 (Perrey, *Dijon Acad.* 1845–46, p. 416).

A slight shock at Comrie.

375. **1846** FEB. 3, 8 *p.m.* Cat. No. 672 (Perrey, *Dijon Acad.* 1845–46, p. 428).

A very slight shock at Comrie.

376. **1846** FEB. 4, 7 *a.m.* Cat. No. 673 (Perrey, *Dijon Acad.* 1845–46, p. 428).

A very slight shock at Comrie.

377. **1846** FEB. 16, 10.45 *a.m.* Cat. No. 674 (Perrey, *Dijon Acad.* 1845–46, p. 429).

A rather strong shock (int. 3) at Comrie.

378. **1846** MAR. 2, *about* 11 *a.m.* Cat. No. 675 (Perrey, *Dijon Acad.* 1845–46, p. 430).

A shock at Comrie.

1846 Nov. 24, some minutes before midnight, a shock of intensity 5 was felt at Perth and Crieff, preceded and followed at Crieff by a sound like distant thunder. At Dundee, the shock consisted for a few seconds of a slight tremor ending with a strong vibration, and accompanied by a slight noise. The shock was felt also at Dollar, Cupar, Inverness, and in the Carse of Gowrie (Cat. No. 676, Perrey, *Dijon Acad.* 1845–46, p. 463).

1846 Nov. 25, Mr Macfarlane (Perrey, *Dijon Acad.* 1845–46, p. 463) states that 30 shocks were felt at Comrie, the three stronger occurring at 1 a.m., 1.15 a.m. and 4 p.m. Though no reference is made to the earthquake of Nov. 24 at Comrie, it seems probable that the large number of shocks observed on the next day points to an origin in the Comrie district.

379. **1846** NOV. 25, 1 *a.m.* Cat. No. 677 (Perrey, *Dijon Acad.* 1845–46, pp. 463–464).

A shock (int. 3) at Comrie.

380. **1846** NOV. 25, 1.15 *a.m.* Cat. No. 678 (Perrey, *Dijon Acad.* 1845–46, pp. 463–464).

A shock (int. 2) at Comrie.

381. **1846** NOV. 25, 4 *p.m.* Cat. No. 679 (Perrey, *Dijon Acad.* 1845–46, pp. 463–464).

A shock (int. 3) at Comrie.

382. **1846** DEC. 17, 11.45 *p.m.* Cat. No. 680 (Perrey, *Dijon Acad.* 1845–46, p. 464).

A shock (int. 1) at Comrie.

383. **1846** DEC. 18, 3 *a.m.* Cat. No. 681 (Perrey, *Dijon Acad.* 1845–46, p. 464).

A shock (int. 1) at Comrie.

384. **1847** JAN. 20, 3.30 *a.m.* Cat. No. 682 (Perrey, *Bull.* vol. 15, 1848, p. 442).

A slight shock at Comrie.

385. **1847** JAN. 20, 9.30 *a.m.* Cat. No. 683 (Perrey, p. 169).

A shock (int. 2) at Comrie.

386. **1847** JAN. 28, 3 *p.m.* Cat. No. 684 (Perrey, *Bull.* vol. 15, 1848, p. 443).

A shock (int. 1) at Comrie.

387. **1847** JAN. 28, 9 *p.m.* Cat. No. 685 (Perrey, *Bull.* vol. 15, 1848, p. 443).

A shock (int. 1) at Comrie.

388. **1847** FEB. 1. Cat. No. 686 (Perrey, *Bull.* vol. 15, 1848, p. 443).

A slight shock at Comrie.

389. **1847** FEB. 19, 11.15 *p.m.* Cat. No. 687 (Perrey, p. 169).

A very slight shock at Comrie.

390. **1847** MAY 13, 9 *p.m.* Cat. No. 688 (Perrey, p. 169).

A very slight shock at Comrie.

391. **1847** MAY 13, *between* 10.0 *and* 10.15 *p.m.* Cat. No. 689 (Perrey, p. 169).

Three very slight shocks at Comrie.

392. **1847** MAY 20, 11.30 *p.m.* Cat. No. 690 (Perrey, *Bull.* vol. 15, 1848, p. 446).

A very slight shock at Comrie.

393. **1847** OCT. 7, 7 *p.m.* Cat. No. 691 (Perrey, *Bull.* vol. 15, 1848, p. 451).

A slight shock at Comrie.

394. **1847** OCT. 17, 7.30 *p.m.* Cat. No. 692 (Perrey, *Bull.* vol. 17, 1850, p. 226).

A shock felt at Comrie and to a distance of 6 miles.

395. **1848** MAY 7, 7 *p.m.* Cat. No. 693 (Perrey, *Bull.* vol. 17, 1850, p. 227).

A slight shock at Comrie.

396. **1848** JULY 7, *noon.* Cat. No. 694 (Perrey, *Bull.* vol. 17, 1850, p. 227).

A slight shock at Comrie.

397. **1848** JULY 19, 6.30 *p.m.* Cat. No. 695 (Perrey, *Bull.* vol. 17, 1850, p. 227).

A strong shock, followed immediately by two slight shocks, at Comrie.

398. **1848** NOV. 13, 8.30 *a.m.* Cat. No. 697 (Perrey, *Bull.* vol. 17, 1850, p. 235).

A slight shock at Comrie.

399. **1848,** *between* NOV. 13, 8.30 *a.m. and* NOV. 15, *midnight.* Cat. No. 698 (Perrey, *Bull.* vol. 17, 1850, p. 235).

A slight shock at Comrie.

400. **1848** NOV. 15, *midnight.* Cat. No. 699 (Perrey, *Bull.* vol. 17, 1850, p. 235).

A slight shock at Comrie.

401. **1848** NOV. 23, 9.30 *a.m.* Cat. No. 700 (Perrey, *Bull.* vol. 17, 1850, p. 235).

A slight shock at Comrie.

402. **1851** FEB. 15, *between* 1 *and* 2 *a.m.* Cat. No. 701 (Perrey, *Bull.* vol. 19, 1852, p. 356).

Two slight shocks at Comrie.

403. **1851** MAY 17, 11.30 *p.m.* Cat. No. 702 (Perrey, *Bull.* vol. 19, 1852, p. 371).

A slight shock at Comrie.

404. **1864** JULY 3, *night.* Cat. No. 732 (Perrey, *Mém. Cour.* vol. 18, 1866, p. 18).

Two distinct shocks felt at Comrie and in the neighbourhood.

405. **1865** MAY 7, *between* 8 *and* 9 *p.m.* Cat. No. 737; Intensity 4 (Perrey, *Mém. Cour.* vol. 19, 1867, p. 69; etc.).

A strong shock, accompanied by a noise like a heavy peal of thunder or the discharge of cannon, at Comrie; felt also at Ochtertyre, Crieff and other places to the E. of Comrie.

406. **1865** MAY 10, *early morning*. Cat. No. 738 (Perrey, *Mém. Cour.* vol. 19, 1867, p. 69; etc.).

Several slight shocks at an early hour at Comrie.

1867 May 8, about 10 p.m., May 9, between 7 and 8 a.m., and May 10, early in the morning, shocks, accompanied by a noise like thunder or the discharge of distant artillery, were felt in the district including Comrie and Greenloaning (*Times*, May 14; etc.).

407. **1869** JULY 7, *5 p.m.* Cat. No. 752 (Perrey, *Mém. Cour.* vol. 22, 1872, p. 68; etc.).

A slight shock at Comrie. After this earthquake, several slight shocks were felt at night at Comrie, each accompanied by a noise like that of a railway train passing or of distant thunder. In the autumn, there were also several slight shocks at Comrie (*Brit. Ass. Rep.* 1870, pp. 48–49).

408. **1870** APR. 29, *about 11 p.m.* Cat. No. 753 (*Brit. Ass. Rep.* 1870, pp. 48–49; etc.).

Two shocks at Comrie, separated by an interval of about half a minute, accompanied by sound. The disturbed area was about 10 miles long in the direction of the valley of the Earn, and about 5 miles wide.

Shortly after the earthquake, some unexplained noises and tremors were observed at Crieff, and were recorded in a private journal kept by a resident in that town. The dates at which they occurred are 1871 Feb. 2–3; June 20, about 8.50 p.m.; 1872 Jan. 19, 7 a.m.; 1873 May 4, about 7.40 a.m.; June 10, about 10.30 a.m. and 7.35 p.m.; Aug. 12, about 7.30 p.m.; 1876 Oct. 6, about 4.5 p.m.; and 1885 July 29, about 7.50 p.m.

409. **1876** JAN. 14, *morning*. Cat. No. 772 (*Brit. Ass. Rep.* 1876, p. 74).

A very slight shock and noise at Comrie, not strong enough to affect the seismometer or cylinders.

410. **1876** JAN. 16, *about 3 a.m.* Cat. No. 773 (*Nature*, vol. 13, 1876, p. 236).

Two shocks at Comrie.

411. **1876** JAN. 16, *afternoon*. Cat. No. 774 (*Brit. Ass. Rep.* 1876, p. 74).

A very slight shock and noise at Comrie, not strong enough to affect the seismometer or cylinders.

412. **1877** MAY 11, *between 2 and 3 a.m.* Cat. No. 780 (*Nature*, vol. 16, 1877, p. 54).

A shock, accompanied by a rumbling noise, felt at Comrie and in the surrounding district.

413. **1882** OCT. 14, *about 3 p.m.* Cat. No. 789 (*Times*, Oct. 16).

A very distinct shock from the S.W., with a noise like the distant booming of a cannon, at Comrie.

414. **1882** OCT. 14, *about 7.30 p.m.* Cat. No. 790 (*Times*, Oct. 16).

A stronger shock than the preceding, with a noise like the distant booming of a cannon, at Comrie.

415. **1886** APR. 25, *about 0.30 a.m.* Cat. No. 803 (*Nature*, vol. 33, 1886, p. 611).

A tremor felt at Comrie and St Fillans, with a noise like distant thunder.

416. **1894** JULY 12, *about 11 p.m.* Cat. No. 906 (*Geol. Mag.* vol. 7, 1900, p. 114).

A tremor was felt at Comrie and Dalginross (about ¼ mile S.E. of Comrie), accompanied by a dull rumbling sound, and lasted about 6 seconds.

417. **1895** JULY 12, *about 7.40 a.m.* Cat. No. 909; Intensity 3; Centre of isoseismal 3 in lat. 56° 22·3′ N., long. 3° 59·3′ W.; No. of records 21 from 19 places and negative records from 17 places; Fig. 8 (*Geol. Mag.* vol. 3, 1896, pp. 75–78).

In the map, the continuous curve represents the isoseismal of intensity 3. It is 5½ miles long, 4⅛ miles wide and contains 18½ square miles. Its centre lies ¼ mile S.W. of Comrie. The longer axis is directed N. 75° E. Outside this line, the shock was felt at Easter Ballindalloch and Westerton. If the boundary of the disturbed area were coaxial with the isoseismal 3 and passed through those places, it would contain about 40 square miles.

At Comrie, the shock consisted of a slight uninterrupted tremor, lasting 2 or 3 seconds, accompanied by a rumbling sound. The sound was heard at all the places at which the shock was felt, and also at Braefordie, Carroglen, Glenturret Lodge, and Wester and Easter Dundurn, all from ½ to 3 miles outside the isoseismal 3.

418. **1898** AUG. 22, *about 7.15 a.m.* Cat. No. 930; Intensity 3; Centre of sound-area in lat. 56° 22·9′ N., long. 3° 57·9′ W.; No. of records

14 from 8 places and negative records from 12 places; Fig. 8 (*Geol. Mag.* vol. 7, 1900, pp. 170–172).

The shock was felt at Clathick and possibly in the neighbourhood of Lawers. At Clathick, three vibrations were felt, lasting for about 3 seconds. The dotted line in Fig. 8 represents the boundary of the sound-area, which is 3¾ miles long, 3 miles wide, and contains about 9 square miles. Its centre is 1 mile N. 63° E. of Comrie, and must coincide nearly with the epicentres of the earthquakes of 1789 Nov. 5 and 1839 Oct. 23.

419. **1920 JULY 21.** Cat. No. 1186; Intensity 3 (*Nature*, vol. 106, 1920, p. 190).

A shock felt at Comrie.

420. **1920 SEPT. 14,** *early morning.* Cat. No. 1189; Intensity 4 (*Nature*, vol. 106, 1920, p. 190; etc.).

A shock, accompanied by a dull rumbling sound, at Comrie.

421. **1921 APR. 30,** 10.35 *a.m.* Cat. No. 1190; Intensity 4 (*Nature*, vol. 107, 1921, p. 370; etc.).

The strongest shock experienced by the present generation at Comrie, accompanied by a sound like the firing of guns.

CHARACTERISTICS OF THE COMRIE EARTHQUAKES

Number of Earthquakes. No other district has contributed so largely to the catalogue of British earthquakes as that which includes the village of Comrie. But, though it contains the records of 421 earthquakes, the preceding list is far from complete. For instance, from 1789 Sept. 2 to 1791 Sept. 2, there are notices of 6 earthquakes only; but, probably, no other interval of two years in the century that followed has been so prolific in earth-sounds and slight shocks. In his published paper, the Rev. R. Taylor remarks that many noises have been observed within a few hours—as many as 30 within two hours after the shock of 1789 Nov. 5. Most of them, however, were confined to Comrie, Lawers and Glen Lednock, and few of them were audible as far as Ochtertyre. Through the kindness of a great-grand-daughter of Mr Taylor, I have been allowed to see three drafts of the published letter. In the first draft (written 1789 Dec. 21), he speaks of "the multitude of shocks which still continue to be felt," and he adds that "a multitude of slight shocks are felt at Lawers, at Comrie and in Glenleadnick from time to time of which none of us are sensible at Ochtertyre." In the third draft (written 1790 Jan. 19), he says that "more than thirty of them have sometimes been noticed in the space of two or three hours; and on a moderate computation it cannot be reckoned that there have been fewer than eight or nine hundred since the time of their commencement." As these estimates were excluded from the final paper, it is evident that Mr Taylor felt that there was not sufficient foundation for the figures given. The only inference that we are entitled to draw from them is that, at Comrie and in the immediate neighbourhood, earth-sounds and slight shocks were extraordinarily numerous during the two years which succeeded the beginning of September 1789.

Again, from October 1839 to the end of December 1841, the above catalogue contains notices of 203 earthquakes. During the same interval, Mr P. Macfarlane recorded 258 shocks at Comrie, while Mr J. Drummond felt 54 shocks and heard 276 earth-sounds[1]. There can thus be little doubt, I think, that the number of earthquakes observed in the Comrie district (including both shocks and sounds) must amount to considerably over one thousand.

Intensity of Earthquakes. The earthquakes of the Comrie district may be divided into three classes:

(i) In the first are the principal earthquakes of 1789 Nov. 11 (no. 5, int. probably 6); 1795 Mar. 12 (no. 25); 1801 Jan. 11 (no. 57) and Sept. 7 (no. 60, int. 7); 1839 Oct. 23 (no. 107, int. 8); 1841 July 30 (no. 248, int. 8); and 1844 Jan. 14 (no. 332). They disturbed areas ranging from about 500 to about 26,500 square miles; the average being about 4800 square miles.

(ii) The second class also contains seven earthquakes, namely, those of 1789 Nov. 5 (no. 3, int. 5) and Nov. 10 (no. 4, int. 5); 1795 Jan. 2 (no. 23); 1839 Oct. 10 (no. 89, int. about 4), Oct. 14 (no. 99, int. 5) and Oct. 16 (no. 101, int. 5); and 1843 Aug. 25 (no. 319). The disturbed areas range from about 80 to 300 square miles, the average being about 180 square miles.

(iii) In the third class are the remaining 407 earthquakes. Of these 336 (or 82 per cent.) are recorded at one place only, usually Comrie. Though many of them must have been felt in the surrounding country, it is clear that in most of them the disturbed areas contained only a few square miles.

---

[1] The discrepancy between the above figures is due to two causes: (i) Mr Macfarlane in his lists gives the numbers of shocks each day, but the times of the three stronger only; (ii) many of the tremors and earth-sounds recorded by Mr Drummond were observed by himself alone.

Clustering of Earthquakes. The Comrie earthquakes known to us are grouped in four well-defined series:

(i) The first series was brief in duration, lasting from 1788 Nov. to 1789 Dec. It included the strong shocks of Nov. 5, 10 and 11 (nos. 3, 4 and 5, belonging to classes II, II, and I, respectively), that of Nov. 11 being the principal earthquake. The catalogue (depending mainly on observations made at Ochtertyre) includes only 6 earthquakes, but the number of slight shocks and earth-sounds observed at Comrie probably amounted to several hundred.

This series was followed by an interval, lasting from 1790 Jan. to 1794 May, during which 8 slight shocks (nos. 7–14) were felt.

(ii) The second series lasted from 1794 Sept. to 1796 Mar. It consisted of 8 fore-shocks (nos. 15–22) from Sept. to Dec. 1794, the first principal earthquake (no. 23, class II) on 1795 Jan. 2, one slight shock (no. 24) on 1795 Jan. 22, the second and stronger principal earthquake (no. 25, class I) on 1795 Mar. 12, and 16 after-shocks (nos. 26–41) from 1795 Mar. 13 to 1796 Mar. 16.

During the second interval, from 1796 Apr. to 1800 Nov., 13 slight shocks (nos. 42–54) were felt.

(iii) The third series lasted from 1800 Dec. to 1802 Oct. It contained 2 fore-shocks (nos. 55, 56) in 1800 Dec. and 1801 Jan., the first principal earthquake (no. 57, class I) on 1801 Jan. 11, two slight shocks (nos. 58, 59) on Sept. 6 and 7, the second and stronger principal earthquake (no. 60, class I, int. 7) on 1801 Sept. 7, and 4 after-shocks (nos. 61–64) from 1801 Sept. 18 to 1802 Oct. 8.

The third interval lasted from 1802 Nov. to 1839 Sept., during which 17 slight shocks (nos. 65 to 81) were felt.

(iv) The fourth and most important series lasted from 1839 Oct. to 1848 Nov. It contained 25 fore-shocks (nos. 82–106) from Oct. 3 to Oct. 22, including 3 shocks (nos. 89, 99 and 101, class II) on Oct. 10, 14 and 16; the first principal earthquake and strongest of all Comrie earthquakes (no. 107, class I, int. 8) on Oct. 23; 140 after-shocks (nos. 108–247) from 1839 Oct. 23 to 1841 July 30; the second principal earthquake (no. 248, class I, int. 8) on 1841 July 30; 83 after-shocks (nos. 249–331) from 1841 July 30 to 1844 Jan. 9; the third principal earthquake (no. 332, class I, int. unknown) on 1844 Jan. 14; and 69 after-shocks (nos. 333–401) from 1844 Jan. 14 to 1848 Nov. 23. Altogether in this series, there were 3 principal earthquakes and 317 minor shocks.

The fourth interval began in 1848 Dec. and still continues. It includes 20 slight shocks (nos. 402–421).

Periodicity. The following table gives the number of earthquakes recorded during each half-month, the upper figure referring to the first half (the first 14 days in February and the first 15 days in the other months), and the lower figure to the second half:

Jan.	Feb.	Mar.	Apr.	May	June	July	Aug.	Sept.	Oct.	Nov.	Dec.
25	14	19	16	16	10	14	7	23	34	21	21
21	11	17	15	18	4	14	9	24	29	21	17

The form of harmonic analysis known as the method of overlapping means gives the following results:

	MAX. EPOCH	AMPLITUDE
Annual period	end of Nov.	·38
Semi-annual period	middle of Apr. and Oct.	·32

The accuracy of the results must, however, be affected by the tendency of the earthquakes to cluster in series.

The hourly distribution of the earthquakes is as follows:

	0–1	1–2	2–3	3–4	4–5	5–6	6–7	7–8	8–9	9–10	10–11	11–12
a.m.	12	28	25	19	14	16	11	9	13	16	7	15
p.m.	10	20	16	13	14	8	7	12	10	13	9	21

The double maximum, shortly before and shortly after midnight, is characteristic of all non-instrumental records and is due to the favourable conditions for observation prevailing at these times. The results obtained from the analysis of such figures are consequently of little value. They give apparently the following elements:

	MAX. EPOCH	AMPLITUDE
Diurnal period	4 a.m.	·23
Semi-diurnal period	2 a.m. and p.m.	·37

Instrumental records usually give a diurnal maximum about noon for ordinary earthquakes, and about or shortly after midnight for aftershocks. Possibly the maximum epoch obtained for the Comrie earthquakes is due in part to their prevailing character as after-shocks.

Nature of the Shock and Sound. In the slighter earthquakes of the Comrie district, the shock consists of a brief tremor only; a tremor so weak that it often passes unnoticed by strangers. In earthquakes of moderate strength, a single strong vibration occurs in the midst of the tremors, and this is so characteristic a feature that, among residents in the district, it has become known as the "thud." In the strongest earthquakes (those of class I), two, and even three, such vibrations are felt.

It would seem that all Comrie earthquakes are accompanied by sound, though it is not always recorded. In many, the sound is by far the more prominent feature; very often, it is the only phenomenon observed; so that "hearing the earthquakes" became a common phrase in the valley.

The variation in the nature of the sound with the intensity of the shock is shown in the following table, in which the figures represent percentages of reference to the different types of sound (p. 9):

	TYPE						
	1	2	3	4	5	6	7
Class I	41	23	13	7	4	7	4
Classes II and III	20	38	7	10	4	18	3

Distribution in Space. The most remarkable feature of Comrie earthquakes is their extreme localisation. After studying the earthquake of 1839 Oct. 23, Mr Milne was inclined to conclude that the earthquakes came from some part of the hill between Dunira and Lord Melville's monument, or about a mile N.W. of Comrie, though he suggests that the "central radiating point" had moved a little towards the west. In 1844, he again located the epicentre as a little to the west or north-west of the Melville monument, and he remarks that this position is in agreement with the apparent direction of the shocks at Comrie, Clathick, Crieff, Dalchonzie and Dunira. In six earthquakes (nos. 3, 60, 107, 248, 417 and 418), the epicentre is known with some approach to accuracy, and lies between about $1\frac{3}{4}$ miles east of Comrie and a mile or so to the north-west of the same place, the extreme distance between the epicentres being thus little more than 2 miles.

Of the 421 earthquakes in the above list, the number known to have been felt at more than one place is 86. Of the rest 330 were recorded at Comrie only, two at Crieff, and one each at Lawers, Clathick and Ochtertyre. This seems to imply that, while movements take place along a range of 7 miles, they are confined for the most part to the immediate neighbourhood of Comrie. The frequent vertical motion experienced at Comrie and lateral motion elsewhere is also evidence of the same fact.

ORIGIN OF THE COMRIE EARTHQUAKES

The Comrie earthquakes have for many years been ascribed to movements along the great Highland Border fault which traverses Scotland from Stonehaven and passes about a mile S.E. of Comrie. The seismic evidence, so far as it goes, is clearly in favour of this connexion. In the neighbourhood of Comrie, the mean direction of the fault is N. 69° E. Only one isoseismal line is traced with approximate accuracy, namely, the bounday line of the slight shock of 1895 July 12 (no. 417). The direction of the longer axis of the curve is about N. 75° E. In the still slighter shock of 1898 Aug. 22 (no. 418), the longer axis of the boundary of the sound-area is roughly in the same direction, and its epicentre lies N. 63° E. of that of the earthquake of 1895.

The direction in which the fault hades in the neighbourhood of Comrie is unknown from geological evidence. From Stonehaven to the neighbourhood of Fettercairn (about 15 miles), Dr Campbell has shown that the hade is to the N.W., and Prof. Jehu and Dr Campbell[1] give the same direction of hade at Aberfoyle (20 miles S.W. of Comrie). This direction, assuming it to be maintained at Comrie, is confirmed by the seismic evidence. All the six known epicentres lie on the north side of the fault-line[2]; Crieff lies on the south side, but Comrie and all the other places from which single records come lie on the other side.

It will be noticed from Fig. 8 that, in the neighbourhood of Comrie, the fault is tripled. Movements along one branch would tend to precipitate slips along the others, and it is possible that this peculiarity of the fault-structure may be responsible in part for the abundance of the shocks and their localisation in the neighbourhood of Comrie.

The connexion of the remaining earthquakes (those of Dunoon, Rothesay and Kintyre) with the Highland Border fault cannot be proved, though the epicentres are not far distant from the continuation of the fault towards the south-west. The Dunoon and Rothesay earthquakes may be associated with a fault (apparently a branch of the great fault) which passes just to the south of Dunoon and through Rothesay. Another fault runs from the west end of Loch Rannoch to the head of Loch Fyne and, if continued, would traverse that Loch and pass close to the epicentre of the Kintyre earthquakes.

[1] *Edin. Roy. Soc. Trans.* vol. 48, 1913, pp. 923–960; vol. 52, 1921, pp. 175–212.
[2] *Geol. Mag.* vol. 3, 1896, pp. 77–78.

DUNOON EARTHQUAKES

1. **1756** NOV. 17, 11.50 *p.m.* Cat. No. 127; Intensity about 5 (*Gent. Mag.* vol. 26, 1756, p. 591).

A shock, preceded by a noise like that of distant thunder, was felt at Inverhallan (Argyllshire), also at Rothesay, Kilfinnan and Glendaruel.

Two days later, two other shocks were felt in this district.

2. **1871** APR. 15, 7.55 *p.m.* Cat. No. 763 (*Times*, Apr. 18; etc.).

A strong shock, felt along the coast of Argyllshire, at Dunoon, Blairmore, Kilcreggan and Roseneath. The waters of the Gareloch were disturbed, as if boiling.

3. **1904** SEPT. 18, 4.7 *a.m.* Cat. No. 982; Intensity 5; Centre of isoseismal 5 in lat. 55° 54·5′ N., long. 5° 10·4′ W.; No. of records 79 from 28 places and negative records from 4 places; Fig. 10 (*Geol. Mag.* vol. 5, 1908, pp. 297–298).

The isoseismal lines (continuous lines in Fig. 10) correspond to intensities 5 and 4. The isoseismal 5 is 26 miles long, 15 miles wide, and 300 square miles in area. Its centre is 9 miles W. of Dunoon. The isoseismal 4 is 36 miles long, 21 miles wide and 564 square miles in area. The direction of its longer axis is N. 50° E. The distance between the two curves is 2·1 miles towards the south-east and 3·7 miles towards the north-west, indicating a hade of the originating fault in the latter direction.

The shock consisted of a continuous series of vibrations, 2·5 seconds in mean duration, and gaining in strength towards the close.

The sound was heard by all the observers. It was compared to passing waggons, etc., in 38 per cent. of the records, to thunder in 31, wind in 6, loads of stones falling in 5, the fall of a heavy body in 2, explosions in 17, and miscellaneous sounds in 2, per cent.

4. **1908** JULY 3, 6.15 *a.m.* Cat. No. 1043; Intensity 4; Centre of disturbed area in lat. 56° 6·7′ N., long. 4° 56·5′ W.; No. of records 70 from 15 places and negative records from 1 place; Fig. 10 (*Geol. Mag.* vol. 7, 1910, pp. 315–316).

The approximate boundary of the disturbed area is represented by the broken-line in Fig. 10. It is 25 miles long, 21 miles wide, and contains about 400 square miles. Its centre is 11 miles N. of Dunoon, and the longer axis is directed about N. 41° E.

At most places, the shock consisted of a single series of tremors, the mean duration of which was about 4 seconds. At Strachur, however, two parts were felt by four observers, the average length of the

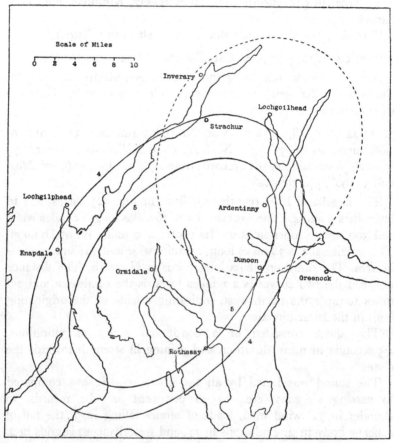

Fig. 10. Dunoon Earthquakes of 1904 Sept. 18 and 1908 July 3.

interval between the two parts being 3 seconds, and the first part according to each observer the stronger.

The sound was heard by all the observers. It was compared to passing waggons, etc., in 27 per cent. of the records, to thunder in 29, wind in 14, loads of stones falling in 15, the fall of a heavy body in 4, explosions in 10, and miscellaneous sounds in 1, per cent.

ROTHESAY EARTHQUAKES

1. **1821** OCT. 15, *early morning*. Cat. No. 305 (*Phil. Mag.* vol. 58, 1821, p. 458).

An earthquake felt over the whole of the island of Bute and also, though slightly, at Greenock. At and near Rothesay, a tremor was felt, which lasted a few seconds, and was accompanied by a sound like the distant rolling of a carriage.

2. **1821**, *end of* OCTOBER. Cat. No. 309 (*Ann. de Chim. et de Phys.* vol. 18, 1821, p. 415).

A noise, like that of a carriage passing in the distance, heard at Rothesay.

3. **1826** NOV. 26, *about* 4 *p.m.* Cat. No. 320; Intensity 4 (*Edin. Journ. of Sci.* vol. 6, 1827, p. 370).

A shock, lasting 3 or 4 seconds, and accompanied by a noise like that of heavy carts passing, in the island of Arran.

KINTYRE EARTHQUAKES

1. **1889** JULY 15, *about* 6 *p.m.* Cat. No. 825; Intensity 5; Centre of disturbed area in lat 55° 42′ N., long. 5° 31′ W.; Records from 8 places and negative records from 2 places; Fig. 11 (*Geol. Mag.* vol. 8. 1891, pp. 365–367).

The boundary of the disturbed area, an isoseismal of intensity 4, is about 25 miles long, 18 miles wide, and contains about 350 square miles. The centre of the area is about 3½ miles S.E. of Clachan, its longer axis being directed about N. 30° E.

The shock consisted of a single vibration, without tremulous motion, either before or after. The sound-area and disturbed area were approximately coincident.

2. **1890** JULY 24, 11.37 *a.m.* Cat. No. 836; Intensity 5; Records from 7 places and negative records from 2 places; Fig. 11 (*Geol. Mag.* vol. 8, 1891, pp. 453–454).

This shock was felt at nearly the same places as the preceding, but not at Bellochantuy, Lochranza or Gigha Island, though in the latter the sound was heard. Towards the north and south, the boundaries of both areas must have coincided approximately; towards the east and west, the area of the later earthquake did not extend quite so far as that of the former. Thus, the epicentres of both earthquakes must have been nearly coincident.

At Clachan, the shock began with a series of tremors, which increased in strength until a prominent vibration was felt, ending with a briefer

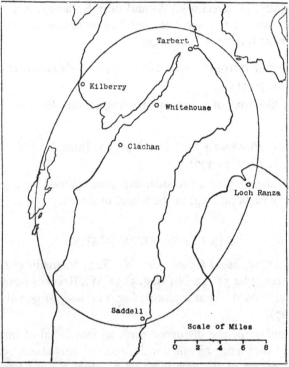

Fig. 11. Kintyre Earthquake of 1889 July 15.

series of tremors. A sound, like the noise of stones falling down a chimney, was heard during the whole time, and, at the moment when the vibration was felt, there was a dull thud as of a suppressed explosion.

July 24, about noon, a second shock was felt at Clachan, but by one observer only.

CHAPTER V

MENSTRIE EARTHQUAKES

THE remarkable series of earthquakes described in this chapter originated in the district lying at the southern foot of the Ochil Hills and to the north of the river Forth. Though the stronger earthquakes disturbed areas of about a thousand square miles, extending across the hills as far as Comrie to the north and to or near Falkirk to the south, the great majority of the shocks were felt only in the "Hillsfoot" towns and villages—especially Airthrey, Blairlogie, Menstrie, Alva and Tillicoultry. The village of Menstrie lies about 4 miles N.E. of Stirling, and its name is here connected with the whole series of earthquakes as the number of shocks recorded in it is about twice as great as at any other place. The record begins with the year 1736, and for a century and a half includes notices of only 9 earthquakes. As the two shocks in 1736 and the two in 1802 were the strongest known in the district, it is probable that they were succeeded by tremors and earth-sounds quite as numerous as those which followed the earthquakes of 1905, 1908 and 1912, but of which no notice has been preserved owing to the lack of interested observers.

1. **1736** APR. 30, *noon*; and 2. **1736** MAY 1, 1 *a.m.* Cat. Nos. 90 and 91; Intensity about 8 (*Gent. Mag.* vol. 6, 1736, p. 289).

At each of the above times there "was a terrible Earthquake along the Ochil Hills in Scotland, which rent several Houses, and put the People to Flight, it was accompanied with a great Noise under Ground."

3 and 4. **1767** APR. 20, 9*h.* and 9*h.* 15*m.* Cat. Nos. 138 and 139 (Milne, vol. 31, p. 104).

"At Stirling and Alloa, at 9 o'clock, and another in a quarter of an hour after."

5 and 6. **1802** AUG. —, 5 *a.m. and* 6 *a.m.* Cat. Nos. 235 and 236 (Milne, vol. 31, p. 114).

The shocks were felt at Alloa, Clackmannan, Kennet (1 mile S.E. of Clackmannan) and Harvieston (midway between Tillicoultry and Dollar). At Clackmannan, chimneys were thrown down. The disturbed area was evidently small, and its centre probably a little more than 1 mile S. of Tillicoultry.

7. **1809** JAN. 17, *between 2 and 3 a.m.* Cat. No. 245; Intensity 4 (*Edin. Adver.*).

A smart shock felt at Bridge of Allan and along the foot of the hills.

1821 Oct. 23, about 3 p.m., a strong shock felt at Comrie, Blackford and near Dunblane (Cat. No. 307; *Ann. de Chim. et de Phys.* vol. 18, 1821, p. 415; *Phil. Mag.* vol. 58, 1821, p. 458).

1846 Nov. 24, about midnight, a shock, with noise, felt at Perth, Comrie, Crieff, Dollar, Cupar, Dundee and across the Ochils as far as Edinburgh. A slight tremor was felt at first, and then, after a few seconds, a stronger shock (Cat. No. 676; Perrey, pp. 168–169).

8. **1872** AUG. 8, 4.10 *p.m.* Cat. No. 764; Intensity 5 (*Brit. Ass. Rep.* 1873, pp. 194–197; *Nature*, vol. 6, 1872, p. 378; etc.).

The shock was strongest at Bridge of Allan, Dunblane, Green-loaning, Braco and Kinbuck; it was also felt at St Fillans, Glen Lednock and Crieff. The boundary of the disturbed area seems to have coincided nearly with the isoseismal 4 of the earthquake of 1905 Sept. 21 (no. 19; Fig. 14); so that the area disturbed must have been about 700 square miles, and the epicentre about 3 miles N.N.E. of Menstrie.

9. **1881** JAN. 12, *about 7 a.m.* Cat. No. 787 (*Nature*, vol. 23, 1881, p. 275).

A smart shock at Bridge of Allan.

On 1900 Sept. 17, a new and remarkable series of earthquakes began, which, after including 191 earthquakes, terminated on 1916 Oct. 24. These earthquakes are described in *Quart. Journ. Geol. Soc.* vol. 63, 1907, pp. 362–374 (to 1907 Apr. 11); *Edin. Roy. Soc. Proc.* vol. 36, 1917, pp. 256–287 (to 1914 Dec. 18).

10. **1900** SEPT. 17, 3.30 *p.m.* Cat. No. 932.

A slight shock at Menstrie.

11. **1900** SEPT. 17, 10.5 *p.m.* Cat. No. 933.

A slight shock at Alva.

12. **1900** SEPT. 17, 10.15 *p.m.* Cat. No. 934; Intensity 4; Centre of disturbed area in lat. 56° 11·5′ N., long. 3° 50·5′ W.; No. of records 56 from 26 places and 13 negative records from 11 places; Fig. 12.

The boundary of the disturbed area, represented by the continuous line in Fig. 12, is an isoseismal of intensity slightly less than 4. It is 15 miles long, 9½ miles wide, and 117 square miles in area. The longer axis is directed N. 77° E., and the centre of the area is 3 miles N. 8° E. of Menstrie.

All over the disturbed area, the shock consisted of a single prominent vibration, followed by a tremor, such as would be caused by a heavy weight falling on the floor and making the building shake. The duration of the shock was not more than 3 seconds. The sound was heard by 94 per cent. of the observers, and was compared to passing waggons, etc., in 29 per cent. of the records, thunder in 10, wind in 5, the fall of a load of stones in 12, the fall of a heavy body in 17, explosions in 12, and miscellaneous sounds in 15, per cent.

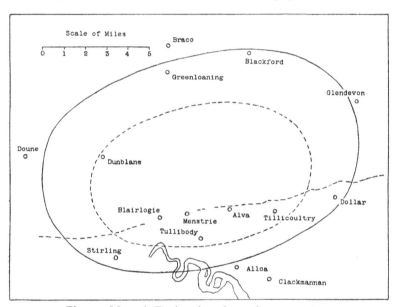

Fig. 12. Menstrie Earthquakes of 1900 Sept. 17 and 22.

13. 1900 SEPT. 22, 4.30 p.m. Cat. No. 935; Intensity 4; Centre of disturbed area in lat. 56° 10·5' N., long. 3° 50·4' W.; No. of records 20 from 13 places and 17 negative records from 15 places; Fig. 12.

The boundary of the disturbed area, represented by the broken-line in Fig. 12, is an isoseismal of intensity slightly less than 4. It is 11 miles long, 7 miles wide, and 60 square miles in area. The longer axis is directed N. 77° E., and the centre of the area is 1¾ miles N. 17° E. of Menstrie.

The shock consisted of a single prominent vibration followed by a tremor. The sound was heard by 87 per cent. of the observers.

Slight shocks were also felt on Sept. 18 at 2 a.m at Alva, and at about 2.55 a.m. at Bridge of Allan, in each case by one observer only.

14. **1903** MAY 15, 6.15 *p.m.* Cat. No. 967.

A distinct shock felt generally at Menstrie.

15. **1905** APR. 23, 0.15 *a.m.* Cat. No. 985.

A distinct shock felt at Red Carr (Blairlogie).

16. **1905** JULY 23, 0.15 *a.m.* Cat. No. 988; Intensity 5; Centre of isoseismal 4 in lat. 56° 11·3′ N., long. 3° 47·6′ W.; No. of records 33 from 16 places and negative records from 4 places; Fig. 13.

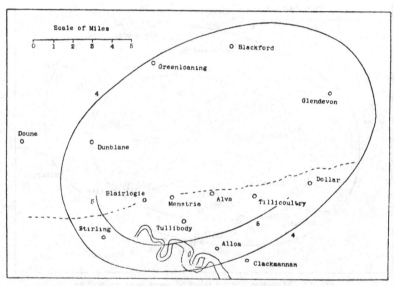

Fig. 13. Menstrie Earthquake of 1905 July 23.

The boundary of the disturbed area is an isoseismal of intensity 4. It is 16½ miles long, 10¼ miles wide, and 136 square miles in area. The direction of the longer axis is N. 63° E., and the centre is 3½ miles N.E. of Menstrie. On the same map, the portion of the isoseismal 5, lying on the south side of the Ochil Hills, is shown. Its distance from the isoseismal 4 is 1½ miles.

Within the isoseismal 5, the shock consisted of one or more prominent vibrations followed by a series of tremors. Its mean duration was 3 seconds. The sound-area coincided with the disturbed area in all directions, except perhaps towards the east. The sound was heard by 89 per cent. of the observers. It was compared to passing waggons, etc., in 17 per cent. of the records, thunder in 10, wind in 27, the fall of a heavy body in 40, explosions in 3, and miscellaneous sounds in 3, per cent.

17. **1905** JULY 26, 6.3 *p.m.* Cat. No. 989; Intensity 4; No. of records 9 from 7 places.

The only records of this shock are from places in the Hillsfoot district, namely, Alva, Gogar, Logie, Menstrie, Sauchie, Tillicoultry and Tullibody. The disturbed area was much smaller than that of the preceding shock, and probably it did not extend so far as Dunblane, Greenloaning, Blackford and Glendevon, and must have contained less than 80 square miles.

The shock was as if a huge body had fallen on the floor, followed by the quivering which such a fall would produce. Its duration was 1½ or 2 seconds. At Menstrie, the noise was as loud as thunder over-head.

18. **1905** AUG. 3, *about* 6 *p.m.* Cat. No. 990.

A very perceptible shock, felt by several observers, at Red Carr.

19. **1905** SEPT. 21, 11.33 *p.m.* Cat. No. 991; Intensity 6; Centre of isoseismal 5 in lat. 56° 11·8′ N., long. 3° 49·1′ W.; No. of records 139 from 57 places and 22 negative records from 19 places; Fig. 14.

This may be regarded as the first principal earthquake of the 1900–1916 series. On the map are shown three isoseismal lines of intensities 6, 5 and 4. The isoseismal 6 is 13 miles long, 8 miles wide, and 82 square miles in area. Towards the north-east, the course of the line is uncertain. The isoseismal 5, which is determined by a much larger number of observations, is 20½ miles long, 14 miles wide, and 227 square miles in area. Its longer axis is directed N. 61° E., and its centre is 3 miles N. 25° E. of Menstrie. The distance between the isoseismals 6 and 5 is 3¼ miles on the north side and 2¾ miles on the south. The isoseismal 4 is 33½ miles long, 26½ miles wide, and 700 square miles in area. Its longer axis is directed N. 63° E. The distance between the isoseismals 5 and 4 is 7½ miles on the north side and 5 miles on the south. Outside this isoseismal, the shock was felt at St Fillans and Balmeanach, 3 and 4½ miles to the north-west, and at Falkirk, 1 mile to the south. The disturbed area must therefore contain about 1000 square miles.

The shock was similar to its predecessors except that, near the epicentre, the principal vibrations were preceded, as well as followed, by tremors. The mean of 54 estimates of the duration of the shock is 3·4 seconds. The sound-area is co-extensive with the disturbed area. The percentage of audibility is 84 for the whole area, 89 within the isoseismal 6, 82 between the isoseismals 6 and 5, and 79 between

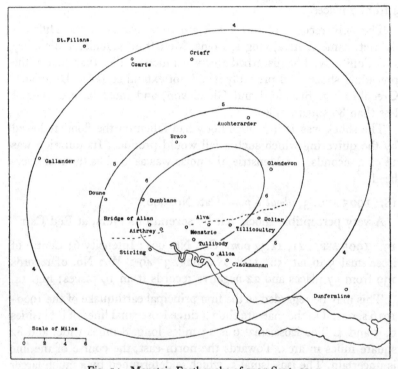

Fig. 14. Menstrie Earthquake of 1905 Sept. 21.

the isoseismals 5 and 4. Towards the latter isoseismal, there was a marked decline in audibility. The sound was compared to passing waggons, etc., in 24 per cent. of the records, thunder in 16, wind in 8, loads of stones falling in 10, the fall of a heavy body in 24, explosions in 16, and miscellaneous sounds in 2, per cent.

20. 1905 SEPT. 22, *about* 1.30 *a.m.* Cat. No. 992.

A tremor felt at Alloa and Bridge of Allan.

21. 1905 SEPT. 25, *early morning.* Cat. No. 993.

A very slight tremor felt at Alloa and Cambus.

22. **1905** SEPT. 30, 9.45 *p.m.* Cat. No. 994.

A slight shock, accompanied by a rumbling noise, at Menstrie.

23. **1905** OCT. 29, 10.53 *a.m.* Cat. No. 995.

A sound like distant thunder, without any accompanying tremor, at Menstrie.

24. **1905** DEC. 22, 9.15 *p.m.* Cat. No. 996.

A slight shock felt throughout the village of Menstrie.

25. **1906** JULY 3, 2.15 *p.m.* Cat. No. 1000; Intensity 3.

A slight shock felt at Menstrie, accompanied by a sound like the fall of a heavy body.

26. **1906** JULY 4, 3.45 *a.m.* Cat. No. 1001; Intensity 4.

The shock was felt at Airthrey, Alva, Blairlogie, Menstrie, Red Carr and Tullibody. The shock is described as a quivering thud, strongest at the beginning and lasting 2 seconds. The sound was heard by all the observers.

27. **1906** JULY 7, 5.29 *a.m.* Cat. No. 1002; Intensity 4.

The shock was felt at Alva and Menstrie. The sound was heard by all the observers.

28. **1906** AUG. 24, 5.25 *p.m.* Cat. No. 1003.

A very slight shock at Menstrie.

29. **1906** SEPT. 28, 0.25 *p.m.* Cat. No. 1005.

A slight shock, with sound, observed at Blairlogie, Menstrie and Red Carr.

30. **1906** OCT. 3, 4.32 *a.m.* Cat. No. 1006.

A shock, strong enough to waken several observers, felt at Blairlogie and Menstrie; the accompanying sound was like a cannon-shot.

31. **1906** OCT. 8, 7.24 *a.m.* Cat. No. 1007; Intensity 5; Centre of isoseismal 4 in lat. 56° 10·9′ N., long. 3° 51·3′ W.; No. of records 23 from 14 places and 8 negative records from 6 places; Fig. 15.

The disturbed area is bounded by an isoseismal of intensity 4, represented by the continuous line in Fig. 15. It is 12 miles long, 9½ miles wide, and 90 square miles in area. Its centre is 2 miles N. of Menstrie. At Logie, Menstrie and Tillicoultry, the shock consisted as a rule of one prominent vibration, its intensity being 5. The sound was heard by 86 per cent. of the observers, and, in 57 per cent. of the records, was compared to the dull concussion caused by the fall of a heavy body.

32. **1906** OCT. 8, 8.16 *a.m.* Cat. No. 1008; Intensity 4.

The shock was felt at Airthrey, Alva, Blairlogie, Bridge of Allan, Dunblane, Logie, Menstrie and Red Carr. The disturbed area was probably about 9½ miles long, 6 miles wide, and 45 square miles in

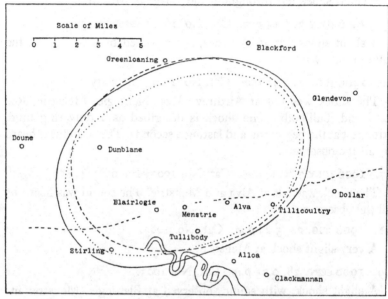

Fig. 15. Menstrie Earthquakes of 1906 Oct. 8, Dec. 28 and 30.

area. The shock consisted of two vibrations, lasting not more than 2 seconds, and was accompanied by a sound resembling that of a muffled explosion.

33. **1906** OCT. 12, 7.20 *a.m.* Cat. No. 1009.

A distinct shock at Menstrie.

34. **1906** OCT. 20, 7.15 *a.m.* Cat. No. 1010.

A slight tremor, with sound, at Airthrey.

35. **1906** OCT. 24, 7.11 *p.m.* Cat. No. 1011.

A report, like that of a distant cannon, heard generally at Menstrie.

36. **1906** OCT. 26, 7.15 *p.m.* Cat. No. 1012.

A slight shock, felt at Airthrey (with sound) and Menstrie.

37. **1906** OCT. 30, 0.15 *p.m.* Cat. No. 1013.

A slight shock and sound at Menstrie.

38. **1906** DEC. 28, 4.12 *p.m.* Cat. No. 1014; Intensity 6; Centre of isoseismal 4 in lat. 56° 11·3′ N., long. 3° 51·0′ W.; No. of records 10 from 7 places and negative records from 5 places; Fig. 15.

The shock was felt at Airthrey, Alva, Braco, Dunblane, Menstrie (int. 6), Red Carr and Tillicoultry. The disturbed area, the boundary of which is represented by the broken-line in Fig. 15, was probably about 11 miles long, 8½ miles wide, and about 74 square miles in area, its centre being 2½ miles N. of Menstrie. The sound was heard by 78 per cent. of the observers.

39. **1906** DEC. 29, 1.30 *p.m.* Cat. No. 1015.

A slight shock, accompanied by a noise like the fall of a heavy body, at Airthrey.

40. **1906** DEC. 30, *about* 1 *a.m.* Cat. No. 1016.

A shock at Menstrie.

41. **1906** DEC. 30, 2.10 *p.m.* Cat. No. 1017; Intensity 3.

A shock felt at Airthrey, Alva, Dunblane, Menstrie and Red Carr. The sound was like the fall of a heavy body.

42. **1906** DEC. 30, 4.15 *p.m.* Cat. No. 1018; Intensity 6; Centre of isoseismal 4 in lat. 56° 10·8′ N., long. 3° 51·3′ W.; No. of records 13 from 8 places and 6 negative records from 5 places; Fig. 15.

The disturbed area, the boundary of which is represented by the dotted line in Fig. 15, is about 11 miles long, 8½ miles wide, and about 82 square miles in area, its centre being about 2 miles N. of Menstrie. The intensity of the shock was 6 at Menstrie. The sound was heard by all the observers.

43. **1906** DEC. 31, 1 *a.m.* Cat. No. 1019.

A shock, accompanied by sound, at Menstrie.

44. **1907** FEB. 10, 5.40 *p.m.* Cat. No. 1021.

A slight shock at Menstrie.

45. **1907** MAR. 19, 7.33 *p.m.* Cat. No. 1022; Intensity 5.

A shock at Airthrey, Alva, Bridge of Allan and Menstrie. It consisted of two prominent vibrations, and was accompanied by a loud report like that of an explosion.

46. **1907** APR. 7, 11.11 *p.m.* Cat. No. 1023.
A very slight shock, accompanied by sound, at Menstrie.

47. **1907** APR. 7, 11.19 *p.m.* Cat. No. 1024.
A slight shock, consisting of two vibrations, at Menstrie.

48. **1907** APR. 8, 6.45 *a.m.* Cat. No. 1025.
A slight shock at Menstrie.

49. **1907** APR. 11, 5.30 *a.m.* Cat. No. 1026; Intensity 4.
A shock, accompanied by sound, at Menstrie and Red Carr.

50. **1907** APR. 11, 5.40 *a.m.* Cat. No. 1027; Intensity 4.
A shock at Airthrey, Menstrie and Red Carr.

51. **1907** APR. 11, 6.5 *a.m.* Cat. No. 1028.
A shock, with very slight noise, at Menstrie.

52. **1907** JUNE 14, 1.59 *a.m.* Cat. No. 1029.
A slight shock, consisting of two vibrations each accompanied by sound, at Menstrie.

53. **1907** JUNE 30, 3.36 *p.m.* Cat. No. 1030; Intensity 4.
A shock, preceded and accompanied by a rumbling sound, at Alva and Menstrie.

54. **1907** JULY 5, 9.48 *p.m.* Cat. No. 1032.
A slight shock, consisting of a single vibration, at Menstrie.

55. **1907** JULY 21 *or* 28, *between 5.0 and 7.30 p.m.* Cat. No. 1033.
A slight shock at Menstrie on one of the last two Sundays in July.

56. **1907** SEPT. 18, *about 5.30 p.m.* Cat. No. 1034.
A very slight shock at Menstrie.

57. **1908** JAN. 19, 1.27 *a.m.* Cat. No. 1036.
A distinct shock at Menstrie.
Feb. 9, 4.6 a.m., a shock, stronger than the preceding, felt at Menstrie, but only, so far as known, by one observer.

58. **1908** MAY 1, 6.54 *p.m.* Cat. No. 1037; Intensity 4.
A shock felt at Airthrey, Alva, Dunblane, Menstrie and Tilli-coultry. The shock consisted of a single series of vibrations, with one maximum of intensity, and lasted 2 seconds. It was accompanied by a loud noise like a muffled explosion.

59. **1908** MAY 2, 7.5 *a.m.* Cat. No. 1038; Intensity 4.

A shock of duration 3 seconds at Airthrey, Alva and Menstrie. At Airthrey, it consisted of two concussions, with intervening tremor, the latter concussion being the stronger; it was preceded, accompanied and followed by a rumbling sound.

60. **1908** MAY 10, 0.48 *a.m.* Cat. No. 1039; Intensity 4.

A shock of duration about 4 seconds at Airthrey, Alva, Menstrie, Tillicoultry and Tullibody. At Airthrey and Menstrie it consisted of two concussions, each accompanied by a loud noise like an explosion, the former being much the stronger at Airthrey, and both of about the same intensity at Menstrie.

61. **1908** MAY 10, 0.58 *a.m.* Cat. No. 1040; Intensity 4.

A shock felt at Airthrey, Alva and Menstrie, accompanied by a muffled sound like that of an explosion.

62. **1908** JUNE 21, 3 *a.m.* Cat. No. 1041.

A distinct shock, consisting of a single series of vibrations, at Alva and Menstrie, at Alva being accompanied by a loud noise.

63. **1908** JUNE 21, 4.20 *a.m.* Cat. No. 1042.

A slight but distinct shock, consisting of a single series of vibrations, at Menstrie.

64. **1908** JULY 17, 5.27 *p.m.* Cat. No. 1044.

A slight but distinct tremor, accompanied by noise, at Alva and Menstrie.

65. **1908** SEPT. 2, 8.16 *a.m.* Cat. No. 1045; Intensity 3.

A shock, consisting of a single series of vibrations and accompanied by a rumbling noise, at Menstrie.

66. **1908** SEPT. 2, 8.51 *a.m.* Cat. No. 1046; Intensity 3.

A single series of vibrations, accompanied by a rumbling noise, at Menstrie.

67. **1908** OCT. 16, 9.53 *p.m.* Cat. No. 1047; Intensity 4.

A tremor of duration 2 seconds at Airthrey, Alva, Blair Ochil (Dunblane), Menstrie and Red Carr. At Menstrie, the shock consisted of two prominent vibrations. The sound was compared to an underground explosion, a clap of thunder, or the thud of falling rock.

68. **1908 OCT. 19, 9.18 a.m.** Cat. No. 1048.

A slight tremor accompanied by sound at Airthrey. The shock was also felt at Blair Ochil (Dunblane).

69. **1908 OCT. 19, 9.39 a.m.** Cat. No. 1049.

An earth-sound at Menstrie.

70. **1908 OCT. 20, 4.8 p.m.** Cat. No. 1050; Intensity 7; Centre of isoseismal 6 in lat. 56° 11·4′ N., long. 3° 47·3′ W.; No. of records 59 from 34 places and 20 negative records from 18 places; Fig. 16.

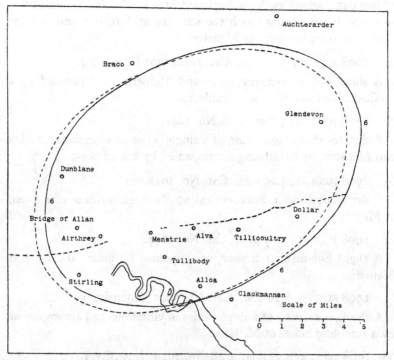

Fig. 16. Menstrie Earthquakes of 1908 Oct. 20.

This shock was stronger than all the earlier recorded shocks in the Ochil district, with the exception perhaps of those of 1736 and 1802 (nos. 1, 2, 5 and 6). The intensity was not less than 7 at Alva and Tillicoultry.

The only isoseismal that can be drawn is that of intensity 6, indicated by the continuous line in Fig. 16. The curve is 15 miles long, 10½ miles wide, and contains 123 square miles. Its centre is 3½ miles N. 42° E. of Menstrie, and the direction of its longer axis is N. 65° E.

The boundary of the disturbed area differs slightly from that of the earthquake of 1905 Sept. 21 (no. 19), for the shock was felt one or two miles farther to the south, at Falkirk, Polmont and Bo'ness, while it was not felt at Comrie, Crieff and Monzie. The disturbed area must contain about 1000 square miles.

At several places within the isoseismal 6 (Alloa, Cambus, Green-loaning, Menstrie and Tillicoultry), the shock consisted of two distinct parts, separated by an interval of about 3 seconds, the first part being much the stronger. At places outside this isoseismal, only one part was observed, the effect resembling that caused by the fall of a heavy weight or the passing of a traction-engine. The average of 15 estimates of the duration of the shock is 2·7 seconds.

The sound was heard by 94 per cent. of the observers within the isoseismal 6, by 87 per cent. of those outside it, and by 92 per cent. of those in the whole disturbed area. It was compared to passing waggons, etc., in 47 per cent. of the records, to thunder in 27, to the fall of a heavy body in 17, and to explosions in 10, per cent.

71. **1908 OCT. 20, 4.13 *p.m.*** Cat. No. 1051; Intensity 5; Centre of disturbed area in lat. 56° 11·3′ N., long. 3° 48·0′ W.; No. of records 18 from 13 places; Fig. 16.

, The boundary of the disturbed area (represented by the broken-line in Fig. 16) coincides nearly with the isoseismal 6 of the preceding earthquake, except that it is displaced about ½ mile to the W.S.W. It is 15 miles long, 10¼ miles wide, and contains about 123 square miles. Its centre is 3 miles N. 40° E. of Menstrie, and its longer axis is directed N. 65° E. The shock as a rule was merely a tremor, lasting about 1½ seconds, without any prominent vibration or "thud." The sound was compared to thunder or an explosion.

72. **1908 OCT. 20, 9.26 *p.m.*** Cat. No. 1052.

A very slight shock at Airthrey and Menstrie. At Airthrey the sound and vibration began together and terminated simultaneously with a thud.

73. **1908 NOV. 6, 4.45 *p.m.*** Cat. No. 1053.

An earth-sound at Menstrie.

74. **1909 JAN. 19, 5.28 *a.m.*** Cat. No. 1054; Intensity 4.

A shock, accompanied by a loud sound, at Airthrey, Alva and Menstrie.

75. **1909** JAN. 19, 5.29 *a.m.* Cat. No. 1055.

A shock at Menstrie. Two tremors were felt on this day after 5.28 a.m. at Alva.

76. **1909** JAN. 23, 0.15 *p.m.* Cat. No. 1056; Intensity 3.

A slight shock, accompanied by a noise like that of an underground explosion, at Airthrey and Menstrie.

77. **1909** JAN. 24, 0.15 *p.m.* Cat. No. 1057; Intensity 3.

A shock at Menstrie.

78. **1909** JAN. 24, 2.28 *p.m.* Cat. No. 1058.

A tremor at Airthrey.

79. **1909** JAN. 24, 3.35 *p.m.* Cat. No. 1059.

A tremor, of about the same intensity as the preceding, at Airthrey.

80. **1909** JAN. 27, 1.40 *p.m.* Cat. No. 1060.

A slight shock at Menstrie.

81. **1909** FEB. 22, 7.26 *p.m.* Cat. No. 1061.

An earth-sound at Menstrie.

82. **1909** MAR. 19, 9.35 *a.m.* Cat. No. 1062.

A shock accompanied by noise at Menstrie.

83. **1909** MAY 22, 3.23 *p.m.* Cat. No. 1063; Intensity 5.

A shock, accompanied by a noise like that of an explosion, at Airthrey and Menstrie.

84. **1909** MAY 22, 5.1 *p.m.* Cat. No. 1064; Intensity 4.

A shock, accompanied by a noise like that of a slight explosion, at Airthrey.

85. **1909** MAY 22, 5.24 *p.m.* Cat. No. 1065.

A slight shock at Menstrie.

86. **1909** MAY 22, 8.23 *p.m.* Cat. No. 1066; Intensity 5.

A shock, accompanied by a sound like that of an explosion, at Airthrey.

87. **1909** OCT. 21, 8.37 *a.m.* Cat. No. 1067; Intensity 4.

A shock, followed by a sound like that of an explosion, at Airthrey.

88. **1909** OCT. 21, 9.53 *a.m.* Cat. No. 1068.

A slight shock at Menstrie.

89. **1909** OCT. 22, 6.55 *a.m.* Cat. No. 1069.

A slight shock at Menstrie.

90. **1909** OCT. 22, 7.57 *a.m.* Cat. No. 1070; Intensity 4.

A shock, preceded and followed by a sound like that of a heavy body falling, at Airthrey.

91. **1909** OCT. 22, 9.8 *p.m.* Cat. No. 1071.

At Airthrey, there were three concussions, separated by intervals of 2 and 5 seconds, preceded and followed by sounds like those of sharp explosions.

92. **1910** FEB. 8, 9.33 *a.m.* Cat. No. 1072; Intensity 4.

A shock at Airthrey, Bridge of Allan, Menstrie and Red Carr. At Airthrey, it consisted of two parts, separated by an interval of 2 seconds, the first part being the stronger; the sound resembled that of an explosion.

93. **1910** FEB. 10, 9.30 *a.m.* Cat. No. 1073.

A loud rumbling noise at Menstrie.

94. **1910** FEB. 10, 9.50 *a.m.* Cat. No. 1074; Intensity 4.

A shock at Airthrey, Bridge of Allan and Menstrie. At Airthrey it consisted of two parts, separated by an interval of 2 seconds, the first part being the stronger; the sound resembled that of an explosion.

95. **1910** FEB. 11, 10.55 *p.m.* Cat. No. 1075; Intensity 4.

A shock, consisting of a single series of vibrations and accompanied by a sound like that of an explosion, at Airthrey.

96. **1910** APR. 17, 1.45 *a.m.* Cat. No. 1076; Intensity 4.

A shock, consisting of a single series of vibrations and accompanied by a noise like that of a heavy waggon passing, at Airthrey.

97. **1910** MAY 19, 9.45 *a.m.* Cat. No. 1077.

A bumping noise, as of a huge rock rolled over and over, at Menstrie.

98. **1910** MAY 20, 2.40 *p.m.* Cat. No. 1078; Intensity 4.

A shock at Airthrey, Alva and Menstrie; at Airthrey being preceded by a sound like that of an explosion.

99. **1910** MAY 20, 2.50 *p.m.* Cat. No. 1079; Intensity 4.

A shock, preceded by a sound like that of an explosion, at Airthrey and Alva.

100. **1910** MAY 20, 3.5 *p.m.* Cat. No. 1080; Intensity 4.

A shock, consisting of a single vibration and preceded by a sound like that of an explosion, at Airthrey and Menstrie.

101. **1910** MAY 20, 3.15 *p.m.* Cat. No. 1081.

A bumping noise at Menstrie.

102. **1910** MAY 27, 0.30 *a.m.* Cat. No. 1082; Intensity 4.

A shock, followed by a sound like the firing of heavy guns, at Airthrey.

103. **1910** JULY 10, 3.7 *p.m.* Cat. No. 1083; Intensity 4.

A shock, preceded and followed by a sound like that of an explosion, at Airthrey; the noise was also heard at Menstrie.

104. **1910** JULY 11, 0.1 *a.m.* Cat. No. 1084; Intensity 4.

A shock, accompanied by a sound like that of an explosion, at Airthrey.

105. **1910** JULY 22, 11.59 *p.m.* Cat. No. 1085; Intensity 4.

The same.

106. **1910** OCT. 25, 8.55 *p.m.* Cat. No. 1086.

A bumping noise at Menstrie.

107. **1910** OCT. 29, 7.41 *p.m.* Cat. No. 1087.

The same.

108. **1910** NOV. 20, 1.57 *p.m.* Cat. No. 1088.

A bumping noise, consisting of two loud reports, at Menstrie.

109. **1910** DEC. 7, 5.54 *p.m.* Cat. No. 1089.

A slight shock, with a bumping noise, at Menstrie.

110. **1910** DEC. 8, 3.45 *p.m.* Cat. No. 1090.

The same.

111. **1911** JULY 27, 8.10 *a.m.* Cat. No. 1094.

A bumping noise at Menstrie.

112. **1911** JULY 27, 5.15 *p.m.* Cat. No. 1095.

The same.

113. **1911** JULY 28, 7.46 *p.m.* Cat. No. 1096.

The same.

114. **1911** OCT. 12, 1.50 *a.m.* Cat. No. 1097; Intensity 4.

A shock, accompanied by a loud sound, at Alva and Menstrie.

115. **1911** OCT. 12, *about* 2.20 *a.m.* Cat. No. 1098.

A shock, slighter than the preceding and accompanied by a loud sound, at Alva.

116. **1911** OCT. 12, 3.45 *a.m.* Cat. No. 1099.

A shock, accompanied by a bumping noise, felt generally at Menstrie.

117. **1911** OCT. 17, 6.35 *a.m.* Cat. No. 1100.

A slight shock, accompanied by a bumping noise, felt generally at Menstrie.

118. **1911** OCT. 21, 1.25 *p.m.* Cat. No. 1101; Intensity 5.

A shock, accompanied by a loud sound, at Alva; the sound, consisting of three thuds, was heard at Menstrie.

119. **1912** JAN. 20, *about* 4 *a.m.* Cat. No. 1102.

A single slight vibration, lasting about 2 seconds, at Bridge of Allan.

120. **1912** JAN. 26, 3.59 *a.m.* Cat. No. 1103; Intensity 5.

A shock, consisting of a single series of vibrations, at Alva, Dunblane and Menstrie. The sound was compared to dull thunder, the fall of a heavy body, and an approaching train in a tunnel ending with a detonation as of a deep explosion.

121. **1912** JAN. 28, 4 *a.m.* Cat. No. 1104; Intensity 4.

A shock felt at Airthrey, Coalsnaughton, Dunblane, Kinbuck and Tillicoultry. The disturbed area contained about 80 square miles. The shock consisted of a single series of vibrations lasting 2 seconds. The sound resembled that of a load of coals falling.

122. **1912** JAN. 28, 4.40 *p.m.* Cat. No. 1105.

A single series of vibrations, accompanied by sounds like those of wind and a heavy body falling, at Red Carr.

123. **1912** JAN. 28, 5.25 *p.m.* Cat. No. 1106.

A rather strong shock, accompanied by a loud noise, at Red Carr.

124. **1912** JAN. 30, 5.25 *p.m.* Cat. No. 1107; Intensity 4.

A shock, consisting of a single series of vibrations and lasting 2 or 3 seconds, at Menstrie; a sound like loud thunder followed the shock.

125. **1912** JAN. 31, 5.29 *p.m.* Cat. No. 1108; Intensity 3.

A slight shock, accompanied by a noise like that of a prolonged explosion, at Dunblane. A very loud sound was heard at Menstrie.

126. **1912** FEB. 1, 5.25 *p.m.* Cat. No. 1109; Intensity 4.

A shock, accompanied by a loud sound, at Alva.

127. **1912** FEB. 9, 1.23 *p.m.* Cat. No. 1110; Intensity 4.

Two vibrations, followed by tremulous motion, at Alva. The sound was heard at Alva and Menstrie.

128. **1912** FEB. 9, 1.32 *p.m.* Cat. No. 1111; Intensity 5.

The shock was felt at Alva, Coalsnaughton, Dunblane, Logie and Menstrie. The disturbed area towards the south and west was probably bounded by a curve coinciding nearly with the isoseismal 6 of the earthquake of 1905 Sept. 21 (no. 19), but not extending quite so far to the east as that isoseismal. It must therefore have contained somewhat less than 80 square miles. The shock consisted of a single series of vibrations and lasted 2 or 3 seconds. The sound was compared to dull thunder, a heavy body falling, and an explosion in a quarry.

129. **1912** FEB. 9, 1.39 *p.m.* Cat. No. 1112.

A very slight bumping noise, like an echo of that which accompanied the preceding shock, at Menstrie.

130. **1912** FEB. 11, 3.45 *a.m.* Cat. No. 1113; Intensity 3.

A slight tremor, accompanied by a sound like that of the fall of a heavy body, at Alva.

131. **1912** FEB. 24, 3.15 *a.m.* Cat. No. 1114.

A shock, lasting about 1 second, at Alva; the sound, like a double knock, was heard at Alva and Menstrie.

132. **1912** FEB. 24, 4.53 *a.m.* Cat. No. 1115.

A very loud bumping noise at Menstrie.

133. **1912** MAR. 3, 2.24 *p.m.* Cat. No. 1116.

A slight noise at Menstrie.

134. **1912** MAR. 20, 7.50 *a.m.* Cat. No. 1117.

A slight tremor at Alva.

135. **1912** MAR. 21, 10.20 *p.m.* Cat. No. 1118.

A very distinct double tremor at Alva.

136. **1912** MAR. 21, 10.25 *p.m.* Cat. No. 1119.

A slight tremor at Alva.

137. **1912** MAR. 21–22, *midnight.* Cat. No. 1120.

A slight tremor at Alva.

138. **1912** MAR. 22, 5.30 *a.m.* Cat. No. 1121.

A slight tremor at Alva.

139. **1912** MAR. 22, 7.50 *a.m.* Cat. No. 1122; Intensity 5.

A shock felt at Airthrey, Alva, Dunblane and Menstrie. The sound was compared to the fall of a load of stones, the fall of a heavy body, and the sound of cannon firing.

140. **1912** MAR. 22, 10.15 *p.m.* Cat. No. 1123; Intensity 5.

A shock, consisting of a single series of vibrations, felt at Airthrey, Alva, Dunblane and Menstrie. The sound resembled thunder or an explosion.

141. **1912** MAR. 22, 11.15 *p.m.* Cat. No. 1124.

A noise, as of distant thunder, at Dunblane and Menstrie.

142. **1912** MAR. 23, 5.33 *a.m.* Cat. No. 1125; Intensity 5.

A shock, consisting of a single series of vibrations and accompanied by a noise like that of wind or heavy waggons passing, at Airthrey, Alva, Dunblane and Menstrie.

143. **1912** MAR. 23, 6.8 *a.m.* Cat. No. 1126.

A bumping noise at Menstrie.

144. **1912** APR. 3, 10.33 *a.m.* Cat. No. 1127.

An earth-sound heard at Airthrey and Menstrie, at Airthrey like that of an engine passing.

145. **1912** APR. 3, 5.20 *p.m.* Cat. No. 1128.

A bumping noise at Menstrie.

146. **1912** APR. 3, 5.35 *p.m.* Cat. No. 1129.

The same.

147. **1912** APR. 18, 8.14 *p.m.* Cat. No. 1130.

The same.

148. **1912** APR. 18, 10.15 *p.m.* Cat. No. 1131.

The same.

149. **1912** APR. 19, 8.11 *p.m.* Cat. No. 1132; Intensity 4.

A shock (of duration about 3 seconds), increasing in strength to a maximum and then dying away, at Airthrey, Alva, Doune and Dunblane. The sound resembled a loud rumble of thunder.

150. **1912** APR. 19, 9.10 *p.m.* Cat. No. 1133.

A rather strong shock, accompanied by sound and lasting 4 seconds, at Airthrey.

151. **1912** APR. 19, 10.10 *p.m.* Cat. No. 1134.

A shock, accompanied by the usual noise and lasting about 3 seconds, at Alva and Dunblane.

152. **1912** APR. 19, 10.30 *p.m.* Cat. No. 1135.

A slight shock at Doune and Tullibody.

153. **1912** APR. 19–20, *midnight*. Cat. No. 1136.

An earth-sound at Doune.

154. **1912** APR. 20, 2 *a.m.* Cat. No. 1137.

A rather strong shock at Airthrey and Dunblane; at Airthrey it consisted of two parts, the first lasting 4 seconds, the second 2 seconds, with an interval of 2 seconds between.

155. **1912** APR. 20, 2.5 *a.m.* Cat. No. 1138.

A shock, of less intensity than the preceding, at Airthrey, Doune and Tullibody.

156. **1912** APR. 20, 2.8 *a.m.* Cat. No. 1139; Intensity 4.

A shock at Dunblane and in the surrounding district.

157. **1912** APR. 20, 2.10 *a.m.* Cat. No. 1140.

A shock, of less intensity than the preceding but strong enough to awaken people, at Alva, Doune, Dunblane and Menstrie. A bumping noise was heard with it at Menstrie.

158. **1912** APR. 20, 2.12 *a.m.* Cat. No. 1141.

A bumping noise at Menstrie.

159. **1912** APR. 20, 2.14 *a.m.* Cat. No. 1142.

A slight shock was felt at Dunblane and a noise heard at Menstrie.

160. **1912** APR. 20, 2.18 *a.m.* Cat. No. 1143.

A bumping noise at Menstrie.

161. **1912** APR. 20, 4.15 *a.m.* Cat. No. 1144.

The same.

162. **1912** APR. 21, 1.5 *a.m.* Cat. No. 1145.

A slight tremor at Alva, Dunblane and Menstrie, and accompanied by the usual bumping noise at Menstrie.

163. **1912** APR. 21, 2.47 *p.m.* Cat. No. 1146.

A bumping noise at Menstrie.

164. **1912** APR. 22, 7.17 *a.m.* Cat. No. 1147.

A slight tremor and bumping noise at Menstrie.

165. **1912** APR. 23, 6.10 *a.m.* Cat. No. 1148.

The shock was felt as a thud with sound at Airthrey and as a slight tremor at Alva.

166. **1912** APR. 26, *about* 4 *a.m.* Cat. No. 1149; Intensity 4.

A shock at Bridge of Allan.

167. **1912** MAY 3, 6.20 *a.m.* Cat. No. 1150.

A slight tremor at Airthrey.

168. **1912** MAY 3, *about* 2.15 *p.m.* Cat. No. 1151.

A slight shock, accompanied by a sound like distant thunder, at Alva.

169. **1912** MAY 3, 4.13 *p.m.* Cat. No. 1152; Intensity 7; Centre of isoseismal 6 in lat. 56° 10·5′ N., long. 3° 52·5′ W.; No. of records 91 from 53 places and negative records from 11 places; Fig. 17.

Though the disturbed area of this earthquake is less than that of the earthquakes of 1905 Sept. 21 and 1908 Oct. 20 (nos. 19 and 70), the shock within the central isoseismal was stronger than that of any other member of this series.

On the map (Fig. 17) are shown four isoseismal lines, corresponding to intensities 7, 6, 5 and 4. The isoseismal 7 is 8½ miles long, 4¾ miles wide, and 32 square miles in area. The isoseismal 6, which is the most accurately drawn of the series, is 14 miles long, 8¼ miles wide, and contains an area of 92 square miles. Its centre is about 2 miles N.N.W. of Menstrie, and the direction of its longer axis is N. 82½° E. The isoseismal 5 is 19 miles long, 12¼ miles wide, and 182 square miles in area. The isoseismal 4, which may be regarded as the boundary of the disturbed area, is 30 miles long, 29½ miles wide, and contains 605 square miles. The distance between the isoseismals 7 and 6 is 2¼ miles on the north side and 1¼ miles on the south; between the isoseismals 6 and 5, 2¼ miles on the north side and 1¼ miles on the south; and between the isoseismals 5 and 4, 9 miles on the north side and 4¼ miles on the south. Outside the isoseismal 4, the earthquake was felt by persons lying down at Dunfermline (17 miles from the centre of the isoseismal 6), Glasgow (26 miles), Rutherglen (27½ miles) and Musselburgh (35 miles). If the disturbed area were regarded as bounded by a circle of radius 35 miles, it would contain about 3850 square miles.

The shock is usually described as a single series of about four prominent vibrations, followed, and occasionally preceded, by a tremulous motion. A few observers (at Alloa, Blair Drummond, Greenloaning, Logie and Stirling) felt two series of vibrations, the first of which (except at Logie) was the stronger, the mean duration of the interval between the series being 2 seconds. The average of 38 estimates of the total duration of the shock is 3·0 seconds.

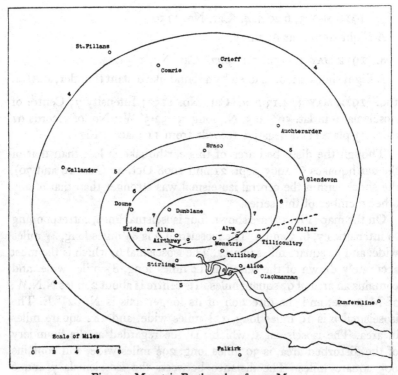

Fig. 17. Menstrie Earthquake of 1912 May 3.

The sound-area is co-extensive with the disturbed area, but towards the boundary there is a marked decline in audibility. While the percentage of audibility is 92 for the whole area, it is 100 within the isoseismal 7, 96 between the isoseismals 7 and 5, and 72 between the isoseismals 5 and 4. The sound was compared to passing waggons, etc., in 20 per cent. of the records, thunder in 22, wind in 4, loads of stones falling in 6, a heavy body falling in 26, and explosions in 22, per cent.

170. **1912** MAY 3, 11.45 *p.m.* Cat. No. 1153.

A shock, accompanied by a loud thud with rumbling before and after, at Airthrey.

A shock was also felt during the evening of May 3 at Tillicoultry.

171. **1912** MAY 4, 0.10 *a.m.* Cat. No. 1154.

A slight shock, consisting of a single series of vibrations, at Alva, Blackford, Dunblane and Kinbuck. The sound was compared to that of a heavy body falling and an explosion.

172. **1912** MAY 4, 2 *a.m.* Cat. No. 1155; Intensity 3.

A shock, accompanied by a noise like that of a heavy body falling, at Alva.

May 4 or 5, at about 11.50 p.m., a slight shock felt by one observer only at Auchendoune (Doune).

173. **1912** MAY 9, 11.28 *p.m.* Cat. No. 1156.

A shock at Airthrey, Alva and Dunblane; at Airthrey it consisted of a thud with noise preceded and followed by tremors.

174. **1912** MAY 10, 9.20 *p.m.* Cat. No. 1157.

A shock, consisting of a thud preceded and followed by tremors (duration 2 or 2½ seconds), at Airthrey, Alva, Doune and Dunblane. The sound resembled that of an explosion.

175. **1912** MAY 11, *about* 11.23 *p.m.* Cat. No. 1158.

A slight vibration and noise, increasing to a maximum and then dying away, at Dunblane.

176. **1912** MAY 14, 6.54 *a.m.* Cat. No. 1159; Intensity 3.

A shock at Airthrey, Alva, Dunblane and Logie; a sound, as of a heavy body falling, was heard at Airthrey.

177. **1912** MAY 14, 6.58 *a.m.* Cat. No. 1160; Intensity 3.

A shock (duration 2 seconds) at Airthrey, Alva, Dunblane, Greenloaning and Logie.

178. **1912** MAY 18, 4 *a.m.* Cat. No. 1161.

A slight shock at Dunblane.

179. **1912** MAY 19, 3.50 *a.m.* Cat. No. 1162.

A thud with sound, preceded and followed by tremors, at Airthrey.

180. **1912** MAY 19, 11.50 *p.m.* Cat. No. 1163.

A slight shock with sound at Airthrey, Bridge of Allan and Dunblane.

May 20–21, a slight shock was felt during the night by one observer at Bridge of Allan.

181. **1912** JULY 5, 1.50 *a.m.* Cat. No. 1164.

A shock, followed after 1 or 2 seconds by another less pronounced, at Greenloaning. Each part was accompanied by a sound as of a heavy body falling, the former sound being the more intense.

182. **1912** JULY 6, 1.52 *a.m.* Cat. No. 1165.

A slight shock at Alva, Doune and Dunblane. The sound resembled that of a falling body at Alva and of an explosion at Doune.

183. **1912** JULY 6, *about* 2.7 *a.m.* Cat. No. 1166.

A very slight tremor at Dunblane.

184. **1912** JULY 7, 10.20 *a.m.* Cat. No. 1167.

A slight vibration and rumbling noise at Alva.

185. **1912** JULY 9, 1.25 *p.m.* Cat. No. 1168.

A shock, lasting 1 or 2 seconds and accompanied by very little noise, at Bridge of Allan.

186. **1912** JULY 17, 2.25 *a.m.* Cat. No. 1169; Intensity 4.

A shock, consisting of a single series of vibrations, and accompanied by a sound like that of wind blowing on a door, at Greenloaning.

187. **1912** JULY 17, *about* 3.20 *a.m.* Cat. No. 1170.

A shock, lasting 4 or 5 seconds and strong enough to awaken people, at Dunblane.

188. **1912** JULY 18, 3.15 *a.m.* Cat. No. 1171.

A slight shock, accompanied by a noise like that of a falling body, at Alva.

189. **1912** JULY 29, 9.22 *a.m.* Cat. No. 1172.

The same.

190. **1912** SEPT. —, *about midnight.* Cat. No. 1173.

On a day about the middle of the month, a slight tremor at Dunblane.

191. **1912** OCT. 18, *about* 4.15 *p.m.* Cat. No. 1174.

A slight shock at Dunblane.

192. **1912 NOV.** 6, 5.15 *a.m.* Cat. No. 1175.

A slight noise, accompanied by a sound like that of a falling body, at Alva.

193. **1913 AUG.** 27, 4.30 *a.m.* Cat. No. 1176.

A thud followed by a tremor, and succeeded immediately by a second thud and tremor, the whole lasting 4 seconds, at Airthrey.

194. **1913 OCT.** 28, 4.30 *a.m.* Cat. No. 1177; Intensity 4.

A shock, accompanied by a sound like that of a falling body, at Alva.

195. **1914 DEC.** 18, 0.54 *a.m.* Cat. No. 1178.

A slight shock at Airthrey, Alva, Dunblane and Menstrie; at Airthrey there were two shocks in rapid succession, like two shots fired from a cannon in the immediate neighbourhood.

196. **1916 OCT.** 21, 2.55 *p.m.* Cat. No. 1181.

A slight shock at Airthrey.

197. **1916 OCT.** 21, 3.50 *p.m.* Cat. No. 1182.

An undulatory movement at Airthrey.

198. **1916 OCT.** 21, 4.10 *p.m.* Cat. No. 1183.

A slight shock at Airthrey.

199. **1916 OCT.** 21, 4.15 *p.m.* Cat. No. 1184.

A slight shock at Airthrey.

200. **1916 OCT.** 24, 3.35 *a.m.* Cat. No. 1185.

A tremor, lasting 4 seconds, at Airthrey.

CHARACTERISTICS OF THE MENSTRIE EARTHQUAKES

Intensity of Earthquakes. The earthquakes of the Ochil district may be divided into three classes:

(i) The first class contains the principal earthquakes of 1736 Apr. 30 and May 1 (nos. 1 and 2, intensity 8); 1802 Aug. — (nos. 5 and 6, intensity about 8); 1872 Aug. 8 (no. 8, intensity 5); 1905 Sept. 21 (no. 19, intensity 6); 1908 Oct. 20 (no. 70, intensity 7); and 1912 May 3 (no. 169, intensity 7). The last four disturbed areas of about 700, 1000, about 1000, and at least 605, square miles respectively.

(ii) The second class contains 10 earthquakes, namely, those of 1900 Sept. 17 (no. 12, intensity 4); 1905 July 23 (no. 16, intensity 5) and July 26 (no. 17, intensity 4); 1906 Oct. 8 (no. 31, intensity 5), Dec. 28 (no. 38, intensity 6) and Dec. 30 (no. 42, intensity 6); 1908 Oct. 20 (no. 71, intensity 5); 1912 Jan. 28 (no. 121, intensity 4), Feb. 9 (no. 128, intensity 5) and May 4 (no. 171, intensity probably 3). The disturbed areas range from 74 to 136 square miles, the average for the ten earthquakes being about 94 square miles.

(iii) The third class contains the remaining 182 earthquakes. In 24 of these, it is doubtful whether any tremor was felt. Of the remainder, 8 (and probably also nos. 3, 4, 7 and 9) were of intensity 5, 34 of intensity 4, and 106 of intensity 3 or about 3, while 6 were earth-sounds without any accompanying tremor.

Periodicity. The following table gives the number of earthquakes recorded during each half-month, the upper figure referring to the first half (the first 14 days in February and the first 15 days in the other months), and the lower figure to the second half:

Jan.	Feb.	Mar.	Apr.	May	June	July	Aug.	Sept.	Oct.	Nov.	Dec.
1	10	1	12	17	1	11	2	2	7	2	2
16	3	12	22	13	3	14	2	10	27	1	8

The method of overlapping means gives the following results:

	MAX. EPOCH	AMPLITUDE
Annual period	end of Apr.	·32
Semi-annual period	middle of Apr. and Oct.	·43

The hourly distribution of the earthquakes is as follows:

	0–1	1–2	2–3	3–4	4–5	5–6	6–7	7–8	8–9	9–10	10–11	11–12
a.m.	8	9	10	9	7	9	10	9	5	9	3	0
p.m.	5	8	9	9	11	14	3	6	4	8	9	9

The double maximum, shortly before and shortly after midnight, is less marked than usual, but there is a distinct maximum about 4 or 5 p.m., due no doubt to the greater quiet of the afternoon tea-hour.

Nature of the Shock and Sound. When the shock is of intensity 3, it consists as a rule of a series of tremors, the separate vibrations of which do not vary much in strength. Sometimes, however, only one vibration is felt; and, in a few cases, there is one prominent vibration, usually called a "thud," followed by slight tremulous motion, the effect being like that caused by the fall of a heavy weight on the floor, succeeded by the quivering which such a fall would produce. When the intensity of the shock attains the degree 4 or 5, the thud, followed by the tremor, is the normal form. Shocks of intensity 6 or 7 contain from 2 to 4 prominent vibrations, preceded and followed by tremulous motion. In a few cases, the shock consists of two parts, separated by an interval of about 2 seconds, but there is no reason for regarding them as twin earthquakes.

In earthquakes of the first class, the sound was heard by 88 per cent. of the observers; in those of the second class by 92 per cent.; and in those of the third class by 98 per cent.

The variation in the nature of the sound in the three classes of earthquakes is shown in the following table, in which the figures represent percentages for each class of earthquake to the different types of sound (p. 9):

	TYPE						
	1	2	3	4	5	6	7
Class I	28	19	5	7	24	15	2
,, II	21	11	11	7	30	14	6
,, III	9	13	5	5	25	42	1

Omitting the miscellaneous sounds in the last column, the first three types are of long, and the others of short, duration. In Class I, 47 per cent. of the comparisons are of short duration; in Class II 54 per cent.; and in Class III 73 per cent.

Length of Focus. Taking the difference between the longer and shorter axes of the inner isoseismal lines as a measure of the length of the seismic focus, the average length of the focus is 5 miles in earthquakes of the first class, and 4 miles in those of the second. For the third class, the corresponding figure cannot be ascertained. It is, however, clear from the nature of the shock and sound that the focus in earthquakes of this class was usually of very small dimensions.

ORIGIN OF THE MENSTRIE EARTHQUAKES

The seismic evidence gives the following elements for the originating fault: (i) The direction of the longer axis of the isoseismal lines ranges between N. 61° E. and N. 82½° E., the average of ten directions (in nos. 12, 13, 16, 19, 31, 38, 42, 70, 71 and 169) being N. 77° E.

(ii) The direction of the hade of the fault is indicated by the relative positions of the isoseismal lines of the earthquakes of 1905 Sept. 21 and 1912 May 3 (nos. 19 and 169). As these lines are farther apart on the north side than on the south, the fault should hade to the north. (iii) The fault-line should therefore pass through a point a short distance to the south of the epicentres of these earthquakes.

That it appears to traverse the district in the immediate neighbourhood of the Hillsfoot villages is further clear from the unusual intensity of some of the shocks at Alva, Menstrie, Airthrey, Bridge of Allan, etc., and from the fact that so many of the slighter shocks are felt only at one or a few of these places. Again, that the fault should

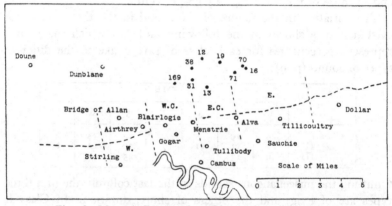

Fig. 18. Menstrie Earthquakes: Distribution of Epicentres.

hade to the north is shown by the distribution of the known epicentres (Fig. 18), which all lie on the north side of the indicated fault-line and are grouped in a band which is roughly parallel to it.

An important feature of the stronger earthquakes is their great intensity considering the smallness of their disturbed areas. None of them, so far as known, disturbed an area greater than about 1000 square miles; while, in other British earthquakes of intensity 7, the disturbed areas range from 12,000 to 63,600 square miles, with an average of 31,000 square miles. Thus, the seismic foci of the stronger Menstrie earthquakes must have been at a comparatively small depth below the surface.

On the maps of the earthquakes (Figs. 12–17) is shown the course of the great Ochil fault in the epicentral districts. Its mean direction there is N. 77° E., and the fault-line passes through or near the Hillsfoot villages. Thus, it satisfies two of the conditions required by the seismic evidence. On the other hand, as Dr. J. Horne, F.R.S., has kindly informed me, the fault in the district including the Clackmannan coalfield is exposed in two or three places, in which it appears as a nearly vertical line of fracture, the hade being to the south. This is, I believe, the only case in which the seismic evidence and the geological evidence seem to conflict. It may be that, owing to the presence of the Ochil Hills, the isoseismal lines are distorted to an unusual extent towards the north, but the seismic evidence in favour of a northerly hade is so strong that I do not think that the discrepancy can be thus explained.

The portion of the fault affected by the recent movements extends from a mile or two east of Tillicoultry to a short distance west of Bridge of Allan, that is, over a length of about 9 miles. This portion may be divided into four regions. That in the neighbourhood of Dunblane, Bridge of Allan and Airthrey may be called the western region; that between Airthrey and Menstrie the west-central region. The region between Alva and Tillicoultry may be called the eastern region; that from Menstrie to Alva the east-central region.

Of the earthquakes of the first class, those of 1802 (nos. 5 and 6) originated in the eastern region, those of 1872 and 1905 (nos. 8 and 19) in the east-central region, that of 1908 (no. 70) in the eastern region, and that of 1912 (no. 169) in the west-central region. One effect of the principal slip of 1905 was evidently to cause a sudden increase of stress within and just beyond the lateral margins of the focus, the stress near the eastern margin being relieved by the principal slip of 1908, and that near the western margin by the principal slip of 1912.

The distribution of the epicentres of the earthquakes of the second and third classes from 1900 is shown in the following table, the figures being percentages of the total number for each interval:

	W.	W.C.	E.C.	E.
Before 1905 Sept. 21	33	—	67	—
Between 1905 Sept. 21 and 1908 Oct. 20	8	24	68	—
Between 1908 Oct. 20 and 1912 May 3	25	22	53	—
After 1912 May 3	58	16	26	—

Thus, during the first 12½ years, seismic activity predominated in the east-central region, and during the last 4½ years in the western region, no minor epicentre throughout the whole 17 years being confined to the eastern region.

CHAPTER VI

EARTHQUAKES OF MISCELLANEOUS AND UNKNOWN CENTRES IN SCOTLAND

ULLAPOOL EARTHQUAKES

1. **1867** AUG. 20, *between 1 and 2 a.m.* Cat. No. 742; Intensity 5 (*Times*, Aug. 31).

In the Loch Broom district, a noise like the approach of a carriage over pavement was first heard, growing louder until it resembled the rumble of thunder, followed immediately by the shock which shook the house as if a great rock had fallen near it.

2. **1887** DEC. 18, *between 5 and 6 p.m.* Cat. No. 806 (C. A. Stevenson, *Edin. Roy. Soc. Proc.* vol. 15, 1889, p. 260).

A very slight shock in the Loch Broom district.

3. **1892** MAR. 4, *about 7.30 a.m.* Cat. No. 874; Intensity 5; Centre of isoseismal 4 in lat. 57° 58·5' N., long. 5° 13·5' W.; No. of records 37 from 33 places; Fig. 19 (*Geol. Mag.* vol. 10, 1893, pp. 293–296).

The boundary of the disturbed area is an isoseismal of intensity 4. It is 38 miles long, 17 miles wide, and contains about 364 square miles. Its centre is 6 miles N. 26° W. of Ullapool, and the direction of its longer axis about N. 30° W. The intensity of the shock was 5 at Isle Martin.

As a rule, the shock consisted of one prominent vibration — a thud, as if some heavy body had fallen—the sound at the same moment being like the boom of a cannon, followed by a tremor lasting for 3 or 4 seconds with a rumbling noise like distant thunder.

Fig. 19. Ullapool Earthquake of 1892 Mar. 4.

The sound-area coincided approximately with the disturbed area, except that it may have extended a short distance farther on the north-east side. The sound was heard by all the observers, and was compared to passing waggons, etc., in 50 per cent. of the records, thunder in 25, the fall of a heavy body in 8, and explosions in 17, per cent.

INVERGARRY EARTHQUAKES

From 1888 to 1899, a series of slight earthquakes was felt in Inverness-shire, in a district in which houses are few and widely scattered. The places at which the earthquakes were felt—Invergarry, Feddan, Ardochy, Glenkingie, Glenquoich and Lochhourn Head—are shown in Fig. 20 (*Geol. Mag.* vol. 8, 1891, pp. 365, 454; vol. 9, 1892, pp. 304–305; vol. 10, 1893, pp. 292–293; vol. 7, 1900, pp. 108, 113, 114, 122).

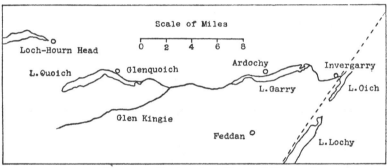

Fig. 20. Invergarry Earthquakes: Places of Observation.

1. **1888** JAN. 5, 5.30 *a.m.* Cat. No. 807; Intensity 4.
Four vibrations at Invergarry.

2. **1888** JAN. 12. Cat. No. 808 (C. A. Stevenson, *Edin. Roy. Soc. Proc.* vol. 15, 1889, p. 260).
A slight shock at Glenquoich.

3. **1888** FEB. 29, 8.10 *p.m.* Cat. No. 811.
One vibration, like a carriage passing, at Invergarry.

4. **1888** MAR. 1, 9.15 *a.m.* Cat. No. 812.
The same.

5. **1888** APR. 4, 9 *a.m.* Cat. No. 813.
A slight shock, like that produced by a carriage passing, at Invergarry.

6. **1888** APR. 4, 11 *a.m.* Cat. No. 814.
The same.

7. **1888** MAY 20, 6.10 *p.m.* Cat. No. 816.
The same.

8. **1888** JULY 3, 2.30 *p.m.* Cat. No. 817.
The same.

9. **1888** OCT. 22, 1.25 *p.m.* Cat. No. 819.
The same.

10. **1889** JUNE 19, 7.40 *a.m.* Cat. No. 824.
The same.

11. **1890** JAN. 5, 2.30 *a.m.* Cat. No. 827; Intensity 4.
A shock at Invergarry.

12. **1890** JAN. 5, 4.35 *p.m.* Cat. No. 828.
The same.

13. **1890** JAN. 5, 4.40 *p.m.* Cat. No. 829.
The same.

14. **1890** JAN. 5, 4.47 *p.m.* Cat. No. 830.
Vibrations, like those produced by a carriage passing, at Invergarry.

15. **1890** JAN. 19, 4.55 *p.m.* Cat. No. 831.
The same.

16. **1890** MAR. 15, 8.45 *a.m.* Cat. No. 832.
A noise, like that of a heavy train, at Invergarry.

17. **1890** MAY 29, 4.45 *p.m.* Cat. No. 833.
A noise, like that of a heavy carriage passing, at Invergarry.

18. **1890** AUG. 8, 3.35 *p.m.* Cat. No. 837.
The same.

19. **1890** NOV. 16, 8.30 *p.m.* Cat. No. 842.
A noise, like that of a passing train, at Invergarry.

20. **1890** NOV. 19, 1.33 *a.m.* Cat. No. 844.
A very slight shock, preceded by a low rumbling noise, at Feddan.

21. **1890** DEC. 1, 10.10 *a.m.* Cat. No. 849.
A slight shock at Invergarry.

22. 1890 DEC. 26, 6.10 *p.m.* Cat. No. 852.

A loud noise, resembling thunder, followed by a tremor, at Feddan. Almost immediately afterwards, another and louder sound was heard, without any accompanying tremor.

23. 1891 FEB. 24, 10.55 *p.m.* Cat. No. 853.

A sound, like that of a carriage, at Invergarry.

24. 1891 FEB. 24, 11.20 *p.m.* Cat. No. 854.

A rumbling noise, with a slight movement, lasting 10 seconds, at Lochhourn Head.

25. 1891 FEB. 25, 1.15 *a.m.* Cat. No. 855.

A noise, like that of a heavy carriage passing, at Invergarry.

26. 1891 MAR. 1, 3.15 *p.m.* Cat. No. 856.

A shock, lasting 8 seconds, at Ardochy.

27. 1891 MAR. 1, 9.25 *p.m.* Cat. No. 857.

A stronger shock than the preceding, lasting 4 seconds, and preceded by a rumbling noise, at Ardochy.

28. 1891 MAR. 2, 10.15 *p.m.* Cat. No. 858.

A noise, like that of a heavy carriage passing, at Invergarry.

29. 1891 APR. 24, 2.30 *p.m.* Cat. No. 860; Intensity 4.

A shock, followed by a noise like thunder, at Ardochy.

30. 1891 AUG. 27, 4.30 *a.m.* Cat. No. 861.

A noise, like that of a heavy carriage passing, at Invergarry.

31. 1891 AUG. 27, 6 *a.m.* Cat. No. 862.

The same.

32. 1891 AUG. 30, 4.15 *p.m.* Cat. No. 863.

The same.

33. 1891 NOV. 16, 10.25 *a.m.* Cat. No. 864.

A slight shock, without any accompanying noise, at Ardochy.

34. 1891 NOV. 16, 2.15 *p.m.* Cat. No. 865.

The same.

35. 1891 NOV. 16, 8.45 *p.m.* Cat. No. 866.

The same.

36. 1891 DEC. 6, 9.55 *a.m.* Cat. No. 867.

A noise, like that of a heavy carriage passing, at Invergarry.

37. **1891** DEC. 26, 1.20 *a.m.* Cat. No. 868.
A very slight noise and trembling at Lochhourn Head.

38. **1891** DEC. 28, 8.20 *p.m.* Cat. No. 869.
A noise, like that of a heavy carriage passing, at Invergarry.

39. **1891** DEC. 28, 9.24 *p.m.* Cat. No. 870.
The same.

40. **1891** DEC. 30, 9.45 *a.m.* Cat. No. 871.
A noise, like that of a heavy train crossing a bridge, at Invergarry.

41. **1892** FEB. 29, 10.35 *p.m.* Cat. No. 872.
A slight shock, accompanied by a low rumbling noise, at Ardochy.

42. **1892** FEB. 29, 11.15 *p.m.* Cat. No. 873; Intensity 4.
A shock, preceded by a low rumbling sound, at Ardochy.

43. **1892** APR. 3, 7.35 *a.m.* Cat. No. 875.
A slight shock, like that caused by a carriage passing, at Ardochy.

44. **1892** SEPT. 25, 8.13 *a.m.* Cat. No. 888.
A rumbling noise, lasting about 4 seconds, at Invergarry.

45. **1892** OCT. 24, 10.10 *a.m.* Cat. No. 889; Intensity 4.
A shock, lasting several seconds, at Ardochy, preceded, accompanied, and followed, by three distinct rumbling noises, of which the first was the loudest.

46. **1892** NOV. 18, 2.17—. Cat. No. 890.
A slight shock, followed by a slight rumbling noise, at Ardochy.

47. **1893** JAN. 2, 7.20 *p.m.* Cat. No. 891.
A slight shock, accompanied by a noise as of a carriage passing very rapidly, at Glenquoich. The noise was heard by another observer about 3 miles higher up the glen.

48. **1893** DEC. 11, *about 3 p.m.* Cat. No. 897.
A low rumbling sound, without any perceptible movement, heard at Glenkingie, Glenquoich and Lochhourn Head.

49. **1894** JAN. 25, 1.7 *p.m.* Cat. No. 903; Intensity 4.
A shock, like that produced by a carriage passing over a wooden bridge, followed by a distant roar, at Ardochy.

50. **1894** SEPT. 18, 10.10 *a.m.* Cat. No. 907.

A shock, weaker than the preceding and followed by a loud noise, at Ardochy.

51. **1899** DEC. 18, *about* 6.50 *a.m.* Cat. No. 931.

The shock was felt at Glenquoich, Glenkingie, Lochhourn Head, Carnack, and Corran (Arnisdale, in Glenelg). On the south side of Loch Quoich, the intensity was about 5. At Glenquoich, a sound was heard like distant thunder, growing louder until the house trembled, and then the sound died away as if a carriage had passed.

STRONTIAN EARTHQUAKES

1. **1809** JAN. 31. Cat. No. 248 (Howard, vol. 1, 1818, after table xxviii; Milne, vol. 31, p. 115; etc.).

For several nights before, considerable noises were heard in the Strontian district. On this day, five shocks, accompanied by a noise like distant thunder, were felt at Strontian and in the surrounding country. One shock, between 10 and 11 a.m., was of intensity 5. Another, about 10 p.m., was felt strongly at Dalclea and Kinlochmoidart, 7 and 9½ miles respectively N.W. of Strontian.

2. **1809** FEB. 1. Cat. No. 249.

A shock at Strontian.

3. **1809** FEB. 4. Cat. No. 250.

Two shocks at Strontian. The earlier shock was the strongest of the series, and was felt in Arisaig (the village of this name being 19 miles N.W. of Strontian). Both were felt in the mines.

4. **1809** FEB. 5. Cat. No. 251.

Two shocks at Strontian.

5. **1909** FEB. 6. Cat. No. 252.

One shock at Strontian.

1841 July 4, about 9.30 p.m., a slight shock, accompanied by a rumbling noise, at Kinlochmoidart (9½ miles N.W. of Strontian), which may be connected with the Strontian centre.

6. **1902** OCT. 14, *about* 5.15 *p.m.* Cat. No. 962; Intensity 5; Centre of isoseismal 5 in lat. 56° 44·4' N., long. 5° 30·5' W.; No. of records 47 from 38 places and 38 negative records from 33 places; Fig. 21 (*Geol. Mag.* vol. 1, 1904, pp. 537–539).

The curves in Fig. 21 are isoseismal lines of intensities 5 and 4. The isoseismal 5 is 38½ miles long, 22 miles wide, and 670 square miles in area. Its centre lies 4 miles N. 10° E. of Strontian, and its longer axis is directed N. 54° E. The isoseismal 4 is 49 miles long, 31 miles wide, and contains 1180 square miles. Its longer axis is parallel to that of the isoseismal 5; and the distance between the two curves is 5 miles towards the north-west and 4 miles towards the south-

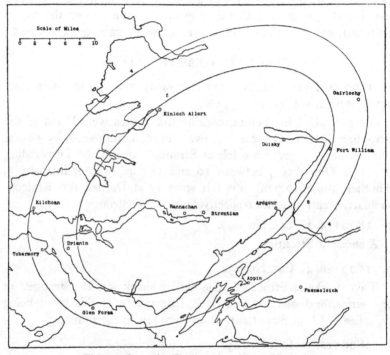

Fig. 21. Strontian Earthquake of 1902 Oct. 14.

east. Outside this isoseismal, the sound was heard at Kinlochhourn, Roy Bridge and Fasnacloich, the shock being also felt at the last-named place. All three places are 4 miles from the isoseismal 4, and lie respectively towards the north, east and south-east.

The shock consisted of a single series of vibrations, which increased in intensity and then died away, the average duration being 3 seconds.

The sound was heard by 98 per cent. of the observers. It was compared to passing waggons, etc., in 43 per cent. of the records, thunder in 43, wind in 3, and miscellaneous sounds in 11, per cent.

MISCELLANEOUS PERTHSHIRE EARTHQUAKES

Besides the important earthquake-centre of Comrie, there are seven minor centres in Perthshire, four of them (Kinloch Rannoch, Pitlochry, Aberfeldy and Amulree) visited by one earthquake only, and the other three (Ardvoirlich, Perth and Dunning) by two earthquakes. Perth and Dunning lie 9 or 10 miles on the south-east side of the Highland Border fault, the other five places on the north-west side. As Amulree is 5 miles, and Ardvoirlich 7 miles, from the fault, it is possible that the earthquakes of these centres may be connected with it.

KINLOCH RANNOCH EARTHQUAKE

1792 NOV. 10. Cat. No. 177 (Lauder, p. 367).

Three smart shocks, accompanied by a noise like distant thunder, felt on the banks of Loch Rannoch.

PITLOCHRY EARTHQUAKE

1842 AUG. 19, *between 7 and 9 p.m.* Cat. No. 583 (*Brit. Ass. Rep.* 1843, p. 121; Milne, reprint, p. 224).

Three tremors, accompanied by a noise like that of a carriage passing, at Pitlochry. The first and third were about equally strong, the second being weaker. Each lasted only a few seconds.

ABERFELDY EARTHQUAKE

About the year 1763, a shock, lasting 1 or 2 seconds, was felt in the parish of Logierait in Perthshire. Logierait lies about 8 miles E.N.E. of Aberfeldy (Lauder, p. 366).

1819 FEB. 11. Cat. No. 294 (*Quart. Journ. Sci.* vol. 7, 1819, p. 191).

A shock felt at Aberfeldy and Ballinloan in Glen Lyon.

AMULREE EARTHQUAKE

1819 DEC. 4, 7.30 *p.m.* Cat. No. 296; Intensity 4 (*Phil. Mag.* vol. 55, 1820, p. 69; *Journ. of Sci.* vol. 9, 1820, p. 205; *Ann. de Chim. et de Phys.* vol. 15, 1820, p. 422).

A shock, lasting 2 or 3 seconds, at Amulree, and dying away with a noise like the slow passing of carts.

ARDVOIRLICH EARTHQUAKES

1. **1826** DEC. 25, 2 *a.m.* Cat. No. 321 (Milne, vol. 31, p. 119).

A shock felt at Ardvoirlich, preceded by a noise like that of a blast in a quarry. This shock was evidently different from one felt at about the same time at Leadhills, 70 miles to the south.

2. **1831** MAR. 1, 11 *p.m.* Cat. No. 326; Intensity 4 (Milne, vol. 31, p. 119).

A shock, accompanied by a sound like a sudden gust of wind, at Ardvoirlich, Killin and Tyndrum.

PERTH EARTHQUAKES

1. **1824** AUG. 10. Cat. No. 316 (*Ann. de Chim. et de Phys.* vol. 33, 1826, p. 408).

A shock, accompanied by a very loud noise, at Perth.

2. **1855** MAY 7, *about 2 a.m.* Cat. No. 713 (Perrey, *Bull.* vol. 24, 1857, p. 86).

A strong shock at Perth, followed almost immediately by a second similar shock.

1855 May 14, between 1 and 2 a.m., Perrey records a similar double shock at Perth, but considers it to be the same as the preceding (*Ibid.* p. 87).

DUNNING EARTHQUAKES

1. **1809** JAN. 18, *about 1 a.m.* Cat. No. 246 (*Edin. Adver.*).

A shock, accompanied by a loud noise like that of thunder, felt at Steeland (near Dunning).

2. **1809** JAN. 18, *about 2 a.m.* Cat. No. 247; Intensity 5 (*Phil. Mag.* vol. 33, 1809, p. 91; etc.).

A sound was heard at Dunning, becoming gradually louder until the ground heaved, with a tremulous motion following, after which the sound died gradually away. The shock was felt also at Steeland and, along the northern base of the Ochil Hills, to Blackford.

GLASGOW EARTHQUAKES

1. **1656** AUG. 17, 4 *a.m.* Cat. No. 62 (R. Chambers, *Domestic Annals of Scotland*, vol. 2, pp. 241–242).

A sensible earthquake in all parts of the town of Glasgow. Another is said to have occurred about five or six years before this.

2. **1732** JULY 11, *between 2 and 3 p.m.* Cat. No. 86 (*Gent. Mag.* vol. 2, 1732, p. 874; *London Mag.* 1732, p. 205; R. Chambers, *Domestic Annals of Scotland*, vol. 3, p. 581).

A shock, which lasted about a second, felt at Glasgow and in the neighbourhood; no damage was done, though plates and cups were seen to move upon the shelves.

3. **1755** DEC. 31, *about* 1 *a.m.* Cat. No. 123; Intensity 5 (Dr R. Whyte, *Phil. Trans.* vol. 49, 1757, pp. 509–511; Milne, vol. 31, p. 102).

A strong shock felt at Glasgow, Greenock, Dumbarton, Kilmalcolm (4 miles S.S.W. of Dumbarton), and Inchinnan (2 miles N. of Paisley).

4. **1787** JAN. 6, *between* 10 *and* 11 *a.m.* Cat. No. 161; Intensity 4 (*Gent. Mag.* vol. 57, 1787, p. 82).

A shock, with sound, at Campsie, Strathblane, New Kilpatrick, Killearn and Fintray. The disturbed area covers a small district (probably containing less than 130 square miles) lying to the north of Glasgow. The epicentre is about 8 miles N. of Glasgow.

Mr W. Creech (*Edinburgh Fugitive Pieces*, 1815, pp. 122–123) and Sir T. D. Lauder (p. 367) give a similar account of an earthquake on Jan. 26 at about 10 a.m.

5. **1820** FEB. 22, 8.30 *a.m.* Cat. No. 297; Intensity 5 (*Ann. de Chim. et de Phys.* vol. 15, 1820, p. 423; *Phil. Mag.* vol. 55, 1820, pp. 231–237; *Journ. of Sci.* vol. 9, 1820, p. 206).

Three strong shocks at Glasgow, separated by very brief intervals, the third strongest, with loud rumbling noise. In one account (*Journ. of Sci.*) the date is given as Jan. 22.

6. **1821** OCT. 29. Cat. No. 308 (Perrey, p. 152).

A shock felt at Glasgow, Greenock, etc.

7. **1839** MAY 24. Cat. No. 352 (Perrey, p. 155).

A shock at Glasgow.

8. **1845** MAR. 9. Cat. No. 657 (Perrey, *Mém. Cour.* vol. 16, 1864, p. 11).

A strong shock at Campsie.

9. **1910** DEC. 14, 8.55 *p.m.* Cat. No. 1091; Intensity 5 or 6; Centre of isoseismal 5 in lat. 55° 53·1′ N., long. 4° 21·4′ W.; No. of records 61 from 22 places and 21 negative records from 16 places; Fig. 22.

Two isoseismals, of intensities 5 and 4, are shown in Fig. 22. The isoseismal 5 is 15¼ miles long, about 8½ miles wide, and 102 square miles in area; the isoseismal 4 is 23¾ miles long, 14 miles wide, and 250 square miles in area; the distance between the two curves being

3 miles on the north side and 2 miles on the south side, but, on the north side, the course of the isoseismal 5 is somewhat uncertain. The shock was also felt at Drymen (5 miles to the N. of the isoseismal 4) and at Mearns (2 miles S.). The disturbed area may thus have contained about 580 square miles. The centre of the isoseismal 5 is about $3\frac{1}{2}$ miles N.E. of Paisley, and the direction of its longer axis N. 83° E.

The shock is usually described as a heavy thud, as if a sack of sand had fallen on the floor, followed by a few shudders or tremors. The

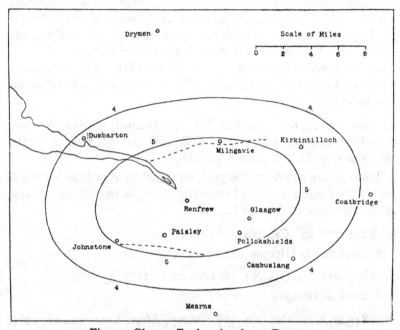

Fig. 22. Glasgow Earthquake of 1910 Dec. 14.

mean of 7 estimates of the duration of the shock is 2·7 seconds. The sound was heard by 96 per cent. of the observers, and was compared to passing waggons, etc., in 18 per cent. of the records, thunder in 5, wind in 5, loads of stones falling in 5, the fall of a heavy body in 52, and explosions in 14, per cent.

10. 1910 DEC. 14, 8.57 p.m. Cat. No. 1092.

A tremor at Bowling and Pollokshields.

Other slight after-shocks are said to have been felt at 9.10 p.m. at Drymen, and 9.25 and 9.50 p.m. at Pollokshields.

ORIGIN OF THE GLASGOW EARTHQUAKES

From the evidence of the earthquake of 1910 (no. 9), it would appear that the mean direction of the originating fault is about N. 83° E. The direction of the hade, and therefore the approximate position of the fault-line, is uncertain. On the map (Fig. 22), the courses of two of the more important east-west faults are indicated by broken-lines. The northern or Queenzieburn fault has a mean direction of N. 83° E. and hades to the south; the southern or Crookston fault has a mean direction of S. 80° E. and hades to the north. Thus, of the two faults, the Queenzieburn fault seems to satisfy the conditions the more closely ("Geology of the Glasgow District," *Mem. Geol. Surv. Scotland*, 1911, pp. 166–172 and Sheet 30).

EDINBURGH EARTHQUAKES

1. 1889 JAN. 18, *about* 4.10 *a.m.* Cat. No. 820; No. of records 7 from 4 places (*Geol. Mag.* vol. 8, 1891, pp. 60–67; R. Richardson, *Scot. Geogr. Mag.* vol. 3, 1889, pp. 135–146).

The shock was felt at Edinburgh, Penicuik and Bellsquarry. A rumbling sound was heard with the shock at Edinburgh; at Broomieknowe a hollow noise was heard without any accompanying shock.

2. 1889 JAN. 18, 6.53 *a.m.* Cat. No. 821; Intensity 6; Centre of disturbed area in lat. 55° 51' N., long. 3° 24' W.; No. of records 93 from 53 places and negative records from 4 places; Fig. 23.

The boundary of the disturbed area is an isoseismal line of intensity greater than 4. Towards the south and west, its course is somewhat uncertain. It is 30 miles long from north to south, 26½ miles wide, and about 830 square miles in area. The centre is about 3 miles S. 48° W. of Balerno.

At most places, the shock consisted of a single vibration, though two vibrations were noticed at Edinburgh, Leith, Ratho, Balerno and Peebles. The mean duration of the shock was 2 seconds.

Towards the east and south, the boundaries of the sound-area and disturbed area coincided approximately; but, towards the north and north-west, there were several places (Leith, Trinity, Burntisland and Polmont) at which no sound was heard. The centre of the sound-area thus lay about 2½ miles to the south or south-east of that of the disturbed area. The sound was compared to passing waggons, etc., in 33 per cent. of the records, thunder in 11, the fall of a heavy body in 39, and explosions in 17, per cent.

ORIGIN OF THE EDINBURGH EARTHQUAKES

The more important faults of the district run from N.E. to S.W., parallel to the axis of the Pentland Hills. "Flanking each side of the anticline, their effect has been to depress the Carboniferous strata against the older rocks of the hills, so that on the west side their

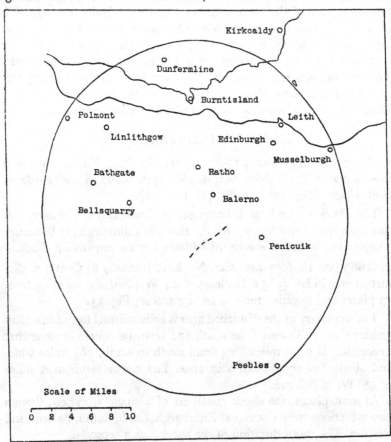

Fig. 23. Edinburgh Earthquake of 1889 Jan. 18.

downthrow is to the west, and on the east side to the east" (Sir A. Geikie, "Geology of the Neighbourhood of Edinburgh," *Mem. Geol. Surv. Scotland*, p. 120). One of these faults is indicated by the broken-line in Fig. 23. Its downthrow is to the north-west, and its distance from the epicentre is $2\frac{1}{2}$ miles. It is probable that the earthquake was caused by a slip near the centre of this fault, and that the sound-vibrations originated principally within the upper margin of the focus.

DUMFRIESSHIRE EARTHQUAKES

LEADHILLS EARTHQUAKES

1. **1748**, *about the end of* OCT. Cat. No. 97 (*London Mag.* 1749, p. 141).

A shock at Leadhills.

2. **1749** FEB. 14, *between* 8 *and* 9 *a.m.* Cat. No. 98; Intensity 5 (*London Mag.* 1749, p. 141).

A shock, accompanied by a noise like the falling of a house, at Leadhills. It was also felt at Wanlockhead (1 mile S.W. of Leadhills) and Penpont (12 miles S.), and also in the mines at Leadhills. The shock was evidently of some duration. Leadhills lies 5½ miles, and Wanlockhead 6 miles, from the Glen App fault on its S.S.E. side.

1812 Oct. 17–18, at night, an instantaneous movement, unaccompanied by any noise, was felt a few miles up the Nith (*Gent. Mag.* vol. 82, 1812, p. 487).

3. **1820** MAY 20. Cat. No. 298 (Milne, vol. 31, p. 118).

A shock at Wanlockhead.

4. **1820** NOV. 28, *about* 8 *a.m.* Cat. No. 300 (*Phil. Mag.* vol. 56, 1820, p. 463; *Edin. Phil. Journ.* vol. 4, 1821, p. 215).

A slight shock, accompanied by a hollow rumbling noise, at Leadhills and Wanlockhead, and felt 8 or 10 miles to the east and 3 or 4 miles to the west of those places. Miners working at a depth of 150 fathoms heard the noise.

5. **1820** NOV. 28, *about* 11 *p.m.* Cat. No. 301.

A shock, stronger than the preceding, felt at Leadhills and Wanlockhead, and accompanied by a rushing noise.

Mr Milne (vol. 31, p. 118) gives the same accounts for 1821 Nov. 28, 8 a.m. and 11 p.m., but the references given above show that the year was 1820.

1820 Nov. 29, 10.30 —, a shock is said to have been felt at Leadhills.

6. **1826** DEC. 25, 2 *p.m.* Cat. No. 322 (Milne, vol. 31, p. 119; Perrey, p. 154).

A shock felt at Leadhills and Crawfordjohn (6 miles N. of Leadhills). At the same time a shock, clearly independent, was felt at Ardvoirlich on the south side of Loch Earn and 70 miles N. of Leadhills.

7. **1828 MAY 20.** Cat. No. 324 (Milne, vol. 31, p. 119).

A shock felt in the mines at Wanlockhead and also at Dumfries (24 miles S. of Wanlockhead).

1865 Jan. 2, 1.30 *a.m.* Twice, at short intervals, windows and doors were shaken at Cargen (Dumfriesshire) as if by a sudden gust of wind, and another observer was awakened by a noise as of a carriage passing. The night was calm and still and no carriage passed (*Times*, Feb. 9; etc.).

8. **1872 DEC. 24.** Cat. No. 765 (*Nature*, vol. 8, 1873, p. 5).

A slight tremor felt in some parts of Upper Nithsdale.

PENPONT EARTHQUAKES

1. **1873 APR.** 16, 9.55 *p.m.* Cat. No. 767; Intensity 4; Centre of isoseismal 4 in lat. 55° 14·2′ N., long. 3° 48·3′ W.; Fig. 24 (*Brit. Ass. Rep.* 1873, pp. 194–197; *Nature*, vol. 8, 1873, p. 5).

A slight tremor, accompanied by a noise like thunder, felt in the parishes of Tynron, Glencairn, Keir, Penpont, Morton, Closeburn and Balmaclellan. The disturbed area, the boundary of which is represented by the continuous line in Fig. 24, is 6 miles long, 4 miles wide, and includes 19 square miles. Its centre is ½ mile W. of Keir, and the direction of its longer axis N. 75° E.

1873 Apr. 17, 2.46 a.m., a shock felt by a single observer at Thornhill.

2. **1876 AUG.** 11, *between* 11 *and* 12 *p.m.* Cat. No. 775 (*Nature*, vol. 14, 1876, p. 369).

Two vibrations felt at Tynron.

3. **1876 AUG.** 12, 3 *a.m.* Cat. No. 776; Centre of disturbed area in lat. 55° 14·2′ N., long. 3° 48·3′ W.; Fig. 24 (*Nature*, vol. 14, 1876, p. 369).

A sharp shock felt in the parishes of Tynron, Glencairn, Keir, Penpont and Morton, and therefore disturbing the same area as the earthquake of 1873 (no. 1). At Penpont a detonating sound was heard.

ECCLEFECHAN EARTHQUAKES

1. **1888 JULY 19**, *shortly before* 4 *a.m.* Cat. No. 818; Intensity 5 (*Times*, July 20).

The shock, which lasted only a few seconds, was felt at Wamphray, Applegarth, St Mungo and Tundergarth, in Annandale; the distance between Wamphray and St Mungo is 10 miles. At Wamphray a loud noise was heard with the shock. The centre of the disturbed area is probably about a mile N. of Applegarth.

2. 1894 MAR. 8, *about noon*. Cat. No. 904 (*Geol. Mag.* vol. 7, 1900, pp. 113–114).

A slight shock, most marked in the central district of Corrie.

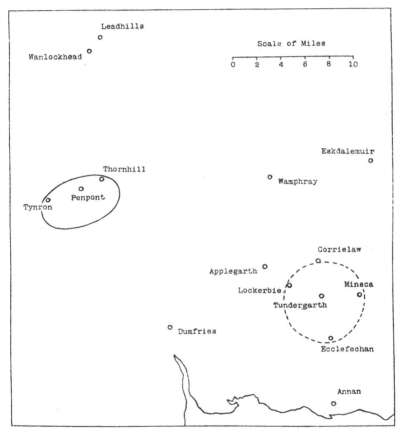

Fig. 24. Penpont Earthquake of 1873 Apr. 16 and
Ecclefechan Earthquake of 1896 May 29.

3. 1894 MAY 14. Cat. No. 905; Intensity 4 (*Geol. Mag.* vol. 7, 1900, p. 114).

A shock felt at Corrie Bridge, Corrielaw and Rosebank, in a small area the centre of which is 2 or 3 miles N.E. of Lockerbie, and about 4 miles east of that of the earthquake of 1888 (no. 1). At Corrie Bridge, the shock was accompanied by a noise like the distant firing of cannon.

4. **1896** MAY 29, 4.47 *a.m.* Cat. No. 911; Intensity 4; Centre of disturbed area in lat. 55° 6′ N., long. 3° 17′ W.; No. of records 15 from 14 places and negative records from 8 places; Fig. 24 (*Geol. Mag.* vol. 7, 1900, pp. 167–168).

The shock was distinctly felt within an area, the boundary of which is represented by the broken-line in Fig. 24, 7 miles long from E. to W., 6 miles wide, and containing 33 square miles. The centre of the area is equally distant (3 miles) from Lockerbie, Ecclefechan and Minsca. The sound alone was observed at several places outside the boundary—at Corsegreen (2½ miles W.), Dormont (1½ miles S.W.) and Craighouse (just outside to the E.). At Castleo'er (6 miles to the N.E.) the sound was heard by two persons, and a faint tremor felt by one of them.

The shock, a mere quiver, lasted about 3 seconds, and was accompanied by a sound like the passing of a waggon or distant thunder.

EARTHQUAKES OF UNKNOWN EPICENTRES IN SCOTLAND

1. **1597** JULY 23, *between* 8 *and* 9 *a.m.* Cat. No. 56 (Calderwood, vol. 5, 1845, p. 655; *Autobiography and Diary of Mr James Melvill*, pp. 420, 525).

A strong shock felt over a large part of the north of Scotland, especially in the counties of Ross and Cromarty, the districts of Kintail (south-west of Ross-shire) and Breadalbane (Perthshire), and as far south as Perth.

2. **1608** NOV. 8, *about* 9 *p.m.* Cat. No. 59 (Calderwood, vol. 6, 1845, p. 819; Lauder, p. 365; etc.).

Though no injury to property is recorded, this must be reckoned as one of the great Scottish earthquakes. It was particularly marked in the county of Fife—at Burntisland, Cupar, Dunfermline, Newburgh and St Andrews—and at Dundee and Edinburgh. The shock must also have been strong at Glasgow, Dumbarton and Aberdeen; for, at Dumbarton, "the people were so affrayed, that they ranne to the kirk..., for they looked presentlie for destructioun" (Calderwood). At Aberdeen, the people were so alarmed that the magistrates and clergy ordered the next day to be set apart for fasting and humiliation. The epicentre was probably in the north of Fifeshire or in the neighbourhood of Perth, and the disturbed area must have been much more than 11,000 square miles.

3. **1621.** Cat. No. 60 (Calderwood, vol. 7, 1845, pp. 461–462).

"About this time" (the last entry being dated May 11 and the next June 28) "there was a great earthquake in the toun of Montrose and thereabout, to the great terrour of the inhabitants, so that manie for fear fledd out of the toun."

4. **1727** MAR. 1, 4 *a.m.* Cat. No. 83 (R. Chambers, *Domestic Annals of Scotland*, vol. 3, p. 543).

A smart shock was felt in Edinburgh and throughout the south of Scotland, especially at Selkirk.

1728 Mar. 2, an earthquake in Scotland (Short, vol. 2, p. 170).

1749, an earthquake in Scotland (Prestwich, p. 544).

5. **1756** OCT. 10. Cat. No. 126 (*Gent. Mag.* vol. 26, 1756, p. 591).

"A violent shock threw down part of a public edifice in the parish of St Andrew...and much terrify'd the neighbourhood."

7. **1789** DEC. 30. Cat. No. 171 (Milne, vol. 31, p. 109).

Three distinct shocks were felt at the house of Parsons Green in the north side of Arthur's Seat near Edinburgh.

Descriptions, in nearly the same terms, are given under other dates —namely, 1782 Sept. 30 by Sir T. D. Lauder (p. 367), and 1789 Sept. 30, about 3 p.m., by W. Creech (*Edinburgh Fugitive Pieces*, 1815, pp. 125–126). As no one of the three mentions either of the other dates, it would seem probable that they refer to the same shocks.

8. **1817** APR. 26, 6.30 *a.m.* Cat. No. 272 (Milne, vol. 31, p. 118).

A smart shock felt at Glasgow, Inverness, Greenock, and (slightly) at Leith.

9. **1818** FEB. 19. Cat. No. 282 (Milne, vol. 31, p. 118).

A shock felt in Aberdeenshire.

10. **1818** JUNE 9, *about* 2.20 *p.m.* Cat. No. 286; Intensity 4 (*Phil. Mag.* vol. 51, 1818, p. 467).

A shock felt at Hayfield (on the N. side of Loch Awe, 10 miles N.N.W. of Inverary) and to a distance of 2 or 3 miles around; accompanied by a report like that of artillery and then of a noise like that of rocks falling down the mountain-side.

The earthquakes of 1821 Oct. 22, 1843 Feb. 25 and Mar. 3 (nos. 12, 16 and 17) may be connected with the same centre as the above in the neighbourhood of Inverary.

11. **1820** DEC. 25. Cat. No. 303 (*Phil. Mag.* vol. 57, 1821, p. 147).

A smart shock felt along the west coast of Kintail (south-west of Ross-shire), about Loch Hourn, the heights of Glenmoriston, and at other places in the central part of Inverness-shire.

12. **1821** OCT. 22, *morning.* Cat. No. 306 (*Ann. de Chim. et de Phys.* vol. 21, 1822, p. 393; Milne, vol. 31, p. 118).

A shock, with a noise like that of several carriages passing, at Inverary; also felt at Comrie, Crieff, near Loch Earn, and at Down (Loch Fyne).

Under the dates 1838 July 30 and Aug. 6, Mallet (1854, p. 278) records shocks at Tureff (Scotland), though he suggests that both entries may refer to the same shock. The place mentioned may be Turriff (in Aberdeenshire, 10 miles S. of Banff) or it may possibly be a misprint for Crieff.

13. **1839** MAR. 20, *about* 3 *a.m.* Cat. No. 350 (Milne, vol. 31, p. 122).

A shock, accompanied by noise, at Kingussie, between 2 and 3 a.m., and in Glengarry at 3.15 a.m.

1839 Oct. 24, about 2 a.m., a shock is said to have been felt at Kinross and in the valley of Glenshee, places which are about 30 miles apart.

1840 Nov. —, an earthquake in Scotland and Ireland (Perrey, p. 176).

1842 June 8, an earthquake in Scotland (Perrey, p. 162).

14. **1841** JULY 4, *about* 9.30 *p.m.* Cat. No. 523 (Milne, vol. 36, pp. 76–77).

A slight shock, with rumbling noise, at Kinlochmoidart (west of Inverness-shire).

15. **1841** DEC. 20, 4 *p.m.* Cat. No. 566 (Milne, vol. 36, pp. 84–85).

A strong shock, with noise, in Kintail (south-west of Ross-shire).

16. **1843** FEB. 25, *about* 8.12 *p.m.* Cat. No. 595; Intensity 5 (*Brit. Ass. Rep.* 1843, p. 121; Milne, reprint, pp. 225–227).

A shock, accompanied by a sound like that of many carts of stones being emptied or several coaches passing, at Lochgilphead, Oban, Dunolly (1 mile N. of Oban) and Castle Toward. Dunolly and Castle Toward are 43 miles apart, so that the disturbed area may have contained about 1450 square miles. The epicentre may be near Loch Fyne about midway between Inverary and Lochgilphead.

17. **1843** MAR. 3, 8.40 *p.m.* Cat. No. 596; Intensity 4 (*Brit. Ass. Rep.* 1843, p. 121; Milne, reprint, pp. 227–228).

A shock, accompanied by a loud noise like that of many carts of stones being emptied, at Lochgilphead and to a distance of 2½ miles from that place.

18. **1844** MAR. 15, *about* 11.20 *p.m.* Cat. No. 638; Intensity 4 (*Ill. London News*, Mar. 30).

A shock, accompanied by a low rumbling sound, at Galashiels.

19. **1855** AUG. 7. Cat. No. 715 (Perrey, *Bull.* vol. 24, 1857, p. 94).

Two slight shocks in the Cuillin Hills district (Skye).

Fuchs (pp. 137, 138) records two shocks of which no further details are known, one on 1871 Apr. 18 in Scotland, the other on 1877 Mar. 20 in the island of Mull.

20. **1878** DEC. 3, 5 *a.m.* Cat. No. 781; Intensity probably 4 (*Times*, Dec. 5).

A very marked shock at Balnacarra (on the north shore of Loch Alsh). A shock was also felt at Plockton (5 miles N. of Balnacarra) between 7 and 8 a.m.

21. **1879** JUNE 16. Cat. No. 783 (*Nature*, vol. 20, 1879, p. 182).

A distinct shock at Tobermory and other places in Mull.

1886 May 11, Roper (p. 41) records (from the *Glasgow News*, May 15) several slight shocks at Gairloch (west of Ross-shire).

CHAPTER VII

EARTHQUAKES OF NORTH WALES

CARNARVON EARTHQUAKES

1782 Oct. 5, 8.40 p.m., a strong shock, accompanied by a loud rumbling noise, felt at Amlwch and many places in Anglesey, and also at Bangor, Mold (intensity 4) and St Asaph. The position of the epicentre is unknown (Cat. No. 156; J. Lloyd, *Phil. Trans.* vol. 73, 1783, pp. 104–105).

1. **1818** DEC. 7, *about* 9 *a.m.* Cat. No. 293 (*Journ. of Sci.* vol. 6, 1819, p. 371; *Ann. de Chim. et de Phys.* vol. 33, 1826, p. 403).

A shock, slight at Bangor, and much more sensible at Pentir (3 miles S. of Bangor).

1818 Dec. 14, 9 a.m., a slight shock at Bangor. This is probably the same as the preceding earthquake (*Ann. de Chim. et de Phys.* vol. 9, 1818, p. 435).

1827 Feb. 9, 7 p.m., a shock, accompanied by a noise like that of a heavily laden cart passing over stones, in north-west Wales and Anglesey, the shock being strong enough to overturn several pieces of furniture (Cat. No. 323; *Phil. Mag.* vol. 3, 1828, p. 463).

1831 Mar. 17, a shock felt throughout Bardsey Island (Cat. No. 328; Mallet, 1854, p. 224).

2. **1842**, *early in* MAY, *between* 6 *and* 7 *a.m.* Cat. No. 574 (Milne, reprint, pp. 223–224).

A shock felt near Carnarvon and at Llanddwyn. Two ships were at the time crossing the bar of Carnarvon harbour when the crews heard a booming noise like thunder, followed by a trembling of the vessels as if they had struck the ground.

3. **1842** AUG. 22, *between* 6 *and* 7 —. Cat. No. 584 (*Brit. Ass. Rep.* 1843, p. 121; Milne, reprint, p. 224).

A shock felt at Bangor and over Anglesey, especially in the south-east of the island.

1874 Nov. 15, 2 *a.m.*, a distinct shock felt in north Carnarvonshire (Cat. No. 770; *Nature*, vol. 11, 1875, p. 57; etc.).

The following earthquakes (nos. 4–11) form a series. In addition to these eight shocks, six shocks and two earth-sounds were observed, in each case by one person. They are distinguished by the absence of a catalogue-number (*Quart. Journ. Geol. Soc.* vol. 60, 1904, pp. 233–242; *Geol. Mag.* vol. 5, 1908, pp. 300–301).

4. **1903** JUNE 19, *about* 4.25 *a.m.* Cat. No. 968; No. of records 2 from 1 place.

A rumbling noise like thunder at Griffiths Crossing (2 miles N.E. of Carnarvon).

5. **1903** JUNE 19, 10.4 *a.m.* Cat. No. 969; Intensity 7; Centre of isoseismal 7 in lat. 53° 3·0′ N., long. 4° 22·9′ W.; No. of records 388 from 206 places and 56 negative records from 44 places; Figs. 25–27.

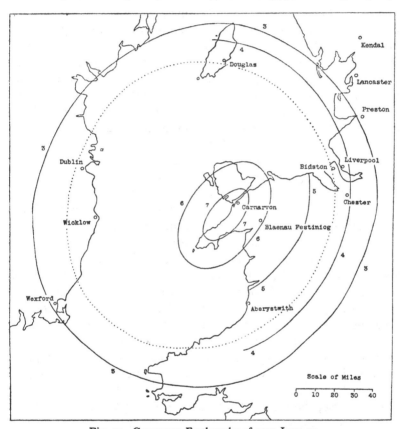

Fig. 25. Carnarvon Earthquake of 1903 June 19.

Isoseismal Lines and Disturbed Area. In Fig. 25 are shown the isoseismals 7 and 6, those portions of the isoseismals 5 and 4 which traverse the land, and the boundary of the disturbed area. In one or two places, buildings were slightly damaged. At Clynnog, a slab of slate, weighing more than a hundredweight, was dislodged from the top of a chimney; and at Penygroes two chimneys were thrown down. Both places are close to the epicentre.

The isoseismal 7 (Figs. 25 and 27) is $33\frac{1}{2}$ miles long, 15 miles wide and 420 square miles in area. The centre is 4 miles W. of Penygroes church, and the longer axis is directed N. 40° E. Of the next isoseismal (intensity 6), little more than half can be drawn with accuracy. The width of the curve is 38 miles, and its distance from the isoseismal 7 is 11·8 miles on the north-west side and 10·6 miles on the south-east. The isoseismal 5 is interrupted by the sea to the north of Flintshire and in Cardigan Bay. Of the isoseismal 4, nearly half can be drawn. It traverses the Isle of Man and the eastern counties of Ireland. The area contained by it is about 15,600 square miles. The outermost isoseismal, which bounds the disturbed area, corresponds to an intensity between 4 and 3. It is 185 miles long from N.E. to S.W., 173 miles wide, and contains 25,000 square miles. The shock was also felt in upstair rooms at four places outside this line—at Dunmore East in Co. Waterford (22 miles from the isoseismal), Ravensdale in Co. Louth (8 miles), Kendal (25 miles) and Didsbury near Manchester (13 miles). A curve passing through Kendal and concentric with the isoseismal 4 would contain about 40,000 square miles.

Nature of the Shock. In its general features, the nature of the shock was nearly uniform throughout the disturbed area, and the following account (from Meyllteyrn, near Nevin) may be regarded as typical for a large part of the area. The shock began with a series of tremors, lasting 4 or 5 seconds, which merged gradually into a single series of principal vibrations of about 3 or 4 seconds' duration, these being succeeded by a brief series of tremors lasting only 1 or 2 seconds. At a few places not far from the central area, two maxima of intensity in the principal vibrations were detected by careful observers and their evidence is confirmed by the Birmingham seismogram (Fig. 26). At a great distance—at Liverpool and Southport and in some parts of Ireland, for instance—the vibrations between these maxima were imperceptible, and the shock seemed to consist of two detached parts. The period of the vibrations also increased with the distance so that, in Lancashire, Ireland and elsewhere, the motion consisted of a gentle swaying several times to and fro. The mean duration of the shock throughout the disturbed area was 6·7 seconds.

Seismographic Records. The earthquake was recorded by a Milne seismograph at Bidston (near Birkenhead) and by an Omori horizontal pendulum at Birmingham.

Bidston is 60 miles from the epicentre in the direction N. 66° E. The first movements there were registered at 10h. 5m. 5s. a.m. The separate oscillations of the pendulum are not shown on the diagram, but there seem to have been two impulses, the second taking place at 10h. 7m. 30s. a.m.

Birmingham lies 111 miles S. 70° E. from the epicentre. The record (enlarged in Fig. 26) gives the E.–W. component of the motion. The whole movement is divisible into three parts: (i) The preliminary tremors (on the right) are first perceptible at 10h. 5m. 56s. and lasted 13 seconds. The enlarged diagram shows hardly any sign of them; but, when the original record is examined under a microscope, they appear as minute notches, 51 in number, on the trace. The average period of the tremors was therefore ·25 second. (ii) The principal vibrations began at

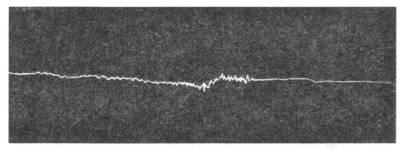

Fig. 26. Seismograph Record at Birmingham of the Carnarvon Earthquake of 1903 June 19.

10h. 6m. 9s., and lasted 26 seconds. The total number of vibrations is 40; but the first 19 of them are, as a rule, of much greater amplitude than the rest. They have an average period of ·63 second, while that of the remaining 21 vibrations is ·67 second. In the 2nd and 19th vibrations, which are the largest of the series, the range (or double amplitude) was ·023 mm. in the E.–W. direction, or ·024 mm. (that is, about one-thousandth of an inch) in the direction of the epicentre. Taking the period of these vibrations at ·63 second, the maximum acceleration would be 1·3 mm. per second per second. (iii) The concluding undulations began at 10h. 6m. 35s. On the enlarged diagram 27 may be seen, with an average period of 1 second; but with the aid of the microscope, they can be detected until 10h. 7m. 40s., though so obscurely in some parts of the trace that their exact number cannot be ascertained. The total duration of the disturbance was thus 1m. 44s.

Sound-Phenomena. The boundary of the sound-area is indicated by the dotted line in Fig. 25. It is 147 miles long from N.E. to S.W., 136 miles wide, and contains about 15,700 square miles. In the whole disturbed area, 88 per cent. of the observers heard the earthquake-sound. The percentage of audibility was 100 within the isoseismal 7, 99 between the isoseismals 7 and 6, 98 between the isoseismals 6 and 5, falling to 48 in the surrounding zone. The sound was compared to passing waggons, etc., in 45 per cent. of the records, thunder in 29, wind in 7, loads of stones falling in 8, the fall of a heavy body in 1, explosions in 7, and miscellaneous sounds in 3, per cent.

Miscellaneous Phenomena. A few observations were made in slate quarries in which the workings are continued underground. At Nantlle, the shock was felt at a depth of from 50 to 70 yards. It was also noticed in underground workings at Blaenau Festiniog (19 miles from the epicentre).

Among the most interesting observations on the earthquake were those made on the movement of the loose material of screes. Owing to the gradual creeping downwards with every change of temperature of all stones free to move, a large part of the material is almost in unstable equilibrium, and a very slight force is necessary to set it in motion (*Quart. Journ. Geol. Soc.* vol. 44, 1888, pp. 232–237, 825–826). At the time of the earthquake, one observer was sitting on a slope of screes 150 yards south of Lleyn dur Arddu and 1 mile N.W. of the summit of Snowdon. The strongest shock sent numbers of stones shuffling and rolling down the surface. Stones of all sizes were involved, blocks of felsite up to 2 feet in diameter among them, the larger moving more quickly than the others, and the resulting noise drowned the rumbling of the earthquake. The screes continued unstable for five minutes, and, at the end of that time, hundreds of newly-fallen blocks were to be seen lying at the base. A somewhat similar observation was made at Blaenau Festiniog, where fragments of slate were seen rolling down the tips of waste slate from the quarries. June 19, 10.7 a.m., a very slight tremor, accompanied by a sound like that of distant thunder, at Meyllteyrn.

6. **1903** JUNE 19, 10.9 *a.m.* Cat. No. 970; Intensity 3; No. of records 4 from 4 places.

A slight tremor was felt at Penygroes and Gaerwen, while a rumbling sound was heard at Gaerwen, Bethesda and Bodfeirig. The boundary of the disturbed area and the position of the epicentre must have coincided nearly with those of the after-shocks of June 19, 11.8 a.m., and June 21, 8.6 a.m. (nos. 9 and 10).

7. **1903 JUNE 19, 10.12 *a.m.*** Cat. No. 971; Intensity 3; No. of records 2 from 2 places.

A slight tremor was felt at Penygroes and a tremulous sound heard at Bethesda. The epicentre was probably near that of the preceding shock.

8. **1903 JUNE 19, 10.16 *a.m.*** Cat. No. 972; Intensity 3; No. of records 2 from 2 places.

A tremulous sound was again heard at Bethesda. At Bettws Garmon, a slight tremor was felt, lasting about 2 seconds, accompanied by a sound like very faint distant thunder.

June 19, 10.23 a.m., a tremulous sound at Bethesda.

June 19, 10.48 a.m., a slight tremor at Penygroes.

9. **1903 JUNE 19, 11.8 *a.m.*** Cat. No. 973; Intensity 3; Centre of disturbed area in lat. 53° 7·6′ N., long. 4° 14.3′ W.; No. of records 7 from 7 places; Fig. 27.

The disturbed area, the boundary of which is represented by the broken-line in Fig. 27, is 20 miles long, 13 miles wide, and 219 square miles in area. The centre of the area is 8 miles N.E. of that of the isoseismal 7 of the principal earthquake (no. 5), and the direction of its longer axis is N. 47° E. A slight tremor was felt at every place, and was accompanied by a faint rumbling sound at Clynnog, Nantlle, Penygroes and Gaerwen.

June 19, 0·5 p.m., a slight shock at Bodfeirig.

June 21, 5.26 a.m., a shock, accompanied by a sound like the tipping of quarry rubbish, at Upper Clynnog.

10. **1903 JUNE 21, 8.6 *a.m.*** Cat. No. 974; Intensity 3; No. of records 5 from 5 places.

The boundary of the disturbed area and the position of the epicentre were nearly the same as those of the preceding after-shock. A slight shock was felt at Nantlle and Penygroes, and a rumbling sound was heard at Bodfeirig, Clynnog and Newborough.

June 21, about 9.6 a.m., an earth-sound at Clynnog.

June 22, 4.26 a.m., a slight shock, accompanied by a rumbling noise, at Penygroes.

June 23, about 5.31 a.m., a very slight shock at Nantlle.

11. **1906** JUNE 29, 3.2 *a.m.* Cat. No. 999; Intensity 4; Centre of disturbed area in lat. 53° 9·2′ N., long. 4° 10·7′ W.; No. of records 23 from 15 places and 17 negative records from 13 places; Fig. 27.

The disturbed area, the boundary of which is represented by the dotted line in Fig. 27, is 16 miles long, 14½ miles wide, and contains about 182 square miles. Its centre is about 4 miles N. 75° E. of Carnarvon,

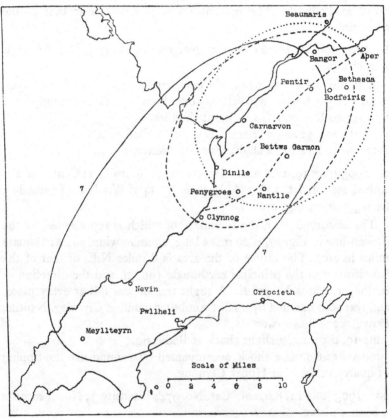

Fig. 27. Carnarvon Earthquakes of 1903 June 19 and 1906 June 29.

11 miles N. 50° E. of the centre of the isoseismal 7 of the principal earthquake (no. 5), and 2¾ miles N.E. of that of the after-shock of 1903 June 19, 11.8 a.m. (no. 9). The shock consisted of a single series of vibrations, lasting as a rule only a few seconds. The sound, which lasted about 5 seconds, was heard by all the observers, and was compared to passing waggons, etc., in 32 per cent. of the records, thunder in 53, wind in 5, loads of stones falling in 5, and miscellaneous sounds in 5, per cent.

The seismic evidence gives the following elements for the position of the originating fault: (i) the mean direction of the fault is N. 40° E.; (ii) the hade is to the N.W.; (iii) the fault-line must pass a short distance on the south-east side of the centre of the isoseismal 7, so that its course may be in part submarine, or it may pass through Clynnog or even farther to the south-east; and (iv) the fault must be of some magnitude, extending about 8 miles N.E. and S.W. of Clynnog.

Two faults, copied from the Geological Survey map (sheets 75 and 78), are shown in Fig. 27, one traced for a distance of 14 miles from Aber to near Dinlle on the coast of Carnarvon Bay, the other for 8 or 9 miles from Bettws Garmon to Clynnog. Of the two, the former satisfies the seismic conditions the more closely. Its average direction at the surface is N. 52° E., it hades to the N.W., and, according to Ramsay[1], the downthrow of the Silurian beds on that side is between 4000 and 5000 feet at Pentir (3 miles S. of Bangor), and between 2000 and 3000 feet at Dinas (4 miles farther to the south-west). If the fault, after leaving Dinlle, is continued under the sea towards Nevin, trending rather more to the south, it would occupy approximately the position assigned to the originating fault.

Assuming it to be so extended, the portion of the fault-surface affected by the above earthquakes reaches from the neighbourhood of Bangor to that of Meyllteyrn, a length of about 32 miles, possibly to Bardsey Island, a total distance of about 45 miles. The focus of the principal earthquake of 1903 was about 16 miles long, extending from near Carnarvon to near Nevin. Thus, the active portion of the fault may be divided into four regions: (i) the northern region, including the part from Bangor to Pentir, connected with the earthquakes of 1818, 1842 Aug., and 1906 (nos. 1, 3 and 11) and two disturbances recorded by single observers in 1903; (ii) the north-central region, near Carnarvon, connected with the earthquakes of 1842 May and 1903 June 19, 4.25, 10.9, 10.12, 10.16 and 11.8 a.m., and June 21, 8.6 a.m. (nos. 2, 4, 6, 7, 8, 9 and 10); (iii) the central region, from Carnarvon to Nevin, in which originated the principal earthquake of 1903 June 19 (no. 5) and 5 disturbances recorded by single observers in 1903; and (iv) the southern region, near Meyllteyrn, connected with one shock recorded by a single observer in 1903, and possibly

[1] A. C. Ramsay, *The Geology of North Wales*, 2nd edition, 1881, pp. 173, 185, 194, 197, 198, 200, 203.

extending to Bardsey Island and connected with the earthquake of 1831 Mar. 17. Denoting these regions by the letters N, NC, C and S, and using small letters for earthquakes recorded by single observers, the migration of the foci from 1818 to 1906 may be represented as follows:

1818 1842   1903   1906

N NC, N   NC, C, s, NC, NC, NC, n, c, NC, n, c, NC, c, c, c   N

BEDDGELERT EARTHQUAKE

1904 OCT. 21, *about* 6.5 *a.m.* Cat. No. 983; Intensity 4; No. of records 5 from 4 places (*Geol. Mag.* vol. 5, 1908, p. 298).

The shock was felt at Beddgelert, Blaenau Festiniog, Croesor (4 miles S.E. of Beddgelert) and Plas Gwynant (3 miles N.E. of Beddgelert). It was accompanied at all these places by a noise like thunder. The disturbed area, which covers a district to the south of Snowdon, was thus at least 8 miles long from E. to W. and 6 miles wide. Its centre may be about 4 miles E. of Beddgelert.

BARMOUTH EARTHQUAKES

1744 Feb. 5, several shocks felt in Merionethshire (Cat. No. 95; Milne, vol. 31, p. 76; *Gent. Mag.* vol. 14, 1744, p. 103).

1. **1769** JUNE 15. Cat. No. 145 (Milne, vol. 31, p. 105).

A shock at Dolgelley.

2. **1820** SEPT. 27, 9 *p.m.* Cat. No. 299 (*Ann. de Chim. et de Phys.* vol. 15, 1820, p. 424).

A shock, accompanied by a sound like that of a gun, at Barmouth.

3. **1820** DEC. 20, *a little after* 9 *p.m.* Cat. No. 302 (*Edin. Adver.*).

A slight shock, accompanied by a noise like that produced by a large ball rolling on a hollow floor, in the district between Barmouth and Harlech.

4. **1851** NOV. —. Cat. No. 703; Intensity 5 (Letter to author).

A shock, accompanied by a loud rumbling noise, at Dolgelley and Caer Ynwch (3 miles E. of Dolgelley).

BALA EARTHQUAKE

1903 JULY 1, 1.16 *a.m.* Cat. No. 975; Intensity 4; No. of records 10 from 5 places and negative records from 3 places (*Geol. Mag.* vol. 1, 1904, p. 539).

The disturbed area lies along the valley of the Dee from near Bala to near Llandrillo, all the places at which the shock was felt being close to the great Bala fault. The area is about 7 miles long, not less than $2\frac{1}{2}$ miles wide, and contains about 14 square miles, its centre being about $\frac{1}{2}$ mile N.W. of Llanderfel.

At Palè, the shock consisted of a single series of horizontal vibrations, lasting 4 or 5 seconds, increasing in strength to a maximum and then dying away. It was accompanied by a sound like thunder or the beating noise of a motor-car.

Though the boundary of the disturbed area cannot be drawn with exactness, it is probable that the earthquake was caused by a shallow slip along the Bala fault or one of its branches.

RHYL EARTHQUAKE

1768 Jan. 18, a shock in Flintshire. Cat. No. 141 (Milne, vol. 31, p. 104).

1773 Jan. 31, between 11 and 12 p.m., a shock in Flintshire recorded by Milne (vol. 31, p. 105), probably due to a landslip.

1805 JAN. *12, about 7 p.m.* Cat. No. 242 (*Gent. Mag.* vol. 75, 1805, p. 173).

A strong shock, lasting 2 or 3 seconds, felt throughout the whole of the lower end of the Vale of Clwyd, in the west of Flintshire.

1810 Oct. 12, 7 p.m., an earthquake at Clwyd (Milne, vol. 31, p. 115).

EARTHQUAKES OF UNKNOWN EPICENTRES
IN NORTH WALES

1768 Feb. 15, Milne records, without details, the occurrence of an earthquake at Llangollen (vol. 31, p. 104).

1. **1780** AUG. *28, about 8.45 a.m.* Cat. No. 154 (*Gent. Mag.* vol. 50, 1780, p. 537).

A strong shock felt in Flintshire and Denbighshire, and also in Anglesey and Carnarvonshire, especially at Llanrwst (in Denbighshire) and Downing and Holywell (in Flintshire).

2. **1780** DEC. 9, *between 4 and 5 p.m.* Cat. No. 155 (*Gent. Mag.* vol. 51, 1781, p. 527).

Two shocks, a slight one followed by one much stronger, at Downing, preceded by the usual noise. It is said to have been felt also at Newcastle, York, Leeds, Whitehaven and other places.

3. **1879** APR. 8, *about 8.35 p.m.* Cat. No. 782; Intensity 4 (*Nature*, vol. 19, 1879, p. 555).

A slight vibration, with a sound like the rumbling of a heavy waggon, felt at Bron Celyn (near Bettws y Coed) and near Bettws Garmon. The earthquake may be connected with the fault that runs from the latter place to near Clynnog (Fig. 27).

4. **1888** APR. 11. Cat. No. 815 (*Nature*, vol. 37, 1888, p. 595; *Times*, Apr. 13).

A shock felt at Llangollen, Corwen, Bala and Dolgelley.

CHAPTER VIII

EARTHQUAKES OF SOUTH WALES

PEMBROKE EARTHQUAKES

1873 Feb. 1, 7 a.m., a double shock (intensity 4) in the west of Pembrokeshire (Cat. No. 766; *Nature*, vol. 7, 1873, p. 283).

An hour later, the double movement seems to have been repeated.

With these possible exceptions, the Pembroke centre seems to have been inactive except during the brief period 1892 Aug. 17–23 (*Quart. Journ. Geol. Soc.* vol. 53, 1897, pp. 157–168, 173–174).

1. 1892 AUG. 17, *about* 11.30 *p.m.* Cat. No. 878; Intensity 3; No. of records 3 from 3 places.

A shock felt at Haverfordwest and Rudbaxton. At the same time, two rumbling noises were heard at Pembroke.

2. 1892 AUG. 18, 0.22 *a.m.* Cat. No. 879; Centre of sound-area in lat. 51° 38·7′ N., long. 4° 50·1′ W.; No. of records 7 from 7 places; Fig. 29.

Two minutes before the principal shock, a distant sound, as of thunder, was heard without any accompanying tremor. The boundary of the sound-area is represented by the incomplete dotted line (Fig. 29). This area is 7 miles wide, its longer axis being directed approximately north. Its centre is either close to the coast, about 1½ miles W. of Manorbier, or else beneath the sea. The land-area over which the sound was heard contains about 33 square miles.

3. 1892 AUG. 18, 0.24 *a.m.* Cat. No. 880; Intensity 7; Centre of isoseismal 7 in about lat. 51° 42·5′ N., long. 4° 54·4′ W.; No. of records 712 from 555 places and 114 negative records from 111 places; Figs. 28, 29.

Isoseismal Lines and Disturbed Area. On the map (Fig. 28) are shown the portions of the isoseismals 7 and 6 which lie on land, and the isoseismals 5 and 4. The isoseismal 7 is also shown on a larger scale in Fig. 29. This curve is in the form of a semi-ellipse, of which only the northern half lies on land. The length of the curve may be about 48 miles, its width is 21 miles, the land-area about 450 square miles and the total area about 790 square miles. The centre of the curve is about 1 mile N. of Pembroke, and the direction of its longer axis is approximately N. The isoseismal 6 includes the whole of Pembrokeshire and parts of the counties of Cardigan, Carmarthen and Glamorgan. Only the eastern part of the curve can be drawn

(Fig. 28), the western part must lie at some distance off the coast; towards the south, it probably passes to the north of Lundy Island. The land-area included within this isoseismal is about 1070 square miles. The isoseismal 5 is 172 miles long from N. to S., 136 miles wide, and contains about 18,660 square miles. The distance between

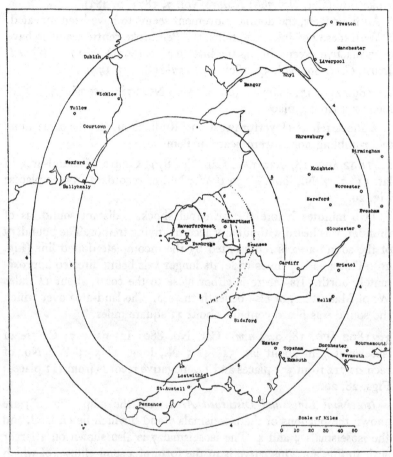

Fig. 28. Pembroke Earthquakes of 1892 Aug. 18.

the isoseismals 7 and 5 is 52 miles on the east side and 62 miles on the west. The isoseismal 4 is 255 miles long from N. to S., 225 miles wide, and about 44,860 square miles in area.

Outside this isoseismal line, the shock was felt at Tresco Abbey (Scilly Isles), Dorchester and South Littleton (near Evesham), the distances of which from the isoseismal are respectively 14, 12 and

8 miles. The shock was probably felt, though very slightly, at Sutton in Surrey, the distance of which from the epicentre is about 200 miles. If we regard the disturbed area as bounded by a circle through Tresco Abbey, it must contain about 56,000 square miles; if by one through Sutton, about 125,000 square miles.

Nature of the Shock. At places near the central area, the shock was continuous and contained two maxima of intensity, the second of which was stronger than the first. For instance, at Haverfordwest, the shock began violently, gradually weakened, and then became stronger than at first, the total duration being about 14 seconds. At some distance from the epicentre, the tremulous motion felt between the two maxima was imperceptible, and the shock then consisted of two detached series of vibrations separated by an interval of 3 seconds in average length, the second series being the stronger. For instance, at Clynderwen (Carmarthenshire), sharp vibrations increasing in strength were felt for about 5 or 6 seconds, ceasing for 2 or 3 seconds, and then renewed with greater intensity for about 10 to 15 seconds.

The double series of vibrations was observed practically throughout the disturbed area,—at Rhyl in the north, Tresco Abbey (Scilly Isles) in the south, Worcester in the east, and Tullow (Co. Carlow) in the west. The relative intensity of the two series was estimated by 30 observers, 22 of who regarded the second as the stronger and 2 the first, while, according to 6, the series were nearly equal in strength. It is probable that the difference between them was not marked, but that all over the disturbed area the second part or maximum was the stronger.

Sound-Phenomena. Towards the north, west and south, the boundary of the sound-area follows closely the course of the isoseismal 4; towards the east and south-east, it extends a few miles outside the isoseismal. The sound was heard by 75 per cent. of the observers in the whole disturbed area, by 100 per cent. within the isoseismal 6, 92 per cent. between the isoseismals 6 and 5, and 69 per cent. between the isoseismals 5 and 4. The sound was compared to passing waggons, etc., in 44 per cent. of the records, thunder in 29, wind in 7, loads of stones falling in 2, the fall of a heavy body in 2, explosions in 5, and miscellaneous sounds in 11, per cent. Close to the epicentre, the sound was observed to change in character. Thus, at Lamphey (near Pembroke), a murmuring sound resembling that of sea-waves at a distance was heard for 3 seconds, changing, when the shock was strongest, to a deep heavy boom like thunder, and continuing after the shock with the same sound as at first. Between the isoseismals

5 and 4, no such variations were perceived; the sound was more monotonous than elsewhere, and resembled frequently the low roll of distant thunder or the moaning of the wind or sea-waves.

Miscellaneous Phenomena. The shock was felt by several persons in boats, the sensation being the usual one of having struck on a rock. At Bulwell, on the southern shore of Milford Haven, two or three waves were seen to run up the shore, the sea both before and after being absolutely still. The shock was felt on a steamboat on the river opposite Langwm. The water, although perfectly calm before, became suddenly swelled as with a heavy breeze; the boat seemed as if it passed over three waves, after which the water was troubled for a few seconds and then became calm as before[1].

The shocks appear to have had some slight effect on underground water. At Green Croft (near Narberth), a well, which contained water shortly before the earthquake, was found to be empty when visited soon afterwards. At Marloes, the springs in the upper part of the village were dry for at least eight weeks after the shock, while those in the lower parts were overflowing.

No traces of the earthquake were detected on the magnetograms of Kew and Greenwich or on the horizontal pendulum records at Strasbourg and Nicolaiew.

4. **1892 AUG. 18, 0.37** *a.m.* Cat. No. 881; Intensity about 4; No. of records 23 from 23 places; Fig. 28.

The disturbed area includes the whole of Pembrokeshire and parts of Cardiganshire, Carmarthenshire and north Devon. The part of the boundary which traverses land is indicated by the broken-line in Fig. 28. From this, it is clear that the epicentre must lie under the sea and several miles to the south of the centre of the isoseismal 7 of the principal earthquake (no. 3). The disturbed area may have contained about 4800 square miles.

The shock consisted of one series of very slight vibrations, the accompanying sound being a low rumble like a distant peal of thunder.

5. **1892 AUG. 18,** *about* **1.5** *a.m.* Cat. No. 882; Intensity 3; No. of records 3 from 2 places.

A very slight shock, felt at Dale and Pembroke Dock, at the latter place being accompanied by a subdued rumbling sound.

---

[1] Soon after the earthquake, one or more so-called tidal waves were observed at various places along the shore of the English Channel. They had no connexion with the earthquake, however, for similar waves were observed on Aug. 16 and 17, and the waves advanced from east to west.

6. **1892** AUG. 18, *about* 1.40 *a.m.* Cat. No. 883; Intensity about 5; No. of records 114 from 111 places; Fig. 29.

This was the strongest of all the after-shocks. The isoseismal 4 (Fig. 29) coincides nearly with the isoseismal 7 of the principal earth-quake, except that it extends a little farther westward. The curve is about 24 miles wide, the longer axis being directed about N., and the land-area enclosed by it about 510 square miles. The epicentre is probably close to the coast almost south of Pembroke, but whether it lies beneath the sea or the land is uncertain. Outside this isoseismal, the places at which the shock was observed are far apart, and it is not possible to trace the boundary of the disturbed area. The most distant places at which the shock was felt are Knighton and Glasbury (Radnorshire), Cantref and Aberclydach (Breconshire), Exmouth and St Austell, and Kyle and Wexford (Co. Wexford). All of these places lie outside the isoseismal 5 of the principal earthquake. The disturbed area must therefore have contained about 29,000 square miles.

The shock was simple in character. At Tenby, for instance, there was a tremulous motion for a few seconds, ending in a sharp shake; at Maenclochog, a thud accompanied by a slight swaying of the ground and tremulous motion. At most places, however, the tremor was not observed, and the shock was simply a jolt or thud.

The sound was much weaker than that which accompanied the principal earthquake. With two exceptions (Llanybyther in Carmarthenshire and Treneglos near Launceston) it was not heard beyond the isoseismal 6 of that earthquake. It was heard by 91 per cent. of all the observers and resembled usually the boom of a gun.

The shock was felt on the river opposite Langwm, and was accompanied by a wave, but not nearly so marked as those seen at the time of the principal earthquake.

7. **1892** AUG. 18, 2.50 *a.m.* Cat. No. 884; Intensity 3; Centre of sound-area in lat. 51° 48·1′ N., long. 4° 58·3′ W.; No. of records 69 from 67 places; Figs. 28 and 29.

The disturbed area includes the whole of Pembrokeshire, most of Carmarthenshire, and parts of the counties of Cardigan and Glamorgan; in all a land-area of about 1680 square miles. The boundary of the area is shown approximately by the dotted line in Fig. 28. The centre of the area obviously lies on land and some miles farther north than the centre of the isoseismal 7 of the principal earthquake.

The shock was a slight tremulous motion in all parts of the disturbed area.

The sound was heard by only 44 per cent. of the observers. It was described as a distant faint sound or as resembling very distant thunder.

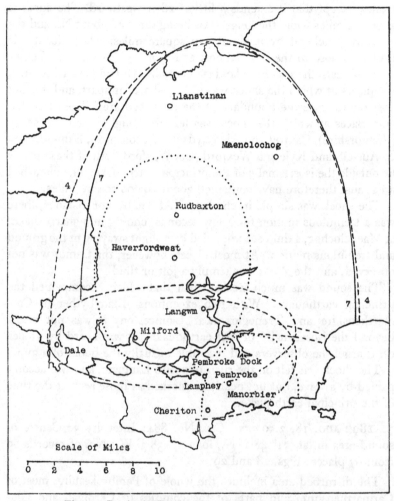

Fig. 29. Pembroke Earthquakes of 1892 Aug. 18 and 22.

The sound-area, the boundary of which is represented by the complete dotted line in Fig. 29, is 16 miles long, 13 miles wide and contains about 175 square miles. The centre coincides nearly with the town of Haverfordwest, and its longer axis is directed a few degrees

E. of N. This axis lies $3\frac{1}{2}$ miles west of that of the isoseismal 7 of the principal earthquake and 6 miles west of that of the preliminary earth-sound (no. 2, Fig. 29).

8. **1892** AUG. 18, *about* 4 *a.m.* Cat. No. 885; Intensity 2 or 3; No. of records 10 from 10 places; Fig. 29.

Only the eastern half of the boundary of the disturbed area can be drawn (represented by the incomplete broken-line in Fig. 29), but this is sufficient to show that the longer axis of the curve runs about east and west, coinciding nearly with the axis of Milford Haven. The width of the curve is 9 miles, the length and area probably about 18 miles and 127 square miles. The shock was at all places a slight quiver. There are no records of any accompanying sound.

1892 Aug. 19, 9.30 a.m., an earth-sound like thunder, but more crashing and grating, was heard at St David's, but by one observer only.

9. **1892** AUG. 22, *about* 11.55 *a.m.* Cat. No. 886; Intensity 4; Centre of disturbed area in lat. 51° 42·8' N., long. 4° 58·9' W.; No. of records 21 from 13 places; Fig. 29.

The boundary of the disturbed area is a narrow oval, represented by the complete broken-line in Fig. 29, 19 miles long, $7\frac{1}{2}$ miles wide, and containing about 112 square miles. Its longer axis is directed east, and lies about $1\frac{1}{2}$ miles N. of that of the preceding earthquake. The centre of the curve is about $1\frac{1}{4}$ miles W. of Honeyborough. The shock was a slight tremor. The sound, a slight rumbling like the firing of a distant gun, was heard at only 6 places.

10. **1892** AUG. 23, 4.30 *a.m.* Cat. No. 887.

A slight vibration felt by several observers at Pembroke Dock.

ORIGIN OF THE PEMBROKE EARTHQUAKES

Of the ten earthquakes in this series, two (nos. 5 and 10) are of unknown origin. Two others (nos. 8 and 9) are connected with an E.–W. fault or faults, the remaining six with a N.–S. fault or faults.

A comparison of the intensities and disturbed areas of these earth-quakes throws some light on the relative depth of the foci. For instance, the earthquake no. 8 was of intensity 2 or 3 and disturbed an area of about 127 square miles; no. 9 was of intensity 4 and disturbed an area of about 112 square miles. On the other hand, the earth-quakes nos. 4, 6 and 7 were of intensities about 4, about 5, and 3, respectively, and disturbed areas of approximately 4800, 29,000 and 5000 square miles. The principal earthquake (no. 3) was of intensity 7 and disturbed an area of more than 50,000 square miles. The conclusion to be drawn from these figures is that the earthquakes nos. 8 and 9 originated at a very much less depth than nos. 3, 4, 6 and 7, and that the focus of no. 8 was deeper than that of no. 9. In other words, the earthquakes connected with the E.–W. fault or faults had shallow foci, while those connected with the N.–S. fault or faults had deep foci.

Most of the faults that have been mapped within the epicentral district are strike-faults running nearly east and west. They are no doubt crossed by a series of transverse faults, though few of them have been traced. In Fig. 29, two strike-faults in the neighbourhood of Milford Haven, hading at a high angle to the south, are represented by broken-lines, with one or both of which the earthquakes nos. 8 and 9 may be connected. Thus, the earthquakes with shallow foci are apparently associated with known faults.

On the other hand, the parent-fault of the earthquakes with deep foci is unknown. Possibly it is so deep-seated that the movements along it have not appreciably affected the structure of the surface-rocks. Owing to the incompleteness of the isoseismal 7 of the principal earthquake (no. 3), the positions of the two epicentres cannot be determined. It is probable, however, that they lie along a nearly N.–S. line, between the axes of the sound-areas of the earthquakes nos. 2 and 7, the southern epicentre being beneath the sea to the south of Lamphey and Cheriton, and the other in the neighbourhood of Haverfordwest. Thus, the distance between the two epicentres may be about 12 or 13 miles.

In tracing the sequence of events, it will be assumed that only one E.–W. fault and one N.–S. fault were in action[1]. The earliest movements in 1892 were two small slips, separated by less than an hour, the first in the northern focus, the second in the southern. Two minutes after the latter, occurred the principal earthquake due to almost simultaneous impulses in both foci. The first and weaker took place in the southern focus, and the interval between the two impulses was longer than the time required by the earth-waves to traverse the interfocal region, for the impulse at the southern focus was apparently felt all over the disturbed area before the impulse at the other. Thir-

[1] It is possible that others of either system may have given rise to some of the shocks, but there is no evidence of any migration beyond that stated above.

teen minutes later, the first after-slip occurred in the southern focus. The position of the next slip is unknown, but the shock was of little consequence. The sixth movement, which occurred about an hour after the fourth, evidently took place in the interfocal region of the fault; and this was followed by another slip, a little more than an hour later, in the northern focus. Except for the slight tenth movement on Aug. 23, the series ended with two small and superficial slips along the strike-fault, the earlier movement being at a slightly greater depth than the later.

Though both a transverse fault and a strike-fault were in action, it should be noticed that all the earthquakes were due to the growth of an Armorican fold. The twin-slip along the transverse fault was caused by the growth of the fold along that fault, that is, by the intensification of the Armorican fold.

CARMARTHEN EARTHQUAKES

1. **1690** AUG. 27, *about 8 p.m.* Cat. No. 73 (W. Spurrell, *Carmarthen*, p. 119).

A very sensible earthquake at Carmarthen.

2. **1802** OCT. 21, *about 8.15 p.m.* Cat. No. 238 (W. Spurrell, *Carmarthen*, p. 134; *Gent. Mag.* vol. 72, 1802, p. 1154).

A smart shock at Carmarthen, felt also at Llandilo, Ferryside and Llandebie in Carmarthenshire, and at Narberth in Pembrokeshire.

1840, about the month of Nov., shocks similar to that of 1841 Jan. 24 are said to have been felt in the neighbourhood of Llanstephen, 6 miles S.S.W. of Carmarthen (Milne, vol. 36, p. 76).

3. **1841** JAN. 24, *between 3 and 4 a.m.* Cat. No. 488 (W. Spurrell, *Carmarthen*, p. 147; Milne, vol. 36, p. 76).

A smart shock, accompanied by a rumbling noise, at Carmarthen; also felt at Ferryside and Hereford.

The earthquakes, nos. 4–7, form a series, the first being the principal earthquake (*Quart. Journ. Geol. Soc.* vol. 53, 1897, pp. 168–174).

At Kidwelly, a rumbling noise was heard 25 seconds before the principal shock. Also, at Hafren Hall (in Carmarthenshire), noises like distant thunder, without any accompanying tremor, were heard at intervals during the afternoon.

4. 1893 NOV. 2, 5.45 *p.m.* Cat. No. 893; Intensity 7; Centre of isoseismal 6 in lat. 51° 52·2' N., long. 4° 36·4' W.; No. of records 633 from 494 places and 164 negative records from 161 places; Figs. 30, 31.

Isoseismal Lines and Disturbed Area. In Fig. 30, the isoseismals 6, 4 and 3 are shown and that part of the isoseismal 5 which lies on land.

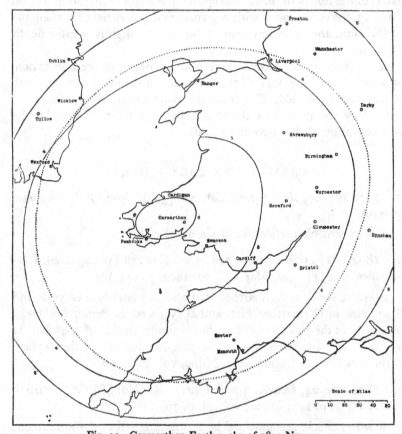

Fig. 30. Carmarthen Earthquake of 1893 Nov. 2.

The isoseismal 6 is elliptical in form, 41 miles long, 28 miles wide and contains about 940 square miles (Figs. 30, 31). Its longer axis is directed N. 75° E., and its centre is 3½ miles N. of Whitland. The isoseismal 5 is confined to Wales and the north-western corner of Devon. If continued in what appears to be its probable form, the length of the curve would be 141 miles, the width 113 miles, and the area included within it about 12,500 square miles. The isoseismal 4 is 233 miles long, 196 miles wide, and about 35,900 square miles in area. The isoseismal 3, which forms the boundary of the disturbed area, is 313 miles long, 269 miles wide, and contains 63,600 square miles, or nearly three-quarters the area of Great Britain.

Nature of the Shock. In the neighbourhood of the epicentre, the shock consisted of two strongly marked maxima connected by slight tremulous motion and rumbling sound. At some distance, this intervening tremor and sound were imperceptible, and the shock consisted of two detached parts, the average duration of the interval between them being 2·3 seconds. In the south-west of Pembrokeshire, the second maximum is always described as the stronger; in other parts of the disturbed area, the first is described as the stronger in most of the records. The bounding curve between the two areas in which the first and second parts were respectively the stronger is clearly concave towards the south-west, passing to the south of Ballyhealy (Co. Wexford) and to the west of Bodmin and Exmouth.

It follows from these observations that there were two distinct foci and that the impulse in the eastern focus was the stronger and occurred first; the greater intensity of the second maximum in the south-west of Pembrokeshire being due to the proximity of this district to the western focus. The positions of the two epicentres cannot be determined with accuracy, but the western epicentre is probably near, if not coincident with, the epicentre of the first after-shock, and, if so, the eastern epicentre would lie about the same distance on the other side of the centre of the isoseismal 6, or about 6 miles W.N.W. of Carmarthen, the distance between the two epicentres being about 13 miles.

Sound-Phenomena. The boundary of the sound-area is represented by the dotted line in Fig. 30. Its length is 231 miles, width 210 miles, and the contained area about 37,700 square miles. The sound-area thus slightly exceeds the area of the isoseismal 4 and is about 59 per cent. of the whole disturbed area. If the isoseismal 4 were shifted about 10 or 12 miles to the north-east, it would coincide roughly with the boundary of the sound-area.

The sound was heard by all the observers within the isoseismal 6 and by 69 per cent. of the observers in the disturbed area. It was compared to passing waggons, etc., in 51 per cent. of the records, thunder in 20, wind in 6, loads of stones falling in 3, the fall of a heavy body in 3, explosions in 9, and miscellaneous sounds in 7, per cent.

5. **1893** NOV. 2, 6.1 *p.m.* Cat. No. 894; Intensity 4; Centre of disturbed area in lat. 51° 50·0' N., long. 4° 45·9' W.; No. of records 64 from 57 places; Fig. 31.

The shock was felt over the greater part of Pembrokeshire and Carmarthenshire and in the south of Cardiganshire. The disturbed area, the boundary of which is represented by the broken-line in Fig. 31, is 43 miles long, 29 miles wide, and contains about 1000 square miles. The longer axis is directed N. 65° E., and the centre

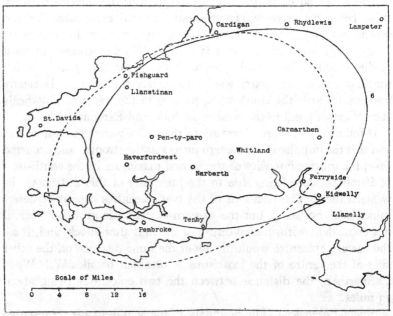

Fig. 31. Carmarthen Earthquakes of 1893 Nov. 2.

of the area lies 2 miles N. of Narberth and about 7 miles S. 70° W. of the centre of the isoseismal 6 of the principal earthquake. The line joining the two centres is thus nearly parallel to the longer axes of both curves.

In all parts of the disturbed area, the shock was a tremulous motion, as a rule extremely slight. The rumbling sound was also very faint, and was unnoticed at many places, though the sound-area and disturbed area are approximately coincident.

6. **1893** NOV. 2, *about* 6.15 *or* 6.30 *p.m.* Cat. No. 895.

An exceedingly slight shock, felt at Llandyssul and Blaendyffryn in Cardiganshire, possibly also at Llanstinan and Pen-ty-parc in

Pembrokeshire. The epicentre was probably near the eastern epi-centre of the principal earthquake.

7. **1893** NOV. 3, *about* 1 *a.m.* Cat. No. 896.

A slight shock felt at Llanilar and Rhydlewis in Cardiganshire, and Newcastle Emlyn on the border of that county and Carmarthenshire. 1913 Oct. 8, 11.26 p.m., a shock (intensity 4) was felt by several persons at Ferryside; it consisted of a series of small but rapid vibrations, like those produced by a road-engine passing, and lasting from 5 to 10 seconds; the accompanying sound resembled a muffled peal of thunder or the boom produced by a stone falling down a very deep well.

ORIGIN OF THE CARMARTHEN EARTHQUAKES

The strongest of the above earthquakes, the first of the 1893 series and its first successor (nos. 4 and 5) are evidently connected with one and the same fault, for the line joining the centres of the isoseismal 6 of the one and of the boundary of the disturbed area of the other is approximately parallel to the longer axes of both curves. The direction of the originating fault must be about N. 75° E. With regard to its hade, the evidence is doubtful.

The faults which have been mapped in the epicentral district are so numerous that the earthquake cannot be assigned to any particular fault. It is possible, moreover, that the parent-fault was too deeply-seated to affect the structure of the surface-rocks; for it is clear, from its large disturbed area, that the principal earthquake of 1893 originated in deep foci. The approxi-mate positions of the corresponding epicentres have been assigned above.

Excluding the earthquake of 1840, the relation of which to the remaining earthquakes is doubtful, the latter seem to be connected with one or both of these two foci. The earthquakes of 1690, 1802 and 1841 were probably due to movements in the eastern focus; the first earthquake of 1893 to a twin movement, the earlier and stronger impulse taking place in the eastern focus and the second, after 2 or 3 seconds, in the western; the first after-slip (no. 5), 16 minutes later, occurred in the western focus; the second and third after-slips (nos. 6 and 7), after intervals of $\frac{1}{4}$ to $\frac{1}{2}$ hour and $6\frac{1}{2}$ hours, probably in the eastern focus.

SWANSEA EARTHQUAKES

1766, an earthquake in Glamorganshire (Cat. No. 137; Walford, p. 58).
1832 Dec. 30, 8.20 —, at Swansea a noise like the distant firing of
artillery was heard, followed in about 2 or 3 seconds by four strong
vibrations, which lasted from 1 to 1½ seconds. The shock was also
felt at Neath, Ferryside and Carmarthen (Cat. No. 329; *Gent. Mag.*
vol. 102, 1832, p. 640; W. Spurrell, *Carmarthen*, p. 145; *Notes and
Queries*, vol. 6, 1906, pp. 30 and 74).

Mallet records a second earthquake the following morning felt at
Swansea, Neath, Llandovery, Carmarthen and Castle Bridge (Co.
Wexford). He suggests, and no doubt correctly, that this is the same
as the preceding (*Brit. Ass. Rep.* 1854, p. 235).

1. **1852** JUNE 1, *about* 7.45 *a.m.* Cat. No. 707 (*Times*, June 8; *Morning
Herald*, Oct. 8, 1863).

A shock, accompanied by a noise like the falling of furniture,
felt at Swansea, Sketty, the Mumbles, · Neath, Aberavon and
Bridgend.

1868 Mar. 26, a slight earthquake in South Wales (*Brit. Rainfall*,
1868, p. 66).

1869 Feb. 15, a slight shock in Glamorganshire (Cat. No. 747; *Brit.
Rainfall*, 1869, p. 61).

1887 Oct. 12, 6.20 a.m., a slight shock, lasting 2 or 3 seconds, at
Merthyr Tydvil; throughout the Rhondda valley, a tremor lasting
several seconds was felt, accompanied by a rumbling noise.

The epicentres of the preceding earthquakes are all unknown,
though the earthquake of 1832 may be connected with the same foci
as that of 1906. In the latter year, only one earthquake was certainly
felt, but five others, depending on the evidence of single observers,
are recorded, all before the strong earthquake (*Quart. Journ. Geol.
Soc.* vol. 63, 1907, pp. 351–361; *Geol. Mag.* vol. 5, 1908, pp. 304–305).

1906 June 26, about 8 p.m., a slight shock at Aberavon.

June 26, about 11.30 p.m., an earth-sound at Ferryside.

June 27, about 0.45 a.m., the same.

June 27, about 1 a.m., a slight shock at Aberavon.

June 27, about 2 a.m., a slight shock, with a rumbling noise like
thunder, at Llandilo.

2. **1906** JUNE 27, 9.45 *a.m.* Cat. No. 998; Intensity 8; Centre of
isoseismal 8 in lat. 51° 38·0′ N., long. 4° 0·3′ W.; Epicentre of secondary

focus in lat. 51° 37·9′ N., long. 3° 28·8′ W.; No. of records 1168
from 607 places, 89 negative records from 84 places, and 53 records
from 39 mines; Figs. 32, 33.

Isoseismal Lines and Disturbed Area. On the map of the earthquake
(Fig. 32) are shown six isoseismal lines, corresponding to intensities
8 to 3. The isoseismal 8 (Fig. 33) is an elongated ellipse, 26¼ miles long,
14 miles wide, and 299 square miles in area. The centre of the curve

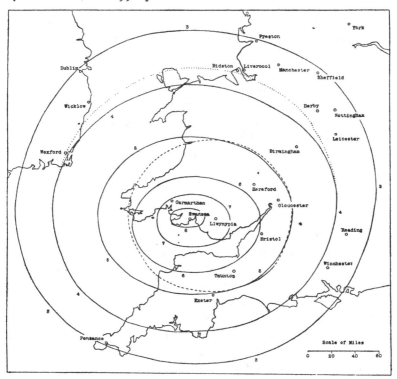

Fig. 32. Swansea Earthquake of 1906 June 27.

is about 3 miles W. of Swansea, and its longer axis is directed N. 85° E.
Damage to buildings (chiefly the fall of chimneys) is reported from
40 towns and villages within the isoseismal line[1]. The southern half
of the isoseismal 7 traverses the Bristol Channel, but the trend of the
curve just before leaving the land and the close approach to intensity 7
at several places in north Devon show that the course of this curve
as drawn on the map cannot err greatly from its true position. The

[1] At Swansea, the number of chimneys thrown down is estimated at not less than
120, at Neath 30, etc.

isoseismal is 62 miles long, 43 miles wide, and 2120 square miles in area. Its distance from the isoseismal 8 is 12 miles on the north side and 17 miles on the south, 13 miles on the west side and 23 miles on the east. The inequality between the former pair of figures indicates that the originating fault hades to the south, while that between the latter pair suggests the existence of a secondary focus lying to the east of the principal focus near Swansea.

The isoseismal 6 is 117 miles long from E. to W., 79 miles wide, and contains 6700 square miles, and is separated from the isoseismal 7 by 16 miles on the north side and 21 miles on the south. The isoseismal 5 is 163 miles long from E. to W., 130 miles wide, and 16,600 square miles in area. Its distance from the isoseismal 6 is 32 miles on the north side and 19 miles on the south. The isoseismal 4 is 237 miles long from E. to W., 203 miles wide, and contains 37,800 square miles. On the north side it is 42 miles from the isoseismal 5, and on the south side 31 miles. The isoseismal 3 is 303 miles long from E. to W., 274 miles wide, and 66,700 square miles in area. Its distance from the isoseismal 4 is 45 miles towards the north and 26 miles towards the south.

Nature of the Shock. To most observers, the shock appeared to consist of a single series of vibrations, increasing gradually in strength to a maximum and then dying away. Others noticed two distinct parts, differing considerably in strength, and separated by a brief interval of rest and quiet. The mean duration of this interval is almost constant, being 2·2 seconds within the isoseismal 8, and 2·1, 2·0 and 2·3 seconds between successive pairs of isoseismals. The mean duration of the interval within the isoseismal 5 was 2·1 seconds.

The inequality in strength between the two parts of the shock is manifest from the fact that both parts were noticed by only 14 per cent. of the observers within the isoseismal 8, and 16 per cent. of those within the isoseismal 5. It is clear that it was the weaker part of the shock, and not the discontinuity of the motion, that escaped observation, for the mean duration of the shock within the isoseismal 5 was 6·1 seconds according to those who observed the two parts, and 3·7 seconds according to those who noticed only one.

The first part of the shock was regarded as the stronger within a nearly circular area near the east end of the isoseismal 8. The boundary of this area is represented by the broken-line in Fig. 33, and is 23 miles long from E. to W. and 22 miles wide. Outside this curve, the second part of the shock was stronger than the first. From this law of variation in the relative intensity, it follows that (i) a second focus must be situated not far from the western boundary of the circular area, (ii) the intensity of the disturbance within this focus was less than that within the Swansea focus, (iii) the greater intensity of the first part of the shock within the circular area was due to the proximity of the area to the second focus, and (iv) the interval between the impulses was so long that the vibrations from the eastern focus at all places preceded those from the western focus. Thus, the area over which the double shock was felt is the disturbed area of the impulse within the eastern focus. The boundary of this area is represented by the broken-line in Fig. 32. This curve coincides roughly with the isoseismal 5 towards the north, east and south; but, towards the west, it falls short of that isoseismal by 27 miles. It is 135 miles long from E. to W., 127 miles wide, and contains about 13,500 square miles. Its centre is situated about 5 miles N.E. of Maesteg and 20 miles E. of the centre of the isoseismal 8. Owing to the size of the area disturbed by the weaker part of the shock, this point may be distant several miles from the secondary epicentre.

Seismographic Records. The seismographic records of the earthquake do not add to our knowledge of its nature. It was recorded at 9h. 45·0m., a.m., by Milne pendulums at Shide and Bidston, which are 135 and 130 miles respectively from the Swansea epicentre. At Shide, the maximum amplitude was 0·2 mm. and the duration of the disturbance 3 minutes; at Bidston, the maximum amplitude was 0·5 mm. and the duration 6·8 minutes. The register of the Omori horizontal pendulum at Birmingham (110 miles) shows, with the aid of the lens, a series of minute serrations lasting for about ¼ minute.

Sound-Phenomena. Towards the south and east, the boundary of the sound-area cannot be distinguished from the isoseismal 4. The two curves probably coincide approximately towards the west, but, towards the north, the boundary of the sound-area (as shown by the dotted line in Fig. 32) extends about 15 miles beyond the isoseismal. The sound-area is thus about 237 miles long from E. to W., 218 miles wide, and includes about 40,600 square miles.

For a considerable distance from the epicentre, the sound was heard by nearly all the observers, the percentage of audibility being 97 within the isoseismal 8, 90 between the isoseismals 8 and 7, and 92 between the isoseismals 7 and 6. Outside the isoseismal 6, the audibility diminished rapidly, the percentage being 75 between the isoseismals 6 and 5, 60 between the isoseismals 5 and 4, and 25 between the isoseismal 4 and the boundary of the sound-area. Throughout the whole area, 85 per cent. of the observers heard the earthquake-sound. The uniformly high percentage of audibility in the region surrounding the epicentres prevents the construction of isacoustic lines. It is clear, however, that the isacoustic lines, if drawn, would not present those marked distortions which are noticeable when the two parts of the shock coalesce along a synkinetic band (see Figs. 51, 56 and 59).

The sound was compared to passing waggons, etc., in 47 per cent. of the records, thunder in 20, wind in 7, loads of stones falling in 5, the fall of a heavy body in 7, explosions in 13, and miscellaneous sounds in 1, per cent. Within the isoseismal 8 and in the surrounding zone, the percentages of comparison to sounds of an explosive character are 21 and 15 respectively, these unusually high figures being due no doubt to the deep crashes which accompanied the strongest vibrations in the neighbourhood of the origin.

Observations in Mines. The country included within the isoseismal 7 is one of the most important mining districts in Great Britain. The earthquake was observed in 39 pits distributed over an area 49 miles long from near Kidwelly to near Pontypool, but chiefly concentrated within and near the Rhondda valleys.

The observations in this district throw an unexpected light on the position of the eastern epicentre. In eight pits, lying within the area bounded by the dotted line in Fig. 33, the heaving of the floor was distinctly perceived; while, in those situated in the surrounding district, the shock was a mere tremor. The area in question is 9 miles long from E. to W., 8 miles wide, and contains about 56 square miles. Its centre is situated at a point 1 mile W. of Llwynypia and $22\frac{1}{2}$ miles due E. of the centre of the isoseismal 8. As this point is close to the centre of the area within which the double shock was perceptible, and also lies within, and near the western margin of, the area in which the first part of the shock was stronger than the second, it must coincide very nearly with the epicentre of the eastern focus.

The sound observed in the mines differed slightly from that heard on the surface. It was compared to passing waggons, etc., in 54 per cent. of the records, thunder in 12, wind in 2, loads of stones falling in 5, the fall of a heavy body in 5, explosions in 17, and miscellaneous sounds in 5, per cent. The sound was evidently more uniform and monotonous underground than on the surface, although the loud explosive crashes which accompanied the strongest vibrations were occasionally heard.

The observations made in the mines lead to the following conclusions: (i) the shock and sound were observed underground over the same area; (ii) in pits not more than 5 miles from the nearest epicentre, the sound seemed to pass below the workings, in those at a greater distance, as a rule, to pass overhead or to travel along the workings; and (iii) the shock appears to have been felt more severely in the lower than in the upper workings of a pit.

3. 1907 JULY 3, 3.40 *a.m.* Cat. No. 1031; Intensity 4; Centre of isoseismal 4 in lat. 51° 38·1′ N., long. 4° 2·8′ W.; No. of records 21 from 14 places and 13 negative records from 12 places; Fig. 33.

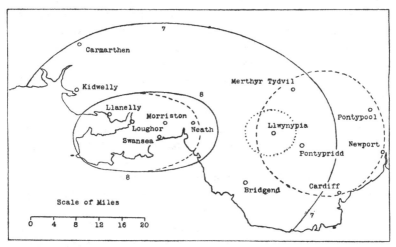

Fig. 33. Swansea Earthquakes of 1906 June 27 and 1907 July 3.

The boundary of the disturbed area is an isoseismal of intensity 4 and is 23 miles long, 14 miles wide, and about 250 square miles in area. Except towards the east, it coincides with the isoseismal 8 of the preceding earthquake. Its centre is 5 miles W. of Swansea, and 2 miles W. of that of the isoseismal 8 referred to. The direction of its longer axis is N. 86° E.

The shock consisted of a single series of vibrations, with an average duration of 3·7 seconds. The sound was heard by all the observers, and was compared to passing waggons, etc., in 27 per cent. of the records, thunder in 45, the fall of a heavy body in 9, and miscellaneous sounds in 18, per cent.

ORIGIN OF THE SWANSEA EARTHQUAKES

For the earthquake of 1906, the elements of the originating fault, as determined by the seismic evidence, are: (i) the mean direction of the fault in the neighbourhood of Swansea is N. 85° E.; (ii) its hade in the same district is towards the S.; and (iii) the fault must pass a short distance to the north of Swansea, and probably not far from the line passing through Llanelly, Loughor, Morriston and Neath. For the earthquake of 1907, the mean direction of the originating fault is N. 86° E. As the line joining the Swansea epicentre of the earlier earthquake and the epicentre of the later is very nearly parallel to this direction, it follows that both earthquakes were connected with the same fault.

There is no known fault in the position assigned by the seismic evidence. The great pre-Triassic E.–W. fault under Swansea Bay, though nearly in the required direction, passes about 5 or 6 miles to the south of the line indicated. Unless it changes considerably in the amount of its hade as it approaches the surface, the Swansea earthquakes cannot have been caused by displacements along this fault. In addition, however, to the faults mapped on the surface, there are many others which have been discovered from surveys in the mines; and it is possible that there are still more deeply-seated faults which have not affected the surface-beds, and that it is to one such fault that the Swansea earthquakes are due.

The total length of the fault affected by the earthquake of 1906 was about 30 or 32 miles. The principal or western focus, with its centre below a point about 3 miles W. of Swansea, must have been about 10 or 12 miles in length, and extended from near Llanelly to the neighbourhood of Neath. The secondary or eastern focus, with its centre below a point 1 mile W. of Llwynypia, was not more than 1 or 1½ miles in length. Thus, as the distance between the two epicentres was about 22½ miles, the length of the interfocal region was about 15 or 16 miles. There may have been a few creeps within the western focus during the 24 hours that preceded the earthquake of 1906; but the first important movement (which gave rise to a shock of intensity 7) occurred in the eastern focus, and this was followed 2 or 3 seconds later by a much greater slip in the western focus. A little more than a year afterwards, another, though much less marked, displacement occurred over a length of 8 or 9 miles in the western focus. As its centre was about 2 miles W. of that of the Swansea focus in 1906, the western margins of the two foci must have coincided approximately.

CONCLUSION

Before the year 1892, the south of Wales was singularly free from earthquakes. In the Pembroke district, one or two shocks of intensity 4 occurred in 1873. In the neighbourhood of Carmarthen, shocks were felt in 1690, 1802 and 1841. In Glamorganshire, there were four earthquakes—in 1766, 1832, 1852 and 1869. It is possible, though not certain, that the Swansea earthquake of 1832 was felt across the Irish Channel in Co. Wexford, and the Carmarthen earthquake of 1841 at Hereford. If so, the disturbed area of the former earthquake may have been 38,000 square miles, and of the latter about 13,000. The remaining shocks were of little consequence.

Thus, only one of these early shocks was comparable in strength with the remarkable earthquakes of Pembroke in 1892, Carmarthen in 1893, and Swansea in 1906. All three were twin earthquakes of the second class (see Chap. XXIII). They belong to the front rank of British earthquakes—their intensities were 7, 7 and 8 respectively, while the areas included within the isoseismal 4 were 44,860, 35,900 and 37,800 square miles. The distance between the centres of the innermost isoseismals of the Pembroke and Carmarthen earthquakes was 19 miles, and between those of the Carmarthen and Swansea earthquakes 26 miles; and the intervals between their times of occurrence about $1\frac{1}{4}$ and $12\frac{3}{4}$ years. If we take into account the brevity of these intervals, it is difficult to resist the impression that the earthquakes were closely connected, the whole region being subjected to compression in two directions which are approximately N.–S. and E.–W.

EARTHQUAKES OF UNKNOWN EPICENTRES
IN SOUTH WALES

1. **1248** FEB. 19. Cat. No. 33; Intensity 8 (*Annal. Cambriae*, p. 87).

A great earthquake in Britain and Ireland, by which the cathedral of St David's was injured. It "probably started or aggravated the curious outward inclination of the nave arcade" (P. A. Robson, *The Cath. Church of St David's*, p. 4).

2. **1255** SEPT. —. Cat. No. 35 (*Brut y Tywysogion*, p. 363).

"In...the octave of the feast of St Mary in September, there was an earthquake in Wales, about the hour of evening tide."

CHAPTER IX

EARTHQUAKES OF THE NORTH-WEST OF ENGLAND

THE counties included in this district are Cumberland, Westmorland and Lancashire and the extreme western portion of Yorkshire. The district contains five well-marked seismic centres, namely, (i) the Carlisle centre, about 7 miles S.S.W. of Carlisle, connected with a secondary centre about 2 miles N. of Wythburn (near Grasmere); (ii) the Grasmere centre, about 3 miles N.N.W. of Grasmere, in all probability distinct from the preceding and less deeply seated; (iii) the Kendal centre, about 12 miles E. of Kendal and 4 miles E. of Sedbergh; (iv) the Rochdale centre, about 1 mile N. of Rochdale; and (v) the Bolton centre, about 2 miles N.N.E. of Bolton. In addition, there are minor and as a rule somewhat ill-defined centres near Clitheroe, Dalton-in-Furness and Halifax, with each of which three or four earthquakes are connected; one near Settle with two earthquakes; and others, visited by a single earthquake, at Cartmel, Everton (near Liverpool) and Shap. The list of earthquakes also includes six shocks, the epicentres of which are unknown.

The seismic history of the district begins with the year 1738. Besides this, Dr Short records earthquakes in Cumberland in the years 1000 and 1014 (vol. 2, pp. 166–167), and also the following:

1649, "an earthquake in Cumberland, Westmorland and the Santorine Islands" (vol. 1, p. 326).

1650 Apr. —, 5 p.m., "Cumberland and Westmorland were so shaken by an earthquake that people left their houses and fled to the fields" (vol. 1, p. 327).

There remain 46 shocks, which seem to me undoubted earthquakes. The majority were of slight intensity, but three earthquakes—in the years 1786, 1843 and 1871—disturbed areas of more than 20,000 square miles.

CARLISLE EARTHQUAKES

1. 1786 AUG. 11, *about* 2.20 *a.m.* Cat. No. 160; Intensity 7; Centre of disturbed area in lat. 54° 41′ N., long. 3° 8′ W.; Fig. 34 (S. Morse, *Phil. Trans.* vol. 77, 1787, pp. 35–36; P. Brydone, *Ibid.* pp. 61–70; Lauder, pp. 366–367; Milne, vol. 31, pp. 107–108; *Gent. Mag.* vol. 56, 1786, pp. 707–708; etc.).

The curve in Fig. 34 bounds all the places at which the earthquake was certainly felt. It is 190 miles long from N. to S., 180 miles wide, and about 27,000 square miles in area. The centre of the curve is 18 miles S.W. of Carlisle and not far from the centre of the isoseismal 5 of the principal earthquake of 1901 (no. 3). Records of the shock also come from isolated places far beyond the limits of the above curve—from as far north as Aberdeen and Argyllshire, and from

Fig. 34. Carlisle Earthquake of 1786 Aug. 11.

Dublin. It is doubtful if the disturbances felt at these places were connected with the earthquake. If they were, the disturbed area must contain about 88,000 square miles. The intensity of the shock must have been nearly 8, for at Whitehaven a chimney was thrown down, at Egremont several chimneys and part of the ruins of the castle, and at Workington the quay was slightly damaged.

The shock was observed to be double at a few places, the interval between the parts being 3 or 4 seconds at Dumfries and 2 or 3 seconds in Nithsdale. The average duration of the shock was 4 seconds.

2. **1787 JULY 6.** Cat. No. 162 (Perrey, p. 149).

A shock felt at Penrith, Threlkeld and Keswick. It originated probably in the region between the two foci of the principal earthquake of 1901 (no. 3).

1787 Aug. 11, about 2 a.m., an earthquake felt at Penrith, Lancaster, Manchester and Lennel (near Coldstream). It seems possible that this earthquake may be the same as that given under the same date in the previous year (Milne, vol. 31, p. 108; *Gent. Mag.* vol. 57, 1787, p. 494).

1901 July 9, about 3 p.m., a slight shock at Loweswater recorded by one observer.

3. **1901 JULY 9, 4.23 *p.m.*** Cat. No. 936; Intensity 5; Centre of isoseismal 5 in lat. 54° 47·8′ N., long. 3° 0·4′ W.; No. of records 267 from 155 places and negative records from 50 places; Fig. 35 (*Quart. Journ. Geol. Soc.* vol. 58, 1902, pp. 371–376).

Isoseismal Lines and Disturbed Area. Two isoseismal lines, of intensities 5 and 4, are shown by continuous lines on the map (Fig. 35). The isoseismal 5 is very nearly a circle, 29 miles in diameter, and 660 square miles in area. Its centre lies 7 miles S.S.W. of Carlisle. The isoseismal 4 is an elongated ellipse, 66 miles long, 46 miles wide, and 2390 square miles in area. Its longer axis is directed N. 5° E. The distance between the isoseismals 5 and 4 is 5 miles on the west side and 10 miles on the east. Outside this isoseismal, the shock was felt at Whitehaven (2 miles to the W.), Westerkirk and Eskdalemuir (4 and 9 miles respectively to the N.), and Barbon (near Kirkby Lonsdale, 5 miles to the S.E.). The disturbed area must therefore have contained about 3700 square miles.

The excentricity of the isoseismal 5 with regard to the isoseismal 4 is the most important feature of these lines; on the north the distance between them is 10 miles, and on the south 27 miles, the distance between their centres being about 9 miles. If, however, an isoseismal line corresponding to an intensity between 4 and 5 could have been drawn, this line would have consisted of two detached portions, one surrounding the isoseismal 5, and the other including Wythburn, Grasmere and the surrounding district, for there is evidence there of a distinct increase in the intensity of the shock. The distance between the two centres would be about 23 miles.

Nature of the Shock. In the central portion of the disturbed area, the shock consisted of a single continuous series of vibrations, containing two maxima of intensity, with intervening weaker motion and fainter sound. At a short distance from the longer axis, the inter-

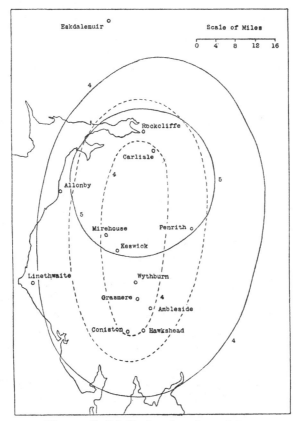

Fig. 35. Carlisle Earthquakes of 1901 July 9.

vening vibrations were imperceptible to many observers, and the shock consisted of two detached series of vibrations, separated by an interval of about 3 seconds' duration. This double series was perceived over an area almost coincident with the boundary of the disturbed area of the earthquake at 4.45 p.m. (no. 5) indicated by the outer broken-line in Fig. 35, except that it is less elongated in form, extending 2 or 3 miles beyond it towards the east, and falling short of it by about the same distance towards the north and south. Outside this area, the shock consisted once more of a single series of vibrations. The average duration of the shock was 4·1 seconds.

Sound-Phenomena. The sound was heard all over the disturbed area and by 90 per cent. of the observers. Though the percentage of audibility varies little on the whole from the central area to near the boundary, it is evident, from the descriptions, that the sound was not of uniform intensity; that there were two regions in which it was of maximum strength. One of them lies within the isoseismal 5; the other includes such places as Ambleside, Coniston, Grasmere, Hawkshead, Rydal and Wythburn, close to the secondary epicentre. The sound was compared to passing waggons, etc., in 33 per cent. of the records, thunder in 35, wind in 6, loads of stones falling in 4, the fall of a heavy body in 7, explosions in 8, and miscellaneous sounds in 6, per cent.

4. **1901 JULY 9**, *about* 4.26 *p.m.* Cat. No. 937; Intensity 4; No. of records 4 from 4 places.

A slight shock, lasting about 2 seconds, felt at Carlisle, Cummersdale, Rockcliffe and Wigton, and at the last two places accompanied by a noise like that of a heavy body falling and of a passing traction-engine. The epicentre was evidently close to the centre of the isoseismal 5 of the principal earthquake (no. 3).

5. **1901 JULY 9**, 4.45 *p.m.* Cat. No. 938; Intensity 4; Centre of disturbed area in lat. 54° 40·2′ N., long. 3° 2·4′ W.; No. of records 64 from 44 places and negative records from 59 places; Fig. 35.

The two broken-lines in Fig. 35 represent isoseismal lines of this earthquake. The inner corresponds to an intensity 4, but the course of the line may be incorrect towards both east and west. As drawn, it is an elongated oval, 37 miles long, 13 miles wide, and including about 380 square miles. The outer line, which forms the boundary of the disturbed area, is more accurately drawn. It is 51 miles long, 28 miles wide, and contains an area of 1130 square miles. The longer axis is directed N. 4° E., and the centre of the curve is 6 miles N.E. of Keswick. The position of the boundary with regard to the isoseismal 4 of the principal shock (no. 3) shows that it must coincide nearly with an isoseismal of intensity slightly greater than 4 of that earthquake.

In all parts of the disturbed area, the shock consisted of a single series of vibrations, which gradually increased in intensity and then died away. There is no evidence of any discontinuity in the shock or of the existence of more than one maximum of intensity. The mean duration of the shock was 3·4 seconds.

The sound was much slighter than that which accompanied the first shock, but was heard all over the disturbed area, and at four places outside it—at Linethwaite and Allonby to the west, and Bentpath and Westerkirk to the north-west. It was heard by 86 per cent. of the observers, the percentage of audibility being the same within and without the isoseismal 4. It was compared to passing waggons, etc., in 46 per cent. of the records, thunder in 33, explosions in 17, and miscellaneous sounds in 4, per cent.

1901 July 10, about 1.30 a.m., a slight shock without sound at Crosthwaite (near Keswick).

6. 1901 JULY 11, *about* 11.10 *p.m.* Cat. No. 939.

A slight tremor at Coniston and Mirehouse (near Keswick). It is probable that the epicentre was near the southern epicentre of the principal earthquake (no. 3).

1901 July 12, about 2 a.m., a slight shock without sound at Crosthwaite.

1901 July 14, about 11 p.m., the same.

ORIGIN OF THE CARLISLE EARTHQUAKES

The Carlisle earthquakes were evidently due to movements along a deep-seated fault directed N. 5° E. The relative position of the isoseismal lines of the first earthquake of 1901 (no. 3) and the overlapping towards the west of the disturbed area by the sound-area of the third earthquake of the same year (no. 5) show that the fault hades towards the east. The uniformity in the percentage of audibility throughout the sound-areas of both earthquakes and the comparatively low value of the percentage in the central areas point to the deep-seatedness of the fault. In the surface-rocks, there is no sign whatever of such a structure.

For the first and strongest earthquake, that of 1786, the evidence we possess points to an origin chiefly, or entirely, within the northern or Carlisle focus. There are two records towards the north of the double character of the shock, the interval between the parts being about the same as in the principal earthquake of 1901. It is therefore possible that the earthquake was a twin. If the other focus coincided with the Grasmere secondary focus, the impulse within it must have

14

been comparatively slight. In less than 11 months, this strong earthquake was followed by one much weaker (no. 2), which may have originated in the interfocal region between the Carlisle and Grasmere foci.

In 1901, the sequence of events is much clearer. The first earthquake was a twin earthquake, the centres of the two foci being about 23 miles apart. The records of the relative intensity of the two parts or maxima of the shock are few in number, but it is probable that the first and stronger impulse occurred in the northern or Carlisle focus, and that the impulse at the southern or Wythburn focus took place after the waves from the Carlisle focus reached it. Throughout the interfocal region, there was a slight displacement. Three minutes afterwards, a small slip occurred in the Carlisle focus, and, about 20 minutes later still, one of greater importance took place in the interfocal region. After a little more than two days, a very slight movement occurred in or near the Wythburn focus, and it is possible, if we may rely on solitary observations, that other small slips may also have occurred in the interfocal region of the fault on July 10, 12 and 14.

GRASMERE EARTHQUAKES

It is somewhat doubtful whether more than one earthquake (that of 1911) should be connected with this centre, the evidence for no. 1 being scanty, while that for no. 2 points to a rather southerly position for the epicentre.

1. **1867 FEB. 23**, *about* 1.15 a.m. Cat. No. 741; Intensity 4 (*Mon. Met. Mag.* vol. 2, 1867, p. 21).

A slight shock at Ambleside (accompanied by a sound like a prolonged clap of thunder), Grasmere, Windermere and Greenside mines (Ullswater).

2. **1885 JUNE 30**, 5.45 a.m. Cat. No. 800; Centre of disturbed area in lat. 54° 24·1′ N., long. 3° 6·2′ W.; Fig. 36 (*Mon. Met. Mag.* vol. 20, 1885, p. 95; *Ann. Reg.* 1885, p. 40).

A rather strong shock, accompanied by a noise like thunder, felt in the southern portion of the Lake district. The boundary of the disturbed area (represented by the broken-line in Fig. 36) is nearly circular, being 15 miles long from N. to S., 14 miles wide, and containing about 165 square miles. Its centre is about 5 miles S.W. of Grasmere.

1886 Jan. 5, 8.40 *a.m.*, Roper (on the authority of the *Standard* for Jan. 9) records an earthquake at Hawkshead Hill (near Coniston) without details.

3. 1911 MAY 16, 8.50 *a.m.* Cat. No. 1093; Intensity 6; Centre of isoseismal 5 in lat. 54° 29·7′ N., long. 2° 59·0′ W.; No. of records 77 from 27 places; Fig. 36.

Two isoseismal lines, of intensities 5 and 4, are shown (by continuous lines) on the map in Fig. 36. The isoseismal 5 is 15 miles

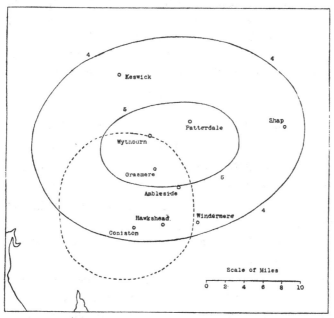

Fig. 36. Grasmere Earthquakes of 1885 June 30 and 1911 May 16.

long, 9 miles wide, and contains 106 square miles. Its centre lies 3 miles N.N.W. of Grasmere and its longer axis is directed N. 80° E. The isoseismal 4 is 30 miles long and 21 miles wide, the disturbed area bounded by it being 495 square miles. The distance between the isoseismals is about 7 miles to the north and about 5½ miles to the south.

At most places, the shock appeared to be continuous, but a double shock was observed at Hawkshead and Newlands, the first part being the stronger and separated from the other by an interval of 3 seconds. At Wythburn, the sound was double, the second part being the louder and following the other after an interval of about 3 seconds.

The sound-area coincided with the disturbed area. The sound was heard by 99 per cent. of the observers, and was compared to passing waggons, etc., in 22 per cent. of the records, thunder in 59, wind in 3, loads of stones falling in 6, the fall of a heavy body in 3, and explosions in 7, per cent.

The evidence of this earthquake points to its association with a fault directed N. 80° E.; the hade is uncertain, though it may be towards the N. In that case, the fault-line should pass through Grasmere or a short distance to the north. From the intensity of the shock, and the elongated forms and closeness together of the isoseismals, it would seem that the focus was situated at a comparatively small depth. In the neighbourhood of the epicentre, there are several short faults with an E.–W. direction; but the district is cut up by such numerous faults that it is impossible to select any one as the parent-fault of the earthquake.

KENDAL EARTHQUAKES

1.   **1817** NOV. 9. Cat. No. 280 (*Gent. Mag.* vol. 88, 1818, p. 557).
A slight shock felt near Dent, at Kendal and Coniston.

2.   **1840** APR. 11. Cat. No. 453 (C. Nicholson, *Annals of Kendal*, p. 300).
A slight shock at Kendal.

3.   **1843** MAR. 16, *about* 10.30 *p.m.* Cat. No. 598 (Milne, reprint, pp. 228–242).
A slight shock at Kendal, also at Fleetwood but so faintly as almost to be imperceptible.

4.   **1843** MAR. 17, 0.55 *a.m.* Cat. No. 599; Intensity probably 7; No. of records 68 from 45 places; Fig. 37 (Milne, reprint, pp. 228–242; *Brit. Ass. Rep.* 1843, pp. 121–122; J. Mayhall, *Annals of Yorkshire*, vol. 1, p. 488; C. Nicholson, *Annals of Kendal*, p. 300; *Times*, Mar. 18, 20, 21; etc.).

The disturbed area, the boundary of which is represented by the continuous line in Fig. 37, does not differ much from that of the earthquake of 1871 Mar. 17, 11.5 p.m. (no. 7), except that it extends slightly beyond it to the north and west and not quite so far to the east. It is about 180 miles long from N. to S., 170 miles wide, and contains about 24,000 square miles. It is also said to have been felt at, and in the neighbourhood of, Belfast, which lies about 36 miles west of the curve.

The shock consisted of two distinct parts at places so near the boundary of the disturbed area as Dumfries, York and Liverpool. The first part was regarded as the stronger at Whitehaven, and the second at Dumfries and Liverpool. The earthquake probably originated in two distinct foci, of which one at least was not far distant from the focus of the earthquake of 1871 Mar. 17 (no. 5).

The sound was heard by 90 per cent. of the observers and over an area which coincides nearly with the disturbed area, the sound having been

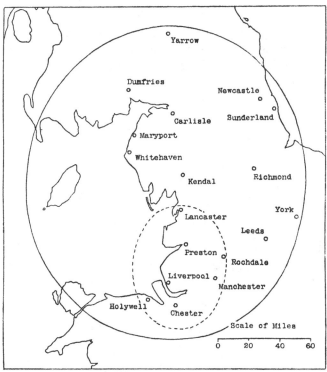

Fig. 37. Kendal Earthquake of 1843 Mar. 17 and
Lancashire Earthquake of 1750 Apr. 13.

heard at Yarrow, Douglas and Castletown (Isle of Man) and Manchester. It was compared to passing waggons, etc., in 40 per cent. of the records, thunder in 13, wind in 27, and the fall of a heavy body in 20, per cent.

5. 1871 MAR. 17, 6.20 p.m. Cat. No. 756; Intensity 4; Centre of disturbed area in lat. 54° 19·8′ N., long. 2° 26·5′ W.; Fig. 38 (*Nature*, vol. 3, 1871, pp. 406, 414; *Brit. Rainfall*, 1871, p. 81; *Mon. Met. Mag.* vol. 6, 1871, pp. 37–44; etc.).

The disturbed area, the boundary of which is represented by the broken-line in Fig. 38, is 52 miles long, 29 miles wide, and contains about 1180 square miles. Its centre lies about 12 miles E. of Kendal. The shock was slight and consisted of one part only.

6. **1871** MAR. 17, 10.56 *p.m.*   Cat. No. 757; Intensity 4 (*Nature*, vol. 3, 1871, p. 414; *Mon. Met. Mag.* vol. 6, 1871, pp. 37–40).

A shock, with accompanying sound, felt at Singleton Brook (near Manchester) and Burton (10 miles S. of Kendal).

7. **1871** MAR. 17, 11.5 *p.m.*   Cat. No. 758; Intensity 7; Centre of disturbed area in lat. 54° 9′ N., long. 2° 24′ W.; Fig. 38 (*Nature*, vol. 3, 1871, pp. 405–406, 414, 426; *Brit. Rainfall*, 1871, p. 81; *Mon. Met. Mag.* vol. 6, 1871, pp. 37–40).

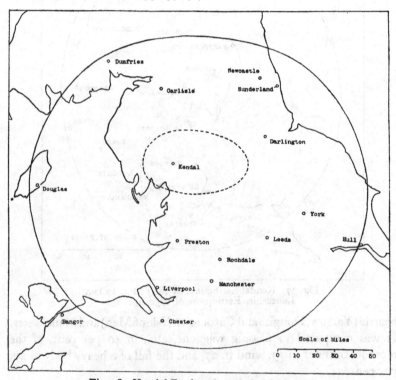

Fig. 38. Kendal Earthquakes of 1871 Mar. 17.

The disturbed area, the boundary of which is represented by the continuous line in Fig. 38, includes the whole of the counties of Lancashire, Cumberland, Westmorland, Durham and Flintshire, the greater part of Northumberland, Yorkshire and Cheshire, and small parts of Lincolnshire, Derbyshire, Staffordshire, Denbighshire, the Isle of Man, and Kirkcudbrightshire and Dumfriesshire. It is bounded by the isoseismal 4, and is 183 miles long from E. to W., 163 miles wide, and about 23,400 square miles in area. Its centre is about 20 miles S.E. of Kendal and about 2 miles N. of Clapham in Yorkshire.

The shock was observed to consist of two parts at Leeds, Nent Head (on Alston Moor), Singleton Brook (near Manchester) and Walton (near Liverpool), the second part being the stronger at Leeds, Nent Head and Singleton Brook. The mean duration of the shock was 8·3 seconds.

The sound-area extended to within a few miles of the isoseismal 4, and contains probably about 18,000 square miles. The sound was heard by 97 per cent. of the observers and was compared to passing waggons, etc., in 75 per cent. of the records, thunder in 8, explosions in 8, and miscellaneous sounds in 8, per cent.

The earthquake appears to have originated in two detached foci, one of which was probably the Kendal focus in action five hours earlier.

8. **1871** MAR. 17–18, *after* 11 *p.m.* Cat. No. 759 (*Mon. Met. Mag.* vol. 6, 1871, pp. 37–40).

A shock at Grasmere.

9. **1871** MAR. 18, *about* 3 *a.m.* Cat. No. 760 (*Times*, Mar. 21).

A slight shock at Coniston.

1871 Mar. 18, 6.3 a.m., a slight shock at Coniston (Perrey, *Mém. Cour.* vol. 24, 1875, p. 94).

10. **1875** SEPT. 23, 9.57 *p.m.* Cat. No. 771; Intensity 4 (*Mon. Met. Mag.* vol. 10, 1875, p. 154).

A distinct shock, lasting about 3 seconds and accompanied by sound, was felt at Oughtershaw Hall (5 miles S. of Hawes), Buckden (8½ miles S. of Askrigg), and at various places in Upper Wensleydale.

11. **1886** SEPT. 15. Cat. No. 804 (Roper, p. 41).

A shock felt in Garsdale and at Sedbergh and Dent.

ROCHDALE EARTHQUAKES

1. **1753** JUNE 8, *between* 11 *and* 12 *p.m.* Cat. No. 117; Intensity 5 (Mallet, 1852, p. 157; *London Mag.* 1753, p. 291).

A rather strong shock at Skipton, Manchester and Knutsford and in many of the neighbouring villages. At Skipton, the shock lasted about 3 seconds and was followed by a rushing noise and a sound as of an explosion.

2. **1777** SEPT. 14, *about* 10.55 *a.m.* Cat. No. 153; Intensity 7; Fig. 39 (T. Henry, *Phil. Trans.* vol. 68, 1779, pp. 221–231; Milne, vol. 31, p. 106; *Gent. Mag.* vol. 47, 1777, p. 458; *Notes and Queries*, vol. 12, 1879, p. 296).

The continuous curve in Fig. 39 surrounds all the known places at which the earthquake was observed, with one exception. It is 112 miles

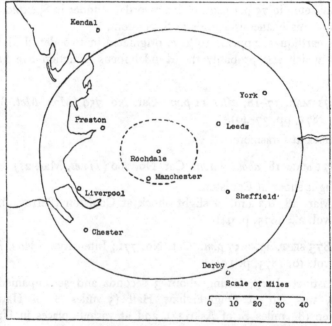

Fig. 39. Rochdale Earthquake of 1777 Sept. 14.

long from E. to W., 104 miles wide, and contains about 9000 square miles. The shock, however, was probably felt far beyond this line. It is said to have been sensible in Birmingham, in which case the disturbed area cannot be less than 22,000 square miles. Its centre coincides very nearly with that of the earthquake of 1869 (no. 6, the boundary of the disturbed area of which is represented by the broken-line in Fig. 39). At places near the centre, the intensity of the shock must have been nearly 8; at Manchester, for instance, several chimneys were thrown down[1].

At Manchester, three concussions were felt within half a minute, each being preceded by a noise as if a large bale of goods or a lofty wall had fallen. In most other places, only two concussions were felt.

3. **1824 MAY 31, 4 p.m.** Cat. No. 314 (*Ann. de Chim. et de Phys.* vol. 27, 1824, p. 378).

A slight shock at Bury.

1833 June 11, a reported earthquake in the district to the north of Manchester (Milne, vol. 31, p. 120).

4. **1843 MAR. 10, about 8 a.m.** Cat. No. 597 (*Brit. Ass. Rep.* 1843, p. 121; Milne, reprint, pp. 227–228; etc.).

A shock felt at Ashton-under-Lyne, Blackburn, Bolton, Cheetham Hill (N. of Manchester), Rochdale, Slaidburn and Turton. With the exception of Slaidburn, these places lie within the area disturbed by the earthquake of 1869 (no. 6). Its epicentre probably coincides with that of this earthquake.

1845 Dec. 5, two shocks are recorded by Roper (without giving any authority) at Stonyhurst at 1.20 a.m. and 2 a.m. (p. 36).

5. **1864 SEPT. 26, 0.35 a.m.** Cat. No. 734; Intensity 5 or 6 (Perrey, *Mém. Cour.* vol. 18, 1866, pp. 78–79).

A shock felt at Broughton, Hebden Bridge, Manchester, Pendlebury, Rochdale and Skipton, all the places being within or near the boundary of the earthquake of 1869 (no. 6).

6. **1869 MAR. 15, 6.6 p.m.** Cat. No. 749; Intensity 6; Centre of disturbed area in lat. 53° 38′ N., long. 2° 9′ W.; No. of records 35 from

---

[1] The shock was felt at Ashbourne (in Derbyshire), where Dr Johnson was staying. "Sir," he said when Boswell told him of the earthquake, "it will be much exaggerated in public talk; for, in the first place, the common people do not accurately adapt their thoughts to the objects; nor, secondly, do they accurately adapt their words to their thoughts; they do not mean to lie; but, taking no pains to be exact, they give you very false accounts. A great part of their language is proverbial. If anything rocks at all, they say *it rocks like a cradle*; and in this way they go on" (*Boswell's Life of Dr Johnson*, ed. by A. Birrell, vol. 4, 1901, p. 138).

26 places; Fig. 40 (*Manchester Lit. and Phil. Soc. Proc.* vol. 8, 1869, pp. 161–163, 175–176; *Brit. Rainfall,* 1869, p. 62; *Mon. Met. Mag.* vol. 4, 1869, p. 37; etc.).

The disturbed area, the boundary of which is represented by the continuous line in Fig. 40, is 31 miles long from E. to W., 24 miles wide, and contains about 580 square miles. Its centre lies 1 mile N. of Rochdale. The shock was a sudden thud accompanied by rapid vibrations. Its mean duration was 4·7 seconds.

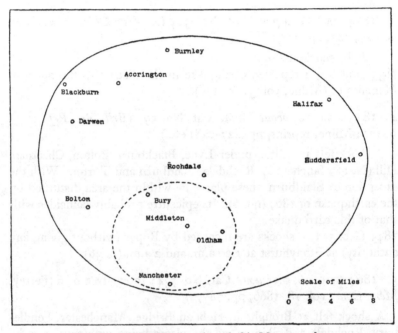

Fig. 40. Rochdale Earthquakes of 1869 Mar. 15 and 25.

7. **1869 MAR. 25, 8.25 *p.m.*** Cat. No. 751; Centre of disturbed area in lat. 53° 32′ N., long. 2° 14′ W.; No. of records 10 from 9 places; Fig. 40 (*Times,* Mar. 29; *Manchester Guardian,* Mar. 27).

The centre of the disturbed area, the boundary of which is represented by the broken-line in Fig. 40, is 2 miles S.W. of Middleton. The shock was a tremulous motion, slight compared with that of the preceding earthquake.

1869 Mar. 28, Perrey records a new trembling in Lancashire (*Mém. Cour.* vol. 24, 1875, p. 13).

BOLTON EARTHQUAKES

1753 June 22, 11.40 *p.m.*, a shock felt at Manchester, Oldham and Ratcliffe (Cat. No. 118; Milne, vol. 31, p. 100).

1. **1887** DEC. 1, *about* 6.50 *a.m.* Cat. No. 805 (*Nature*, vol. 37, 1888, p. 138; etc.).

A slight shock, lasting several seconds, felt at Blackburn, Bolton and Chorley, at the latter place accompanied by a sound like distant thunder.

2. **1889** FEB. 10, 10.36 *p.m.* Cat. No. 822; Intensity 6; Centre of isoseismal 5 in lat. 53° 36·8′ N., long. 2° 24·8′ W.; Records from 156 places; Fig. 41 (*Geol. Mag.* vol. 8, 1891, pp. 306–316).

On the map (Fig. 41), two isoseismal lines are represented by continuous lines. Both are nearly circular in form. The isoseismal 5 is 23 miles long from E. to W., 22 miles wide, and contains about 396 square miles. Its centre is about ½ mile W. of the village of Bradshaw and 2 miles N.N.E. of Bolton. Within this isoseismal, the intensity was 6 at Bolton, Ramsbottom and Tottington, and nearly 6 at Bury, Chesham and Farnworth. The isoseismal 4 extends to within about 3 miles of the outer line, the area included within it being about 1900 square miles. The boundary of the disturbed area is an isoseismal of intensity between 3 and 4. It is 56 miles long from N. to S., 54 miles wide, and contains about 2480 square miles.

The shock usually consisted of one or two prominent vibrations, in some cases followed by tremulous motion. Its mean duration was 3 seconds.

The boundary of the sound-area is represented approximately by the dotted line in Fig. 41. This curve is very nearly a circle 29 miles in diameter and about 636 square miles in area. Its most important feature is its excentricity with reference to the isoseismal lines, its centre being about 3¼ miles S.S.W. of that of the isoseismal 5. The sound was compared to passing waggons, etc., in 45 per cent. of the records, thunder in 5, wind in 11, the fall of a heavy body in 28, explosions in 5, and miscellaneous sounds in 5, per cent.

The shock was felt in a mine at Ince (near Wigan); and the sound was heard (apparently in the roof) in one of the workings of the Pendlebury colliery, and also in a mine at Agecroft.

ORIGIN OF THE BOLTON EARTHQUAKES

The most important fault of the district is the Irwell Valley fault, represented by the broken-line in Fig. 41. It has been traced from a point about 3 miles N.W. of Bolton to near Poynton (in Cheshire), a distance of more than 20 miles. Throughout its whole course, the downthrow is towards the N.E. At Farnworth (near Bolton), the throw is 1050 yards.

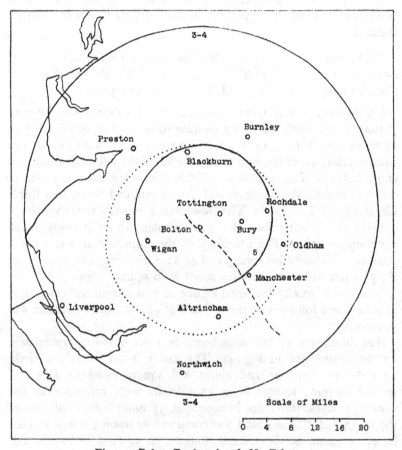

Fig. 41. Bolton Earthquake of 1889 Feb. 10.

If the earthquake of 1887 were connected with this fault, it must have orginated close to its N.W. end. The earthquake of 1889 was probably caused by a slip of the fault a few miles from this end; and (as may be inferred from the approximate circularity of the isoseismal lines and the brevity of the shock) the horizontal length of the focus was short, perhaps not more than a mile. As the inclination of the fault at Bolton is 28° from the vertical towards the N.E., the depth of the principal part of the focus may be about 3¾ miles; but, as shown by the displacement of the sound-area to the S.W., the upper margin of the focus must have extended close up to the surface of the earth.

It may be noticed that several earth-shakes at Pendleton (near Manchester) were probably due to small slips along this fault precipitated by mining operations (see Chap. XIX).

SHAP EARTHQUAKE

1880 DEC. 12. Cat. No. 786 (*Brit. Rainfall*, 1880, p. 30).
A slight shock at Shap.

DALTON-IN-FURNESS EARTHQUAKES

1. **1817** OCT. 31. Cat. No. 279 (*Gent. Mag.* vol. 88, 1818, p. 557).
A shock at Dalton-in-Furness.

2. **1842** JUNE 21, *between* 10 *and* 11 *p.m.* Cat. No. 579; Intensity 4 (Milne, reprint, p. 224).
A shock, accompanied by a rumbling noise, at Cartmel.

3. **1865** JAN. 15, 11 *a.m.* Cat. No. 736 (Perrey, *Mém. Cour.* vol. 21, 1870, pp. 25–26).
A tremor with noise in Furness and near Morecambe Bay.

4. **1867** FEB. 22. Cat. No. 740 (Perrey, *Mém. Cour.* vol. 21, 1870, p. 143).
A shock in Furness and on the north shore of Morecambe Bay.

SETTLE EARTHQUAKE

1688, Roper (p. 16) records a shock at Kettlewell (10 miles N.E. of Settle).

1859 DEC. 15. Cat. No. 722 (Edmonds, p. 93; *Mon. Met. Mag.* vol. 10, 1875, p. 154).
An earthquake felt at Pateley Bridge, near Settle, and by Grassington towards Greenhow Hill.

CLITHEROE EARTHQUAKES

1. **1835** AUG. 19–20, *midnight*. Cat. No. 340 (Milne, vol. 31, p. 121; *Ann. Reg.* 1835, p. 128; etc.).

A slight shock at Clitheroe and Manchester.

2. **1835** AUG. 20, *about* 3.33 *a.m.* Cat. No. 341; Intensity 5; Centre of disturbed area in lat. 53° 56′ N., long. 2° 30′ W.; Records from 14 places.

The disturbed area is 27 miles long, 21 miles wide, and contains about 450 square miles. Its centre lies 2 miles N.W. of Clitheroe, and the direction of its longer axis is about N.W. The shock was a strong tremulous motion accompanied by a sound like that of a heavy piece of artillery drawn rapidly over a pavement, etc.

3. **1884** NOV. 14, 5.10 *p.m.* Cat. No. 796 (*Times*, Nov. 15).

A strong shock at Clitheroe.

HALIFAX EARTHQUAKES

1. **1738** DEC. 30. Cat. No. 94; Intensity not less than 5 (*Gent. Mag.* vol. 9, 1739, p. 45; *London Mag.* 1739, p. 48).

A strong shock, accompanied by a hollow report, at Halifax, Elland, Huddersfield, and other towns in the West Riding.

2. **1757** MAY 17. Cat. No. 130; Intensity 5 (J. Wesley's *Journal*, vol. 2, 1827, pp. 355–356).

A shock, preceded by a hoarse rumbling noise, felt at Heptonstall (7 miles W. of Halifax), Bingley (3 miles E. of Keighley), and, it is said, nearly as far as Preston.

1759 May 17, an earthquake (Roper, p. 23) felt at and near Halifax; probably the same as the preceding.

3. **1871** MAR. 19, *about* 10.20 *a.m.* Cat. No. 761 (*Manchester Guardian*, Mar. 22).

A slight though distinct tremor, accompanied by a low rumbling noise, at Halifax, Hebden Bridge, Cleckheaton and Leeds.

EVERTON (LIVERPOOL) EARTHQUAKE

1863 OCT. 5, 11 *p.m.* Cat. No. 726 (Perrey, *Mém. Cour.* vol. 17, 1865, p. 191).

A very feeble shock at Everton (near Liverpool).

EARTHQUAKES OF UNKNOWN EPICENTRES IN THE NORTH-WEST OF ENGLAND

1. **1750** APR. 13, 10 *p.m.* Cat. No. 108; Intensity 5; Fig. 37 (*Phil. Trans.* vol. 46, 1752, pp. 683, 687, 696–698; *Gent. Mag.* vol. 20, 1750, p. 182; *London Mag.* 1750, p. 186).

The boundary of the disturbed area, which reaches from Lancaster to Wrexham, is 75 miles long from N. to S., 56 miles wide, and contains about 3300 square miles. Its centre lies about 6 miles W. of Wigan. The shock was a strong vibratory motion lasting 2 or 3 seconds.

2. **1755** NOV. 17, *evening.* Cat. No. 122 (Mallet, 1852, p. 173).

A violent shock at Whitehaven and Irton in Cumberland.

3. **1768** MAY 15, 4.15 *p.m.* Cat. No. 142 (Milne, vol. 31, p. 104; Mallet, 1853, p. 162).

A strong shock felt at Manchester, Middleton, Kendal, Skipton, Newcastle and Darlington. At Newcastle, the shock consisted of two parts separated by a brief interval.

1835 Aug. 13, a reported earthquake in Lancashire and Yorkshire (Perrey, p. 155).

1843 Mar. 22, 0.53 a.m., a shock, accompanied by noise, at Liverpool, Prescot and Preston (Perrey, *Mém. Cour.* vol. 24, 1875, p. 7).

1871 June 11, 5.10 p.m., three distinct shocks at St Bees (*Brit. Rainfall*, 1871, p. 87).

4. **1879** OCT. 25, 5.30 *a.m.* Cat. No. 784 (*Nature*, vol. 21, 1880, p. 19).

A strong shock in the west of Cumberland.

CHAPTER X

EARTHQUAKES OF THE NORTH-EAST OF ENGLAND

THE north-east of England, comprising the counties of Northumberland and Durham and the greater part of Yorkshire, is, and has been for several centuries, an almost aseismic district. No true earthquake is known to have originated in either Northumberland or Durham, though, in both counties, there have been local earthshakes due in part to mining operations or the circulation of underground water. In Yorkshire, two reported earthquakes at and near Ripon, in 1796 and 1834, were probably due to subsidences of the land. The most important centre is one—or there may be more than one—near York, and there are minor centres near Malton, Wetherby and Hornsea.

YORK EARTHQUAKES

1. **1573.** Cat. No. 51 (Roper, p. 13).

A severe earthquake at York.

2. **1581,** *beginning of* APRIL, *about* 6 *p.m.* Cat. No. 55 (Sir R. Baker, p. 398).

"An earthquake not far from York, which in some places strook the very stones out of Buildings, and made the Bells in Churches to jangle."

3. **1600.** Cat. No. 57 (Roper, p. 14).

A very serious earthquake at York.

4. **1686** FEB. 18. Cat. No. 71 (Roper, p. 16).

An earthquake at York.

1755 Mar. 25, an earth-sound at York (Mallet, 1852, p. 161).

1755 Mar. 27, between 10 and 11 a.m., a violent tremulous motion at York, the seismic origin of which is doubtful (Mallet, 1852, p. 161).

5. **1822** MAR. 20. Cat. No. 310; Centre of disturbed area in lat. 53° 52′ N., long. 0° 48′ W.; Records from 7 places; Fig. 42 (*Gent. Mag.* vol. 92, 1822, p. 365; *Ann. Reg.* 1822, pp. 61–62).

A slight earthquake felt within an area 7 miles long from N. to S., 6 miles wide, and containing about 33 square miles, with its centre

10 miles E.S.E. of York. The boundary of the disturbed area is represented by the dotted line in Fig. 42.

6. **1869** MAR. 16, *about* 3 *p.m.* Cat. No. 750 (*Times*, Mar. 19; *Manchester Guardian*, Mar. 19).

A slight shock, attended by a rumbling noise, felt at and near Langton (3 miles S. of Malton).

7. **1885** JUNE 18, *about* 10.50 *a.m.* Cat. No. 799; Intensity 5; Centre of disturbed area in lat. 53° 51' N., long. 0° 57' W.; Records from 20 places; Fig. 42 (*Nature*, vol. 32, 1885, pp. 175–176; *Times*, June 22).

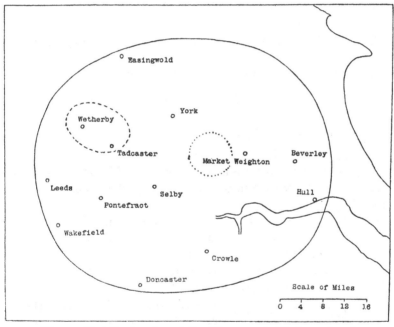

Fig. 42. York Earthquakes of 1822 Mar. 20 and 1885 June 18 and Wetherby Earthquake of 1890 June 25.

The boundary of the disturbed area, which is an isoseismal of intensity about 4, represented by the continuous line in Fig. 42, is 56 miles long, 46 miles wide, and contains about 2000 square miles. Its centre lies 11 miles S.S.E. of York, and its longer axis is directed a few degrees N. of E., passing a short distance to the south of the centre of the earthquake of 1822 (no. 5).

WETHERBY EARTHQUAKES

1. **1890** JUNE 25, *about* 10.30 *p.m.* Cat. No. 834; Intensity 4; Centre of disturbed area in lat. 53° 56′ N., long. 1° 19′ W.; Records from 19 places; Fig. 42 (*Geol. Mag.* vol. 8, 1891, pp. 450–453).

The broken-line in Fig. 42 is an isoseismal of intensity 4. It is 11½ miles long, 7 miles wide, and contains about 60 square miles. Its centre is 3 miles E. of Wetherby, and the longer axis is directed N. 73° W. The shock was also felt at Follifoot and Bolton Percy, 1 and 2 miles respectively from the isoseismal line.

Throughout the disturbed area, the shock consisted of a single vibration, without tremulous motion before or after. At two places near the isoseismal line, no sound was heard.

2. **1890** JUNE 26, *about* 1 *a.m.* Cat. No. 835.

A shock felt at Boston Spa, 2 miles S.W. of the centre of the preceding disturbed area.

1890 June 26, about 4 a.m., a shock felt at Wetherby.

EARTHQUAKES OF UNKNOWN EPICENTRES IN YORKSHIRE

1. **1349.** Cat. No. 40 (*Chron. Monast. de Melsa*, vol. 3, p. 37).

During Lent, there occurred throughout the whole of England an earthquake so great that the monks of Melsa, while at vespers, were thrown so violently from their stalls that they all lay prostrate on the ground.

1602 Feb. —, a severe earthquake, which was felt in different parts of the country and caused much damage, was very sensibly felt in Hull (J. J. Sheahan, *Hist. of Hull*, 1864, p. 87).

1737 Dec. 29, a shock felt at Scarborough, probably due to a landslip (*Phil. Trans.* vol. 41, 1744, pp. 804–806).

2. **1754** APR. 19, *about* 11 *p.m.* Cat. No. 119; Intensity 4 (D. E. Baker, *Phil. Trans.* vol. 48, 1755, pp. 564–566; *Gent. Mag.* vol. 25, 1755, p. 399; *London Mag.* 1754, p. 233).

An earthquake in east Yorkshire, felt at York, Ripon, Stockton, Whitby, Hull and Selby, and at other places between these towns. At York, the shock was a tremulous motion, lasting about 3 seconds and preceded by a rumbling noise.

1755 Mar. 25, J. Wesley (*Journal*, vol. 2, 1827, pp. 282–284) records a shock near Osmotherly (6 miles E.N.E. of Northallerton) evidently due to a landslip.

1872 Jan. 4, Roper (p. 39) records a slight shock at Sheffield.

CHAPTER XI

EARTHQUAKES OF THE WEST
OF ENGLAND

THERE are few seismic centres of importance in the west of England, that is, in the counties of Cheshire, Shropshire, Worcester, Gloucester, Hereford and Monmouth. The earthquakes connected with the twin foci of Herefordshire are described in Chapter XIII. With the exception of these shocks—and they rank among the strongest in the British Isles—there is only one of the first order of magnitude to be chronicled, for the earthquake of 1750 Apr. 13, though felt over the greater part of Cheshire, seems to have originated in Lancashire. The still stronger earthquake of 1775 Sept. 8 was connected with a centre either in the north of Herefordshire or on the borders of that county and Radnorshire. With the single exception of the Malvern earthquake of 1907 Sept. 27, the centres are poorly defined. In Cheshire, only one undoubted earthquake is known to have occurred, and the position of its centre is uncertain. In Monmouthshire, earthquakes are occasionally felt, but their centres lie outside the county.

CHESHIRE EARTHQUAKE

1657 July 8, a shock, preceded by a noise like a clap of thunder, at Bickly, which may perhaps be identified with Bickley 2 miles E. of Malpas (Cat. No. 63; Mallet, 1852, p. 81).

1771 Aug. 24, about 4 a.m., a smart shock, lasting about 3 seconds, at Astbury (1½ miles S.W. of Congleton) (*Gent. Mag.* vol. 41, 1771, p. 422).

1801 JUNE 1, 1.30 *a.m.* Cat. No. 228 (*Gent. Mag.* vol. 71, 1801, p. 565).

A rather strong shock, attended by a rumbling noise, felt at and near Chester, also at Shrewsbury and Salford.

1832 July —, a shock felt at Davenham, Delamere Forest, Sandbach, Tarporley and Waverton (Roper, p. 32, on the authority of the *Chester Chronicle* for 1832 Aug. 3).

1872 Feb. —, a shock at Nantwich (Fuchs, p. 137).

1878 Jan. 11, 2.20 a.m., a shock felt at Edge Hall (2 miles N. of Malpas), the sensation being like that experienced in the upper rooms of a house when rocked by a violent gale.

SHROPSHIRE EARTHQUAKES

1. **1110.** Cat. No. 11 (Florence of Worcester, p. 219; Symeon of Durham, vol. 2, pp. 241–242; R. Fabyan, p. 256; Stow, p. 195; etc.).
A great earthquake at Shrewsbury.

2. **1668** APR. 3. Cat. No. 66 (Mallet, 1852, p. 98).
An earthquake at Middle (8 miles S.E. of Ellesmere)

3. **1712.** Cat. No. 80 (Mallet, 1852, p. 114).
A shock at Boseley (probably Broseley).

4. **1737** MAY 13, *morning.* Cat. No. 92 (*London Mag.* 1737, p. 274).
An earthquake at Ludlow, Bishop's Castle and other places in Shropshire.

1773 May 27, a disturbance, probably a landslip, near Buildwas bridge, described at some length by the Rev. J. Fletcher, who remarks "that all this was owing to an earthquake, there can be no reasonable doubt" (J. Fletcher, *Works*, vol. 5, 1826, pp. 207–223; J. Wesley, *Journal*, vol. 3, 1827, p. 486).

1773 July 3, Milne (vol. 31, p. 106) records, without giving details, an earthquake at Eton in Shropshire. There are two places called Eaton in the county, one 3½ miles S.E. of Church Stretton, the other 7 miles N. of Wellington.

1773 Sept. 8, 9.45 p.m., Milne (vol. 31, p. 106) records an earthquake felt at several places in Shropshire and over a district from Shrewsbury to Bath and from Oxford to Swansea. This is almost certainly the same as the earthquake of 1775 Sept. 8, described below (Herefordshire earthquakes).

1781 Jan. 25, an earthquake at Shrewsbury (Milne, vol. 31, p. 106).

5. **1783** APR. 23, *about* 1.15 *p.m.* Cat. No. 157 (*Gent. Mag.* vol. 53, 1783, p. 442).
A strong shock at Coalbrookdale.

1786 Sept. 26, at night, an earthquake (intensity 4) at Wenlock in Wilts—probably Shropshire (*Gent. Mag.* vol. 59, 1789, p. 947).

1836 Apr. 4, morning, an earthquake in Shropshire (Mallet, 1854, p. 259).

6. **1838** MAR. 17, 1 *p.m.* Cat. No. 346 (Milne, vol. 31, p. 121).
A shock (intensity 4) at Shrewsbury.

7. **1838** MAR. 27, 1 *p.m.* Cat. No. 347 (Milne, vol. 31, p. 121).
A shock felt at Shrewsbury, Hanwood, Dorington, etc.

WORCESTERSHIRE EARTHQUAKES

1. **1048** MAY 1. Cat. No. 2 (Symeon of Durham, vol. 2, p. 164; *Chron. of Melrose*, p. 111; Roger de Hoveden, vol. 1, p. 114; *Anglo-Saxon Chron.* vol. 2, pp. 137, 138; John of Brompton, col. 939; etc.).

"A great earthquake occurred on Sunday, the kalends of May, at Worcester, Wic, Derby, and many other places" (Symeon of Durham).

2. **1119** SEPT. 29. Cat. No. 13 (Symeon of Durham, vol. 2, p. 254; *Anglo-Saxon Chron.* vol. 2, p. 215; *Annal. Monast.* vol. 1, pp. 10, 45; vol. 2, pp. 45, 217; vol. 4, p. 17).

"On the eve of the mass of St Michael was much earth-heaving in some places in this land, though most of all in Gloucestershire and Worcestershire" (*Anglo-Saxon Chron.*).

3. **1768** DEC. 21, *between 5 and 6 p.m.* Cat. No. 144 (*Gent. Mag.* vol. 38, 1768, p. 588).

A shock felt at and near Worcester, and also at Gloucester.

MALVERN EARTHQUAKE

1907 SEPT. 27, 8.12 *a.m.* Cat. No. 1035; Intensity 5; Centre of iso-seismal 5 in lat. 52° 6·6' N., long. 2° 16·8' W.; No. of records 107 from 40 places and negative records from 6 places; Fig. 43 (*Geol. Mag.* vol. 5, 1908, pp. 305–306).

Two isoseismals are represented by the continuous lines in Fig. 43. The isoseismal 5 is 9 miles long, 7 miles wide, and contains 48 square miles. Its centre lies 1½ miles E. of Great Malvern station, and its longer axis is directed N. 6° E. The outer isoseismal, of intensity slightly less than 4, is 18 miles long, 14½ miles wide, and 206 square miles in area. The distance between the two isoseismals is 3½ miles on the west side and 4 miles on the east side. Outside the latter curve, the earthquake was observed at six places, Cheltenham and Charlton Kings being respectively 8 and 9½ miles to the S.E. Thus, the total disturbed area must contain about 800 square miles.

At places near the centre of the disturbed area, the shock consisted of a single violent jerk, like that caused by the fall of a heavy body or an explosion of dynamite, followed by a slight vibration. In the surrounding zone, the jerk only was felt. At places near the boundary of the disturbed area, the shock was a slight tremor. The mean duration of the shock was 2·0 seconds.

The sound-area coincides nearly with that contained by the outer isoseismal, but overlaps it towards the west. The sound was heard by 94 per cent. of all the observers, by 95 per cent. within the isoseismal 5, and by 93 per cent. in the zone between the two isoseismals. The sound was compared to passing waggons, etc., in 35 per cent.

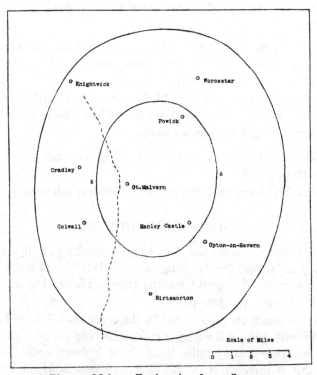

Fig. 43. Malvern Earthquake of 1907 Sept. 27.

of the records, thunder in 10, wind in 4, loads of stones falling in 14, the fall of a heavy body in 24, explosions in 12, and miscellaneous sounds in 1, per cent.

From the seismic evidence, it is seen that the mean direction of the originating fault is N. 6° E., its hade is to the E., and the fault-line must pass a short distance to the W. of the centre of the isoseismal 5, that is, close to Great Malvern. Now, the mean direction in that district of the great fault which skirts the east side of the Malvern Hills is N. 4° E., its hade is to the E., and it passes through Great

Malvern. Thus, the earthquake was probably caused by a small slip of this well-known fault. That the slip took place at a slight depth is shown by the smallness of the disturbed area, the closeness of the isoseismal lines, and by the abrupt nature of the shock near the centre, fading away to a tremor near the boundary of the disturbed area. The most remarkable feature of the earthquake is that it should be the only movement which, so far as known, has occurred in the Malvern fault during the whole of our seismic record.

GLOUCESTERSHIRE EARTHQUAKES

1. **1812** MAY 1. Cat. No. 258 (Milne, vol. 31, p. 115; *Gent. Mag.* vol. 82, 1812, p. 479).

An earthquake in Gloucestershire and South Wales. At Neath, some chimneys are said to have been thrown down. The epicentre may thus have been in Glamorganshire.

2. **1839** SEPT. 1, 1 *a.m.* Cat. No. 353 (Milne, vol. 31, p. 122).

An earthquake of intensity not less than 5 (perhaps as much as 7) felt at Bristol, Newport (Mon.), Cardiff and other places in South Wales, Shrewsbury, etc.

3. **1883** JAN. 16, *about 5 p.m.* Cat. No. 791 (*Nature*, vol. 27, 1883, p. 293; *Times*, Jan. 17).

A slight shock, lasting but a few seconds and accompanied by a rumbling noise, at Monmouth, Clifton and Highnam Court (3 miles W. of Gloucester).

HEREFORDSHIRE EARTHQUAKE

1571 Feb. 17–19, a movement of the ground at Kinnaston (4 miles W.N.W. of Ross), lasting for nearly two days, and attributed at the time to "a prodigious earthquake" (Camden, p. 158; Sir R. Baker, p. 397; Short, vol. 1, p. 245).
1602, an earthquake in Herefordshire (Sir R. Baker, p. 397).
1755 Mar. —, Nov. 17 and 18, Dec. 18, four earthquakes in Herefordshire recorded by Perrey (pp. 136, 142). Two of them, he states, were felt at Glonsow or Glossom in Herefordshire. There is an obvious misprint in one or both names, and I am unable to identify the place referred to. No details are given, so that the seismic nature of the shocks must remain in doubt.

1769 Dec. 29, 8 a.m., a rent occurred in Shobdon's Hill (about 6 miles
W. of Leominster), which was attributed at the time to an earthquake,
of which, as in 1571, there is no evidence (*Gent. Mag.* vol. 39, p. 50;
Milne, vol. 31, p. 105).

1775 SEPT. 8, *about* 9.45 *p.m.* Cat. No. 150; Centre of disturbed area
in about lat. 52° 20′ N., long. 3° 2′ W.; Records from 15 places;
Fig. 44 (*Phil. Trans.* vol. 66, 1776, p. 368; T. Pennant, *Phil. Trans.*
vol. 71, 1781, pp. 193–194; *Gent. Mag.* vol. 45, 1775, pp. 432, 451).

Fig. 44. Herefordshire Earthquake of 1775 Sept. 8.

The disturbed area (Fig. 44) is about 210 miles long (from Preston
to Bristol and Barnstaple), 170 miles wide, and contains about 28,000
square miles. Its centre lies about 10 miles N. of Hereford and close
to the north end of the boundary between Herefordshire and Radnorshire. The longer axis is directed a few degrees E. of N.

CHAPTER XII

EARTHQUAKES OF THE MIDLAND COUNTIES OF ENGLAND

THE following counties are included in this district—those of Stafford, Derby, Nottingham, Leicester, Rutland, Warwick, Northampton, Huntingdon, Oxford, Berkshire, Buckingham and Bedford. The most important earthquakes of the district are the twin-earthquakes described in the next chapter.

STAFFORDSHIRE EARTHQUAKES

1. **1678** NOV. 4, 11 *p.m.*—NOV. 5, 2 *a.m.* Cat. No. 67 (R. Plot, *Hist. of Staffordshire*, 1686, p. 142).

At about 11 p.m. on Nov. 4, a shock, accompanied by a noise like distant thunder, was felt at Brewood (9 miles S. of Stafford), followed by two or three others, the last on Nov. 5 at about 2 a.m. The next night, there was another and slighter shock.
1683 Oct. 9, about 11 p.m., an earthquake in Staffordshire and the adjoining counties (Cat. No. 70; R. Plot, *Hist. of Staffordshire*, p. 143).

2. **1863** OCT. 6, 1.30 *a.m.* Cat. No. 727 (Perrey, *Mém. Cour.* vol. 17, 1865, p. 191; etc.).

A slight shock felt at Stafford and Betley (12 miles N.N.W. of Eccleshall), which occurred two hours before the strong Hereford earthquake.

DERBYSHIRE EARTHQUAKES

1084, an earthquake at Derby (O'Reilly, p. 548).
1678, an earthquake in Staffordshire and Derbyshire (*Gent. Mag.* vol. 20, 1750, p. 56), probably the same as that of 1678 Nov. 4–5 in Staffordshire.

1. **1734** OCT. 31. Cat. No. 89 (S. Glover, *His. and Gaz. of the County of Derby*, 1833, vol. 2, p. 610; *London Mag.* 1734, p. 605).

An earthquake felt at Derby and several places in the neighbourhood.

2. **1738** JULY 10, *between 3 and 4 a.m.* Cat. No. 93 (S. Glover, vol. 2, p. 612).

An earthquake felt by a great many persons in Derby and for several miles around.

1748 June 26, a very smart shock felt at Chapel-en-le-Frith and 1½ miles round (Short, vol. 2, p. 529).

3. **1795** NOV. 18, *about* 11.10 *p.m.*   Cat. No. 209; Intensity 8; No. of records 41 from 33 places; Fig. 45 (E. W. Gray, *Phil. Trans.* 1796, pp. 353–381; *Gent. Mag.* vol. 65, 1795, pp. 891, 964–965).

This was the strongest of all recorded earthquakes in Derbyshire. The position of the epicentre is unknown, but a number of chimneys were thrown down at Derby, Nottingham and Ashover. The boundary of the disturbed area is represented by the line in Fig. 45. It

Fig. 45. Derbyshire Earthquake of 1795 Nov. 18.

includes York, Leeds, Liverpool, Bristol, Oxford, Witney and Norwich, and is about 210 miles long from N.E. to S.W., about 185 miles wide, and contains about 30,000 square miles. At Wollaton (3 miles W. of Nottingham), the shock consisted of two distinct parts. The mean duration of the shock was 3·2 seconds. The shock was felt by men working in lead mines near Chesterfield. In the Gregory mine at Ashover, the noise seemed to pass through the strata above the workings.

1872 Nov. 13, 4.10 p.m., a shock (of intensity 4) was felt at Cavendish Bridge Brewery (near Derby), also at Aston, Castle Donington, Chellaston and Shardlow, lying within a small area, 4 or 5 miles in diameter, with its centre about 4 miles S.E. of Derby (*Nature*, vol. 7, 1873, p. 68).

NOTTINGHAMSHIRE EARTHQUAKES

The most interesting earthquakes in this county are those which occurred in the neighbourhood of Mansfield in 1816, 1817 and 1825. These are described in the next section. The present section is confined to a few shocks of minor interest, which occurred elsewhere in the county.

1. 1180 APR. 25. Cat. No. 22; Intensity 8 (Sir R. Baker, p. 58).

An earthquake at Nottingham and throughout the Midland counties. Many houses were thrown down.

1585 Aug. 4, an earthquake at Nottingham (Perrey, p. 125; Short, vol. 2, p. 69).

2. 1812 SEPT. 3, *between 8 and 9 p.m.* Cat. No. 259 (*Edin. Adver.*).

A slight shock, accompanied by a noise like rushing wind and lasting some seconds, at Hoveringham (8 miles S.W. of Newark) and several other villages.

3. 1857 JAN. 25, 3.20 p.m. Cat. No. 716 (E. J. Lowe, *Times*, Jan. 28).

A trembling, accompanied by a noise like that of a distant heavy luggage train, at Beeston. A pendulum 30 feet long, erected for recording earthquakes (p. 246), moved about ⅛ of an inch from W. to E.

4. 1865 JAN. 2, *between 1 and 2 a.m.* Cat. No. 735 (E. J. Lowe, *Times*, Jan. 3).

Three slight shocks at Beeston, each accompanied by a faint noise not unlike that of a distant railway train, at 1.18, 1.35 and 1.50 a.m. The second shock was the strongest, but was a mere trembling motion; it was apparently felt by two persons in Nottingham. There was a small movement of the pendulum from N. to S.

1865 Feb. 1, 1.42 a.m., a slight shock at Beeston, accompanied by a low rumbling noise, apparently felt by only one observer (E. J. Lowe, *Times*, Feb. 4).

1873 Aug. 28, 6.50 a.m., an earthquake reported at Nottingham (*Nature*, vol. 8, 1873, p. 370).

1873 Sept. 6, an earthquake reported at Nottingham (Fuchs, p. 138).

MANSFIELD EARTHQUAKES

1. **1816** MAR. 17, *between* 0.30 *and* 1 *p.m.* Cat. No. 265; Intensity 7; Centre of disturbed area in lat. 53° 1′ N., long. 1° 0′ W.; No. of records 30 from 23 places; Fig. 46 (Howard, vol. 1, 1818, after table 116; *Gent. Mag.* vol. 86, 1816, p. 366; *Notes and Queries*, vol. 1, 1862, p. 94; *Times*, Mar. 23, 25 and 26).

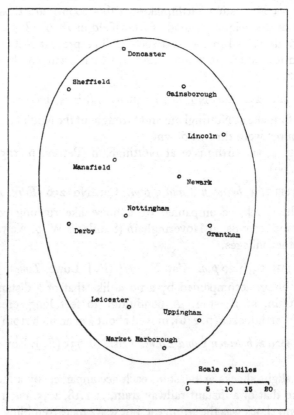

Fig. 46. Mansfield Earthquake of 1816 Mar. 17.

The boundary of the disturbed area is shown in Fig. 46. The area is 75 miles long, 53 miles wide, and contains 3100 square miles. Its centre lies 8 miles N.E. of Nottingham, and its axis is directed about N. There was some slight injury to chimneys, etc., at Mansfield, Kirkby in Ashfield (4 miles S.W. of Mansfield) and Newstead (5 miles S.).

2. **1817** JAN. 27, 11 *p.m.* Cat. No. 271 (*Gent. Mag.* vol. 87, 1817, p. 268). A shock felt at Mansfield and neighbouring villages.

3. **1825.** Cat. No. 318; Intensity 4 (*Notes and Queries*, vol. 12, 1861, pp. 397–398; vol. 1, 1862, p. 16).

A shock at Mansfield and Newstead Abbey.

LEICESTERSHIRE EARTHQUAKE

1660 Dec. —, an earthquake in Leicestershire, Nottinghamshire, etc. (Short, vol. 1, p. 335).

1868 NOV. 19, 7.30 *p.m.* Cat. No. 745 (*Brit. Rainfall*, 1868, p. 66).

A slight earthquake in Leicestershire.

OAKHAM EARTHQUAKE

1898 JAN. 28, *about* 10.5 *p.m.* Cat. No. 926; Intensity 3; Centre of disturbed area in lat. 52° 41·0′ N., long. 0° 40·9′ W.; No. of records 40 from 26 places; Fig. 47 (*Geol. Mag.* vol. 7, 1900, pp. 168–170).

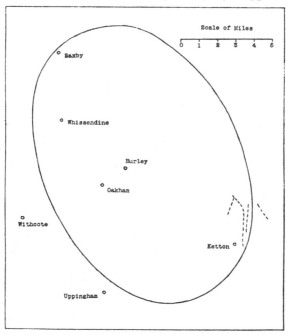

Fig. 47. Oakham Earthquake of 1898 Jan. 28.

This appears to be the only earthquake to which an origin within the county of Rutland can be assigned. Though its intensity was slight, it was felt over nearly the whole of the county and in small portions of the adjoining counties of Lincoln, Leicester and North-ampton. The disturbed area is 15 miles long, 11 miles wide, and con-tains 130 square miles. Its centre lies 2½ miles E.N.E. of Oakham and the longer axis is directed N. 26° W.

At all places, the sound was a much more prominent feature than the shock. The movement, when felt, was merely a faint vibration in the middle of the rumbling noise. The sound was heard by all the observers, and over an area of about 165 square miles which overlaps the disturbed area by from ½ to 1 mile towards the east, south and west. The sound was compared to passing waggons, etc., in 40 per cent. of the records, thunder in 34, wind in 3, loads of stones falling in 11, the fall of a heavy body in 3, explosions in 3, and miscellaneous sounds in 6, per cent.

The mean direction of the originating fault must be about N. 26° W. Owing to the thick covering of clay, which masks a large part of the disturbed area, the faults have not been traced in the central district. In the south-east corner of the area, there are, however, one or two faults (indicated by the broken-lines in Fig. 47) which, if continued, might be responsible for the earthquake (J. W. Judd, "Geology of Rutland," *Geol. Surv. Mem.* 1875, pp. 256–259).

BIRMINGHAM EARTHQUAKES

1. **1769** NOV. 23, 4 *p.m.* Cat. No. 147 (Milne, vol. 31, p. 105).

An earthquake, attended by a rumbling noise like the firing of distant cannon, felt near Birmingham.

2. **1772** NOV. 15, 4 *a.m.* Cat. No. 149 (W. Hutton, *Hist. of Bir-mingham*, 1783, p. 342).

The shock was felt over an area "about eight miles in length, from Hall-green to Erdington, and four in breadth, of which Birmingham was part. The shaking of the earth continued about five seconds, with unequal vibration, sufficient to awake a gentle sleeper, throw down a knife carelessly reared up, or rattle the brass drops of a chest of drawers. A flock of sheep in a field near Yardley, frightened at the trembling, ran away."

1816 Mar. 24, about 1 p.m., a shock felt in Birmingham and neighbour-hood (J. A. Langford, *A Century of Birmingham Life*, vol. 2, 1868, p. 335).

3. **1888** JAN. 31, *about* 2 *a.m.* Cat. No. 809 (*Nature*, vol. 37, 1888, p. 350; *Times*, Feb. 3; *Birmingham Daily Post*, Feb. 3).

A slight shock, accompanied by a rumbling noise, felt at Birming-ham, King's Heath, Penns (near Sutton Coldfield) and Coventry.

NORTHAMPTONSHIRE EARTHQUAKES

The earthquakes of Northamptonshire may be divided into three groups: (i) the twin earthquakes of 1750 Oct. 11 and the associated earthquake of 1768 Jan. 3, described in the next chapter; (ii) a group of five earthquakes (1755, 1792, 1813, 1837 and 1844) connected with a centre about 3 miles S. of Stamford; and (iii) four earthquakes belonging to various or unknown centres within or near the county. The latter are described in the present section.

1. **1563** SEPT. —. Cat. No. 49 (Holinshed, vol. 4, p. 224; *London Mag.* 1750, p. 103).

An earthquake felt in different parts of the country, especially in Lincolnshire and Northamptonshire.

2. **1564** SEPT. *and* NOV. Cat. No. 50 (Holinshed, vol. 4, p. 224; Stow, p. 1112; Short, vol. 2, p. 169).

An earthquake felt in Northamptonshire and Lincolnshire.

3. **1755** JULY 31, *between* 6 *and* 7 *a.m.* Cat. No. 120 (*London Mag.* 1755, p. 394).

A slight earthquake at Rushden (4 miles E. of Wellingborough).

4. **1776** OCT. 27, *about* 9.45 *p.m.* Cat. No. 151 (*Phil. Trans.* 1792, p. 286; *Ann. Reg.* vol. 19, p. 187).

A shock, accompanied by a rumbling noise, at Peterborough, Northampton, Market Harborough and Loughborough.

STAMFORD EARTHQUAKES

1. **1755** AUG. 1. Cat. No. 121 (Mallet, 1852, p. 161).

A shock felt at Stamford.

2. **1792** MAR. 2, *about* 8.45 *p.m.* Cat. No. 174; Intensity 5 (*Gent. Mag.* vol. 62, 1792, p. 272).

A strong shock, accompanied by a loud rumbling noise, felt at Biggleswade, Bottesford, Kettering, Newark, Nottingham and Southwell, and throughout a great part of the counties of Bedford, Leicester, Lincoln, Rutland and Nottingham. The earthquake probably originated in the same centre as the earthquakes of 1813 and 1844, most of the above places being about 35 miles from that centre. The area disturbed was about 4000 square miles.

3. 1813 SEPT. 24, *about* 3.15 *p.m.* Cat. No. 262; Centre of disturbed area in lat. 52° 36′ N., long. 0° 29′ W.; Fig. 48 (*Gent. Mag.* vol. 83, 1813, p. 391; etc.).

The boundary of the disturbed area is represented by the continuous line in Fig. 48. It is 22 miles long from E. to W., 20 miles wide, and contains 350 square miles. Its centre lies 3 miles S. of Stamford. The duration of the shock was about 2 seconds.

4. 1837 DEC. 8, 11.15 *p.m.* Cat. No. 345 (O'Reilly, p. 293).

An earthquake felt at Stamford and to a distance of 20 miles.

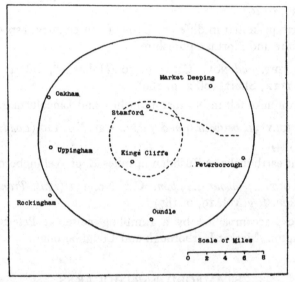

Fig. 48. Stamford Earthquakes of 1813 Sept. 24 and 1844 June 12.

5. 1844 JUNE 12, *about* 7 *p.m.* Cat. No. 645; Centre of disturbed area in lat. 52° 36′ N., long. 0° 29′ W.; Fig. 48 (*Brit. Ass. Rep.* 1844, p. 89).

The boundary of the disturbed area, which is nearly circular, is represented by the broken-line in Fig. 48. It is 8 miles long from N.E. to S.W., 7 miles wide, and contains about 45 square miles. Its centre coincides with that of the earthquake of 1813. The shock was in all parts slight, and was accompanied by a noise like distant thunder.

The movements along a fault, indicated by the broken-line in Fig. 48, may perhaps be responsible for the above earthquakes.

HUNTINGDONSHIRE EARTHQUAKES

1. **1233** NOV., *beginning of.* Cat. No. 28 (R. Fabyan, p. 329; Holinshed, vol. 2, 1807, p. 376).

"An erthquake, to the great fere of the inhabytauntes of Huntyngdon, and nere there aboute" (Fabyan).

2. **1845** MAR. 26, *soon after* 9 *p.m.* Cat. No. 658 (*Ill. London News*, vol. 6, 1845, p. 211).

A slight shock, with a low rumbling noise like that of an explosion, at Huntingdon.

OXFORDSHIRE EARTHQUAKES

Most of the Oxfordshire earthquakes belong to two well-defined centres, one 3 miles S.E. of Banbury, the other 5 miles N.E. of Oxford.

1810 NOV. 18. Cat. No. 255 (Milne, vol. 31, p. 115).

A shock, accompanied by a deep rumbling noise like that of the distant firing of heavy guns, felt in Oxfordshire and the neighbouring counties.

BANBURY EARTHQUAKES

1731 Oct. 19, about 3 a.m., a shock reported at this time from the neighbourhood of Banbury (*Phil. Trans.* vol. 39, 1738, p. 367).

1. **1731** OCT. 21, *about* 4 *a.m.* Cat. No. 85; Centre of disturbed area in lat. 52° 1′ N., long. 1° 17′ W.; Fig. 49 (*Phil. Trans.* vol. 39, 1738, p. 367).

The disturbed area, as shown by the dotted line in Fig. 49, is nearly circular, about 8 miles in diameter and about 50 square miles in area. Its centre is 3 miles S.E. of Banbury. In the north-west corner of the map, a fault is indicated by a broken-line, but it is uncertain whether it can have had any connexion with the earthquake.

1731 Dec. 28, an earthquake felt at and near Banbury (Short, vol. 2, p. 529).

2. **1838** SEPT. 14, 7 *or* 9 *a.m.* Cat. No. 348 (Mallet, 1854, p. 279).

A strong shock, accompanied by a subterranean noise like that of an explosion, at Adderbury, which is 3 miles S.S.E. of Banbury and 1 mile W.S.W. of the epicentre of the earthquake of 1731 Oct. 21.

OXFORD EARTHQUAKES

1. **1666** JAN. 18, *about* 6 *p.m.* Cat. No. 65; Centre of disturbed area in lat. 51° 49′ N., long. 1° 11′ W.; Fig. 49 (J. Wallis, *Phil. Trans.* vol. 1, 1665–66, pp. 166- 169; R. Boyle, *Phil. Trans.* vol. 1, 1665–66, pp. 179–181).

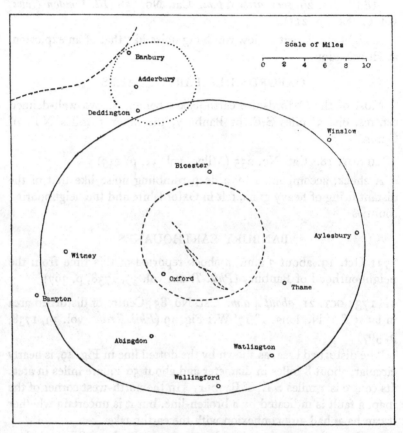

Fig. 49. Banbury Earthquake of 1731 Oct. 21 and
Oxford Earthquakes of 1666 Jan. 18 and 1683 Sept. 17.

The boundary of the disturbed area is represented by the broken-line in Fig. 49. It is nearly circular, 12 miles long from E. to W., 11 miles wide, and contains about 105 square miles. Its centre is 5 miles N.E. of Oxford. In all parts of the disturbed area, the shock was a slight tremulous motion.

2. **1683** SEPT. 17, *about 7 a.m.* Cat. No. 69; Centre of disturbed area in lat. 51° 49′ N., long. 1° 11′ W.; No. of records 27 from 25 places; Fig. 49 (*Phil. Trans.* vol. 12, 1683, pp. 211–321; *Gent. Mag.* vol. 20, 1750, p. 56; *London Mag.* 1750, p. 103).

The boundary of the disturbed area is represented by the continuous line in Fig. 49. It is 35 miles long from E. to W., 28 miles wide, and contains about 770 square miles. Its centre is 5 miles N.E. of Oxford, less than ½ mile from that of the earthquake of 1666. At Oxford, the shock consisted of a rapid trembling motion, and was accompanied by a hollow sound like distant thunder.

3. **1764** NOV. 6, *about 4.15 a.m.* Cat. No. 136; Intensity 5 (*Gent. Mag.* vol. 34, 1764, p. 543).

A strong shock at Oxford, accompanied by a noise as of the fall of a heavy body; a shock, preceded by a rumbling noise, was also felt at Wallingford (14 miles S. of the centre of the preceding earthquakes). 1783, an earthquake at Oxford (Prestwich, p. 544).

4. **1812** JAN. 18. Cat. No. 257; Intensity 4 (*Gent. Mag.* vol. 82, 1812, pp. 80, 283).

A strong shock, accompanied by a rumbling noise like the firing of heavy guns, felt at Bletchingdon, Islip, Radley, Tetsworth, Wolvercot and High Wycombe. The centre was probably the same as that of the earthquakes of 1666 and 1683.

Near this centre, there passes a fault (indicated by broken-lines in Fig. 49), which may be connected with the Oxford earthquakes.

With the Oxfordshire earthquakes may be included the only reported earthquake from Berkshire, of doubtful seismic origin: 1771 Apr. 29, about 5.30 a.m., a strong shock felt throughout Abingdon, so violent as to lift people in their chairs (*Gent. Mag.* vol. 41, 1771, p. 233).

No earthquakes are known to have originated in Buckinghamshire. The only record of earthquakes in Bedfordshire is not free from doubt: 1690 Dec. 18, two shocks in Bedfordshire (Mallet, 1852, p. 99).

CHAPTER XIII

TWIN EARTHQUAKES OF THE WEST AND MIDLAND COUNTIES OF ENGLAND

HEREFORD EARTHQUAKES

THOUGH the seismic record of Herefordshire covers little more than half a century, it includes three earthquakes of great interest—those of 1863, 1868 and 1896. The earthquake of 1896 has been studied in most detail. As will be shown below, it originated almost simultaneously in two detached foci, the epicentre of the more important being about 3 miles S.E. of Hereford and that of the other about 2 or 3 miles N.E. of Ross. Though the evidence is by no means complete, it is probable that the earthquakes of 1863 and 1868 originated in the same foci.

1. 1853 MAR. 27, *about* 11.30 *p.m.* Cat. No. 712; Intensity 5; Fig. 53 (Perrey, p. 37; etc.).

The boundary of the disturbed area is represented by the left-hand dotted line in Fig. 53. It is about 24 miles long, 15 miles wide, and contains about 280 square miles. Its centre lies about 10 miles S.W. of Hereford, and the direction of its longer axis is about N. 65° E. The shock was a sharp shake or thud, preceded and followed by a rumbling sound like the rolling of a heavy carriage on the road.

The seismic nature of the following reported earthquakes is doubtful:

1863 Oct. 6, a little after midnight, a shock felt by a few persons at Oswestry, Wem and Whitchurch (Perrey, *Mém. Cour.* vol. 17, 1865, p. 191).

1863 Oct. 6, about 1.30 a.m., a slight shock at Stafford and Betley (Perrey, *Ibid.*).

2. 1863 OCT. 6, *about* 2.30 *a.m.* Cat. No. 728 (Perrey, *Mém. Cour.* vol. 17, 1865, p. 191).

A rather strong shock with noise at Clifford (16 miles W. of Hereford) and Weston-super-Mare.

3. 1863 OCT. 6, 3.10 *a.m.* Cat. No. 729 (*Times*, Oct. 8).

A shock with sound at Newport (Mon.).

4. 1863 OCT. 6, 3.22 *a.m.* Cat. No. 730; Intensity 8; Centre of isoseismal 5 in lat. 52° 8' N., long. 2° 43' W.; No. of records 210 from 172 places; Fig. 50 (G. B. Airy, *Athenaeum*, 1863, p. 498; T. Godwin-

Austin, *Quart. Journ. Geol. Soc.* vol. 20, 1864, pp. 380–382; E. J. Isbell, *Woolhope Nat. Field Club Trans.* vol. for 1852–1865, pp. 348–350; E. J. Lowe, *Brit. Meteor. Soc. Proc.* vol. 2, 1865, pp. 55–99; R. Mallet, *Quart. Journ. Sci.* vol. 1, 1864, pp. 53–67; Perrey, *Mém. Cour.* vol. 17, 1865, pp. 191–201; etc.).

Isoseismal Lines and Disturbed Area. The continuous lines in Fig. 50 represent the isoseismal 5 and the boundary of the disturbed area.

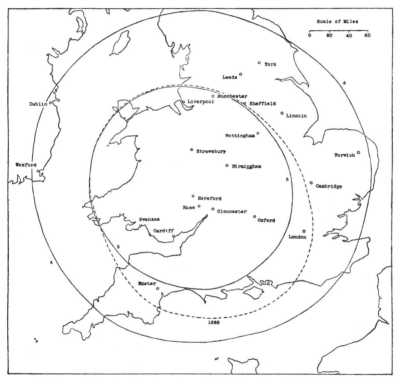

Fig. 50. Hereford Earthquakes of 1863 Oct. 6 and 1868 Oct. 30.

The isoseismal 5 is 233 miles long from N.W. to S.E., 200 miles wide, and contains about 36,000 square miles. Its centre lies 5 miles N. of Hereford. The boundary of the disturbed area is about 365 miles long from E. to W., 295 miles wide, and contains about 85,000 square miles. With more numerous observations, it is possible that the course of this line might be materially altered.

In several places, buildings were slightly damaged. At Sellack (3 miles N.W. of Ross), some houses were injured; at Garway (9 miles

S.W. of Ross), several walls in a house were cracked; and at Hereford, one or two walls were cracked and some bricks fell. The intensity seems to have been slightly greater near Ross than near Hereford.

Nature of the Shock. The shock was recorded as double at nine places—Hereford, Birmingham, Wolverhampton, Woodtown (near Tavistock), Guilsborough (Northamptonshire), Nottingham, Waterloo (near Liverpool), Carmarthen and Ferryside. The second part was distinctly the stronger at Birmingham, Wolverhampton and Wood-town. These places lie close to the minor axis of the isoseismal 5, but are slightly nearer the Ross epicentre; and thus it would seem that the Hereford focus was first in action but that the impulse in the Ross focus was the stronger. The mean duration of the shock was 5·7 seconds.

Instrumental Observations. The records of the three magnetographs at Greenwich show no trace of disturbance by the earthquake. Observations were, however, being made with the altazimuth, and the observer was preparing by a slow motion of the telescope to place one of the horizontal wires on the image of the collimator-mark, when he found himself unable to make a satisfactory bisection. "Before he had actually moved the telescope, the image of the mark moved apparently downwards, remained stationary, or nearly stationary, for a short time, and then returned to its original position, and had no further motion." It was a steady motion, without any quivering, and lasted only a few seconds, the extent of the disturbance being 12″ or 15″. The time must have been either 3.23 a.m. nearly or 3.26 a.m. nearly (G. B. Airy), the former no doubt being the correct time.

At Beeston observatory (near Nottingham), Mr E. J. Lowe had erected a pendulum for recording earthquakes. It consisted of a wooden rod 30 feet long freely hung at the top of a tower. At the lower end of the rod was a solid brass ball, from which there pro-jected downwards an index, the point of which rested on a surface covered with powdered chalk. On this occasion, the movement of the pendulum was from W.N.W. to E.S.E., the pointer having removed the chalk in an elongated oval half an inch in length.

Sound-Phenomena. The boundary of the sound-area falls about 12 miles short of the isoseismal 5 in all directions. It must thus con-tain about 30,000 square miles. The sound was heard by about 95 per cent. of the observers. It was compared to passing waggons, etc., in 23 per cent. of the records, thunder in 32, wind in 9, loads of stones falling in 7, the fall of a heavy body in 4, explosions in 18, and miscellaneous sounds in 7, per cent.

1863 Oct. 6, about 3.25 *a.m.*, a faint rumble was heard at Stretton (8 miles E.N.E. of Hereford) by a single observer (*Times*, Oct. 8).

5. **1868** OCT. 30, 10.35 *p.m.* Cat. No. 744; Centre of disturbed area in lat. 51° 57′ N., long. 2° 24′ W.; No. of records 61 from 48 places; Fig. 50 (Perrey, *Mém. Cour.* vol. 22, 1872, mém. 3, pp. 105–106, and mém. 4, pp. 40–41; *Mon. Met. Mag.* vol. 3, 1868, pp. 153–154, 167, 201–202; Parfitt, I, pp. 653–654; II, p. 281; W. Spurrell, *Carmarthen*, p. 163; *Ann. Reg.* 1868, pp. 139–140).

The boundary of the disturbed area is represented by the broken-line in Fig. 50. It is 125 miles long, 105 miles wide, and contains about 10,000 square miles. Its centre lies about 5 miles S.E. of Ross and the direction of its longer axis is about N.W.

There are very few accounts of the nature of the shock, and only one of a double shock. At Broseley (Shropshire), two distinct parts were felt each lasting 6 or 7 seconds. The mean duration of the shock was 5·3 seconds.

If the earthquake were a twin, it would seem that much the stronger of the two impulses took place in the Ross focus.

1868 Oct. 30, 10.42 p.m., a slight shock at Swansea (Fuchs, p. 136). 1868 Nov. 2, 6.30 p.m., a slight shock in Wales and south-west England (Fuchs, p. 136).

The seismic nature of both these shocks seems doubtful.

6. **1871** MAR. 20, *about* 9.50 *p.m.* Cat. No. 762 (*Nature*, vol. 3, 1871, p. 426; *Times*, Mar. 23).

A slight shock, lasting from 3 to 6 seconds, at Stretton (near Hereford) and Llan Thomas (near Hay). This earthquake may be connected with the same focus as that of 1853 (no. 1).

The most remarkable series of Hereford earthquakes is that which occurred towards the close of 1896. These shocks are fully described in my report on *The Hereford Earthquake of December* 17, 1896 (Birmingham, 1899)[1].

---

[1] Other useful papers, the results of which are incorporated in the above report, are the following: W. Cole, *Essex Nat.* vol. 9, 1896, pp. 258–259; Sir H. G. Fordham, *Hertfordshire Nat. Hist. Soc. Trans.* vol. 9, 1897, pp. 183–208; E. Greenly, *Edin. Geol. Soc. Trans.* vol. 7, 1898, pp. 469–476; J. Lomas, *Liverpool Geol. Soc. Proc.* 1896–97, pp. 91–98; H. C. Moore and others, *Woolhope Nat. Field Club Trans.* 1895–97, pp. 228–235; G. J. Symons, *Mon. Met. Mag.* 1897, pp. 177–185; H. H. Winwood, *Cotteswold Nat. Field Club Proc.* vol. 12, 1897, pp. 187–196.

7. **1896 DEC. 16**, *about* 11.0 *or* 11.30 *p.m.* Cat. No. 913; Intensity probably 4; No. of records 13 from 12 places; Fig. 53.

The disturbed area is about 97 miles long from N.W. to S.E., about 83 miles wide, and contains about 6300 square miles. The boundary coincides nearly with that of the shock at 3 a.m. (no. 11). The epicentre was situated between the two epicentres of the principal earthquake (no. 16) and partly coincided with them. The shock was a slight tremor and was accompanied by a rumbling sound as of stones falling, etc. From the large disturbed area and the comparatively slight intensity of the shock, it may be inferred that the focus was at a considerable depth.

8. **1896 DEC. 17**, *about* 1 *a.m.* Cat. No. 914; Intensity 3; No. of records 4 from 4 places.

A tremor, accompanied by a slight noise, at Stancombe Park and Tetbury (Gloucestershire) and also, it is said, at Birmingham and Leamington.

9. **1896 DEC. 17**, *about* 1.30 *or* 1.45 *a.m.* Cat. No. 915; Intensity 3; No. of records 3 from 3 places.

A slight shock, accompanied by a rushing sound like wind, at Clehonger (3 miles W.S.W. of Hereford), Defford and Great Comberton (in Worcestershire).

10. **1896 DEC. 17**, *about* 2 *a.m.* Cat. No. 916; Intensity 4; No. of records 9 from 9 places.

A slight shock, accompanied by a sound like a sudden gust of wind, at Lugwardine Court (2½ miles E. of Hereford), Evesham, Feckenham, Worcester, Chetwynd Park and Hopesay (Shropshire), Monmouth, The Hendre and Michael Troy (Monmouthshire).

11. **1896 DEC. 17**, *about* 3 *a.m.* Cat. No. 917; Intensity probably 4; Centre of disturbed area in lat. 52° 8′ N., long. 2° 34′ W., No. of records 21 from 20 places; Fig. 53.

The boundary of the disturbed area, represented by the broken-line in Fig. 53, is about 104 miles long, 79 miles wide, and contains about 6400 square miles. It occupies approximately the position of an isoseismal line of the principal earthquake (no. 16) of intensity between 7 and 6. Its centre lies about 10 miles N.E. of Hereford. The shock was slight, there being no evidence of discontinuity in the vibrations, and the noise resembled that of passing waggons. The focus, as in the first shock of the series (no. 7), must have been at a considerable depth, and it probably occupied the interfocal region of the principal earthquake, though it may have included portions of its two foci.

12. 1896 DEC. 17, *about* 3.30 *a.m.* Cat. No. 918; Intensity probably 3; No. of records 7 from 7 places.

A slight tremor, accompanied by a distant rumbling like that of a train passing, at Barnt Green, Defford, Stanbrook and Worcester (in Worcestershire), Grimshill (Shropshire), Newport (Mon.) and Bath.

13. 1896 DEC. 17, *about* 4 *a.m.* Cat. No. 919; Intensity 4; No. of records 8 from 6 places.

A tremor, accompanied by a noise like that of a sudden gust of wind, at Bristol, Droitwich, Great Malvern, Kidderminster, Stourbridge and Hopesay.

14. 1896 DEC. 17, *about* 5 *a.m.* Cat. No. 920; Intensity 4; No. of records 6 from 6 places.

A shock, accompanied by a noise like that of a strong gust of wind, at Hereford, St Michael's Cathedral Priory (3 miles W.S.W. of Hereford), Whitchurch, Oxford, Witney, and Newtown (Montgomeryshire).

15. 1896 DEC. 17, *about* 5.20 *a.m.* Cat. No. 921; No. of records 4 from 4 places.

A slight shock, accompanied by noise, at St Michael's Cathedral Priory, Badgworth and Bentham (Gloucestershire), and Swansea.

16. 1896 DEC. 17, 5.32 *a.m.* Cat. No. 922; Intensity 8; Centre of isoseismal 8 in lat. 52° 1·1' N., long. 2° 32·9' W.; No. of records 2902 from 1943 places; Figs. 51, 52.

Isoseismal Lines and Disturbed Area. The isoseismal 8, which is the most accurately drawn of the series, is 40 miles long, 23 miles wide, and contains an area of 724 square miles. Its longer axis is directed N. 46° W., and its centre lies 8 miles E.S.E. of Hereford. This isoseismal includes within it 73 places at which marked damage occurred to buildings. Of these, 55 are in Herefordshire, 17 in Gloucestershire and 1 in Worcestershire. At Hereford, 217 chimneys required repair and the cathedral was slightly injured. The amount of damage was relatively greatest at Hereford, Dinedor, Fownhope, Dormington, Withington, and other places lying within a small area 8½ miles long, 6¼ miles wide, and containing about 41 square miles. The centre of this curve is 3¼ miles S. 79° E. of Hereford, and occupies approximately the north-west focus of the ellipse formed by the isoseismal 8. For the purpose of measuring distances, this point will afterwards be referred to as the centre.

The isoseismal 7 is 80 miles long, 56 miles wide, and contains 3580 square miles. Its longer axis is directed N. 48° W. The distance between the isoseismals 8 and 7 is 13¼ miles on the south-west side and 20 miles on the north-east. The isoseismal 6 is 141 miles long,

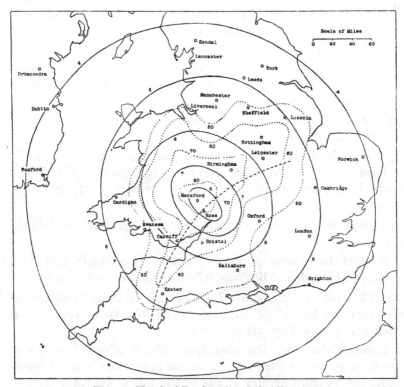

Fig. 51. Hereford Earthquake of 1896 Dec. 17.

116 miles wide, and 13,000 square miles in area. The distance between the isoseismals 7 and 6 is 25 miles on the south-west side and 34 miles on the north-east side. The remaining isoseismals are nearly circular in form. The isoseismal 5 is 233 miles long from N.W. to S.E., 229 miles wide, and contains 41,160 square miles. The distance between the isoseismals 6 and 5 is 60 miles on the south-west side and 55 miles on the north-east. The isoseismal 4 is 357 miles long, 356 miles wide, and includes an area of 98,000 square miles.

Outside the isoseismal 4, the earthquake was felt at the Point of Ayr lighthouse (Isle of Man, 1½ miles from the curve), Flamborough Head lighthouse (7 miles), and Middlesbrough (12½ miles). It is also said to have been felt at Acklington (Northumberland, 59 miles), Graigue (Co. Kilkenny, 5 miles), Drumcondra (Co. Meath, 29 miles), and Killeshandra (Co. Cavan, 65 miles). If we regard the boundary of the disturbed area as a circle concentric with the isoseismal 4 and passing through Middlesbrough, its area would be 115,000 square miles. If it passed through Killeshandra, the disturbed area would contain 185,000 square miles, and a not inconsiderable part of it would lie within the north of France.

Nature of the Shock. The most important feature of the shock was its division into two distinct series of vibrations, separated by a brief interval (of average length 3·6 seconds) of absolute rest and quiet. This twin character of the shock was noticed in nearly all parts of the disturbed area, so near its boundary as at places in Westmorland, the Isle of Man and Ireland. In different parts of the disturbed area, the two parts differed in their relative intensity and duration and in the period of their vibrations. In the south-eastern portion of the disturbed area, a tremulous motion, such as is produced in a house by the passing of a heavy vehicle, was felt for 2 or 3 seconds, followed after about 3 seconds by a stronger series of vibrations which lasted for 4 or 5 seconds. In the north-western portion of the area, the same features were noticed, but in the opposite order. Moreover, the boundary between the two portions was hyperbolic in form, represented by the broken-line in Fig. 51. It also forms the central line of a narrow band (the synkinetic band) within which the shock was continuous, and either gradually increased in strength to a maximum and then died away or contained two maxima of intensity connected by tremulous motion.

The only conclusion that can be drawn from these observations is that the seismic focus consisted of two detached portions arranged along a N.W. and S.E. line (see Chap. XXIII). The epicentre corresponding to the more important focus was evidently near the centre of the oval curve including Hereford and the surrounding villages at which the damage to buildings was distinctly greater than elsewhere, or 3¼ miles E.S.E. of Hereford. The epicentre corresponding to the other focus is known with less accuracy, but it was about 2 or 3 miles N.E. of Ross. Judging from the dimensions of the isoseismal 8, from the durations of the two series of vibrations and of the interval between them, and from the velocity of the earth-waves (about 3000 feet per second), we may estimate the length of the Hereford focus at about 8 miles, of the Ross focus at about 6 miles, and of the interfocal region at about 2 miles. Thus, the total length of the focus was about 16 miles, while the distance between the centres of the epicentral areas was about 9 miles.

As the concavity of the synkinetic band faces the south-east, it follows that the vibrations from the north-west focus travelled farther than those from the south-east focus before the two series coalesced along the synkinetic band, and therefore that the north-west or Hereford focus was first in action, the weaker impulse at the south-east or Ross focus taking place one or a few seconds later. Also, since the two series of vibrations were superposed within the synkinetic band, it is clear that the Ross focus was in action before the vibrations from the Hereford focus had time to reach it.

Besides these variations in the relative strength of the two parts of the shock, there were other variations in the shock due mainly to the distance from the centre. The most important change was an increase in the period with the distance. Close to the epicentre, the shock resembled the rapid jerky movement felt in a carriage without springs. At distances of a hundred miles or more, the movement was of a gently undulating character like that felt in a coach with good springs or like that of a ground-swell at sea.

The mean duration of the shock over the whole disturbed area was 6·2 seconds. It varied, however, with the distance from the origin. Thus, the mean duration was 7·4 seconds within the isoseismal 8, and 6·6, 6·4, 5·8 and 5·4 seconds in the zones bounded by successive pairs of isoseismals.

Direction of the Shock. A large number of observers (469) estimate the direction roughly from personal observations. The distances at which such observations were made are often considerable, the direction having been clearly sensible at Brighton (137 miles from the centre), Maldon (144 miles), Harrogate (147 miles), Douglas (167 miles), Dublin (176 miles) and Baltinglass in Co. Wicklow (180 miles).

In any small district, the apparent directions were very varied. For instance, out of 38 estimates in Birmingham and district, 8 were given as N. and S., 9 as E. and W., 12 as N.W. and S.E., and 5 as N.E. and S.W. As a general rule, however, the apparent direction of the shock was perpendicular or nearly so to the direction of one of the principal walls of the house. Now, the impression of direction must be specially marked in a building with one of the principal walls at right angles to the true direction of the shock; and thus, if a large number of observations were made in a small area at some distance from the centre, the average of all the apparent directions of the shock might be expected to coincide approximately with the true direction of the earth-wave. In Birmingham, the mean direction of the shock was N. 51° E., which differs by only 2° from the direction of this city from the centre. In London, the mean direction was S. 69° E., again differing by only 2° from the direction of the city from the centre.

Sound-Phenomena. The boundary of the sound-area cannot be traced accurately. For the most part, it lies between the isoseismals 5 and 4, coinciding approximately with the isoseismal 4 towards the north-west, but falling short of it towards the south-east by about 33 miles. The sound-area is thus about 320 miles long from N.W. to S.E., about 284 miles wide, and contains about 70,000 square miles.

The dotted lines in Fig. 51 represent isacoustic lines corresponding to percentages of 80, 70, 60 and 50. In drawing them, the county was taken as the unit and the percentage of audibility for the whole county was supposed to represent the percentage at its centre, with the exception of Yorkshire, in which the three Ridings were treated as separate counties. The curves show a marked distortion in both directions along the synkinetic band, due to the coalescence of the sound-vibrations from the two foci. The general decline in audibility with increasing distance from the centre is shown by the fact that 96 per cent. of the observers heard the sound within the isoseismal 8, while in the zones between successive pairs of isoseismals, the percentages were 92, 79, 66 and 67. The audibility-percentage of the whole sound-area is 78.

Throughout the whole sound-area, the sound was compared to passing waggons, etc., in 45 per cent. of the records, thunder in 15, wind in 16, loads of stones falling in 4, the fall of a heavy body in 3, explosions in 7, and miscellaneous sounds in 10, per cent. In the case of passing waggons, etc., it is possible to draw a series of curves of equal percentage of reference. The curves are not shown on the map, but those of highest percentages cling closely to the extremities of the synkinetic band.

Close to the epicentre, heavy crashes were heard by some, but not by all, observers in the midst of the rumbling sound and at the moment when the strongest vibrations were felt. At a distance of 40 or 50 miles from the centre, they were rarely, if ever, heard, but the sound assumed a rougher and more grating character during the strongest part of the shock. Within the isoseismal 8, 47 per cent. of the observers record a change in the nature of the sound. In the next two isoseismal zones, the percentages are 19 and 12. Outside the isoseismal 6, no such change was noticed.

In a large part of the sound-area, two sounds were heard, separated by an interval of complete silence of average duration 2·9 seconds, the louder sound attending the stronger part of the shock. Sometimes, at considerable distances, only one sound was heard while two series of vibrations were felt. As a rule, the single sound accompanied the stronger part of the shock.

Coseismal Lines and Velocity of the Earth-Waves. A series of co-seismal lines, corresponding to the times $5.32\frac{1}{2}$, $5.33\frac{1}{2}$ and $5.34\frac{1}{2}$ a.m., is shown in Fig. 52. In constructing them, each place of observation was indicated on a map by a mark corresponding to the particular minute recorded, only those observations being used in which the time was regarded as accurate to within half a minute. If all the records were correct, there would be a central area occupied by marks corresponding to 5.32, and this would be surrounded by zones in which the times were respectively 5.33, 5.34 and 5.35. The curves separating the marks of different zones would be coseismal lines corresponding to the times $5.32\frac{1}{2}$, $5.33\frac{1}{2}$ and $5.34\frac{1}{2}$. In practice, how-ever, successive zones intrude on each other, and the coseismal lines are drawn through the overlapping regions, special weight being given to the more accurate observations.

The average distance between the first and second coseismal lines is 32 miles, between the second and third $35\frac{1}{8}$ miles, and between the first and third $67\frac{1}{8}$ miles. Taking the latter figure, the mean surface velocity between the coseismals of $5.32\frac{1}{2}$ and $5.34\frac{1}{2}$ is 2955 feet per sec. or ·90 km. per sec.

Observations in Mines. The earthquake was observed in the follow-ing mines: Trafalgar near Mitcheldean (14 miles from the centre), Parkend and Whitecroft near Lydney (19 and 20 miles), Madeley near Ironbridge (42 miles), Wattstown in the Rhondda Valley (44 miles), and Chasetown near Walsall (54 miles). A rather strong shock was felt in the first-named pit, and a tremor in that of Whitecroft. In all cases, a sound was heard, which usually resembled that made by one or several loaded coal-waggons running down an incline. Thus (i) the earthquake was noticed in mines lying within a radius about one-third that of the isoseismal 4, and (ii) in mines more than 20 miles from the centre, the shock was imperceptible and the sound alone was observed.

Effects of the Earthquake on Men and Animals. A feeling of nausea, which as a rule soon passed off, was noticed from Cowarne (7½ miles

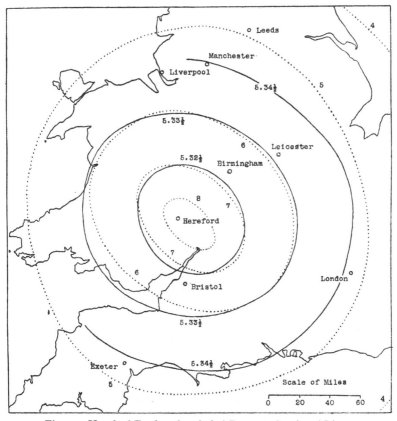

Fig. 52. Hereford Earthquake of 1896 Dec. 17: Coseismal Lines.

from the centre) to such distant places as Weymouth (100 miles), Walton-on-Thames (106 miles), Richmond in Surrey (110 miles), London (113 miles) and Bradford (125 miles).

The effects on animals were usually those of alarm at a sudden and unexpected disturbance. Cats noticed the shock to a distance of 22 miles, cattle to 34 miles, horses to 83 miles, and poultry to 91 miles.

Dogs barked at a distance of 109 miles, small birds were frightened at 108 miles, and pheasants crowed at 145 miles.

At Gloucester (21 miles), more than 30 dead sparrows and pigeons were picked up in the Great Western Railway goods yard; at Norton in Worcestershire (30 miles), small birds were found dead upon the ground. They were probably shaken off the branches on which they were roosting and struck the ground before waking.

Fig. 53. Hereford Earthquakes of 1853 Mar. 27, 1896 Dec. 17 and 1924 Jan. 26.

17. 1896 DEC. 17, *about* 5.40 *or* 5.45 *a.m.* Cat. No. 923; Intensity probably 3; No. of records 7 from 7 places.

A tremulous motion, accompanied by a slight rumbling noise, was felt at Dilwyn, Much Marcle, Putley and Putley Court (Hereford-shire), Bentham and Tortworth (Gloucestershire), and Great Malvern. The focus was probably situated in the interfocal region of the principal earthquake (no. 16).

18. 1896 DEC. 17, *about* 6.15 *a.m.* Cat. No. 924; Intensity probably 3; No. of records 10 from 10 places; Fig. 53.

The boundary of the disturbed area is represented by the broken-line in Fig. 53. It is 41 miles long, 27 miles wide, and contains an area of about 870 square miles. Its centre lies 16 miles S. 70° E. of Ross. The shock was a slight tremor, and was accompanied by a noise as of rushing wind. The earthquake probably originated in the Ross focus of the principal earthquake.

19. 1897 JULY 19, 3.49 *a.m.* Cat. No. 925; Intensity 4; No. of records 10 from 5 places.

The shock was felt at Bridstow (1 mile W. of Ross), Hereford, Ashleworth and Barnwood (Gloucestershire), and Great Malvern. It was a tremor like that produced in a house when a heavy traction-engine passes, and was accompanied by a rumbling sound like distant thunder or the passing of a train or waggon. The earthquake was probably connected with the Ross focus.

20. 1924 JAN. 26, *about* 6.10 *a.m.* Cat. No. 1191; Intensity 4; Centre of disturbed area in lat. 56° 6·3′ N., long. 2° 31·2′ W.; No. of records 13 from 11 places; Fig. 53.

The boundary of the disturbed area is represented by the right-hand dotted line in Fig. 53. It is 30 miles long, 15 miles wide, and contains about 350 square miles. Its centre lies about 6 miles south of Bromyard, and the direction of its longer axis is about N. 65° E. The tremor was accompanied by a sound like thunder or that of snow sliding from a roof and ending in a thud. As will be seen from Fig. 53, the axes of the areas disturbed by this earthquake and that of 1853 (no. 1) are nearly in a straight line which passes close to the north-west or Hereford epicentre of the twin earthquake of 1896 (no. 16).

ORIGIN OF THE HEREFORD EARTHQUAKES

The Hereford earthquakes seem to be connected with two faults. The more important is that in which the strong earthquakes of 1863, 1868 and 1896 originated. The elements of the fault are given by the earthquake of 1896, namely (i) the mean direction of the fault is N. 46° W.; (ii) its hade is to the N.E.; and (iii) the fault-line must pass a short distance on the south-west side of the centre of the iso-seismal 8. The two epicentres are also located by this earthquake, one being 3¼ miles E.S.E. of Hereford, the other 2 or 3 miles N.E. of Ross.

It is possible that, in the interfocal region, the hade of the fault may change, for the distribution of places where damage occurred to buildings seems to imply that the fault in the Ross focus hades to the S.W.

The greater part of the epicentral district is covered by a sheet of Old Red Sandstone, but, just to the north-east of the position suggested for the fault is the well-known Woolhope anticlinal, by which the Silurian beds are brought to the surface. The anticlinal axis runs approximately N.W. and S.E. or roughly parallel to the earthquake-fault. Moreover, a peculiar thinning-out and occasional disappearance of some of the Silurian beds on the south-west side (as compared with those on the north-east side) is suggestive of a N.W. and S.E. fault (hading to the N.E.) or rapid flexure at or near the junction of the Old Red Sandstone and Silurian strata. Again, a few miles to the south-east of the Woolhope anticlinal, and almost in the same line with it, there is a second anticlinal, that of May Hill. This is a triangular area and is bounded on all three sides by faults, of which that on the north-east side has an average N.W. and S.E. direction. The earthquake-fault probably passes between the two anticlinals, forming the south-west boundary of that of Woolhope and the north-east boundary of that of May Hill[1].

The second of the two faults referred to must pass nearly through Abbey Dore in the direction N. 65° E., that is, roughly at right angles to the other fault, and through, or nearly through, the Hereford epicentre.

It is important to notice that all the Hereford earthquakes were due to crust-movements along one system of folds—the twin earthquakes and their minor shocks to the growth of a N.E.–S.W. fold along a fault cutting it transversely, the simple earthquake or earthquakes to a slip or slips along a fault nearly parallel to this direction.

The first earthquake with which we are acquainted, that of 1853, was caused by a slip along the latter fault. Then, after 10½ years of rest, followed the strong earthquake of 1863 (no. 4)[2]. This evidently originated in the two foci of the main fault, the first slip occurring in the

---

[1] The position assigned to the earthquake-fault is a mile or two S.W. of the boundary of the Woolhope anticlinal. The foci, however, must have been very deep-seated, as is shown by the large disturbed areas and slight intensity of some of the minor shocks, especially nos. 7, 11 and 18; and the tangent-plane to the fault-surface in the neighbourhood of the foci might intersect the surface of the earth along a line at some distance from the fault-line.

[2] Preceded by two shocks of little consequence, the origins of which are unknown.

Hereford focus, the second and much stronger in the Ross focus. Five years of rest succeeded, and then, in 1868, a less important earthquake occurred, again, probably, in both foci but with the stronger impulse in the Ross focus. A slight earthquake, which may have originated in the strike-fault, brought this series to a close in 1871.

A quarter of a century later, a more important series of movements occurred, all of them probably confined to the main fault. The first slip (no. 7) seems to have taken place in the interfocal region. The next three apparently originated in or near the Ross focus, and were followed by a fifth (no. 11) in the same region as the first. Then came a series of four small movements that we cannot locate further than by saying that they were more closely connected with the Ross focus than with the other. These preliminary slips prepared the way for the principal earthquake by equalising the effective stress over a large area of the fault-surface. As so many of the preliminary slips occurred within or near the Ross focus, the effective stress in that part of the fault was slightly diminished; and this may be the reason why the first and greatest slip took place in the Hereford focus. The two great slips in the Hereford and Ross foci were, however, so nearly simultaneous that neither can be regarded as a consequence of the other. Both were due to a single generative effort—the growth of a fold along a transverse fault. The result of this double movement would be a sudden increase of stress along the margins of the interfocal region, and thus, after the lapse of ten minutes, there followed a simple movement of the median limb of the fold. Half an hour later, another small slip took place in the Ross focus, and by this, the equilibrium of the displaced masses was almost completely restored; for we have no certain evidence of any further movement until seven months had elapsed when, on 1897 July 19, there was a final slip in or about the same region of the fault. The movement which gave rise to the slight earthquake of 1924 Jan. 26 evidently took place in a continuation of the fault that was in action in 1853.

STAFFORD EARTHQUAKES

With one or two possible exceptions, the only earthquake connected with the Stafford foci is the important twin earthquake of 1916 Jan. 14. The eastern epicentre lies about 2 miles N.E. of Stafford, and the western epicentre about 1½ miles N.W. of Eccleshall.

1. **1916** JAN. 14, 7.29 *p.m.* Cat. No. 1179; Intensity 7; Centre of isoseismal 7 in lat. 52° 52' N., long. 2° 9' W.; No. of records 1241 from 580 places; Figs. 54, 55 (*Geol. Mag.* vol. 6, 1919, pp. 302–312).

Isoseismal Lines and Disturbed Area. On the map of the earthquake (Fig. 54) are shown five isoseismal lines of intensities 7 to 3, the two inner isoseismals being given on a larger scale in Fig. 55. The isoseismal 7 is 12½ miles long, 9 miles wide, and 88 square miles in area. Its centre is 2 miles S.S.W. of Stone, and its longer axis is directed N. 68° W. The isoseismal 6 is 32½ miles long, 25 miles wide, and

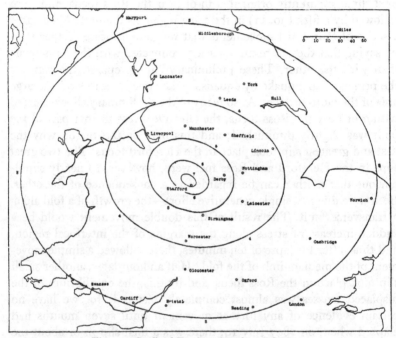

Fig. 54. Stafford Earthquake of 1916 Jan. 14.

contains 637 square miles. Its longer axis is directed N. 62° W., and its distances from the isoseismal 7 are 9½ miles on the north side and 5 miles on the south. The isoseismal 5 is 68 miles long, 55 miles wide, and 2908 square miles in area; its distances from the isoseismal 6 are 18 miles on the north side and 12½ miles on the south. The isoseismal 4 is 144 miles long, 112 miles wide, and 12,500 square miles in area; its distances from the isoseismal 5 are 32 miles on the north side and 12 miles on the south. The isoseismal 3, which bounds the disturbed area, is 281 miles long, 225 miles wide, and contains 50,200 square miles.

Nature of the Shock. The following accounts may be regarded as typical of the nature of the shock in different parts of the disturbed area (see Fig. 55): At Chebsey, the shock consisted of one series of vibrations accompanied by sound which increased in strength to a maximum and then died away. At Coton Hill (near Stone), the shock also consisted of one continuous series of vibrations, lasting about

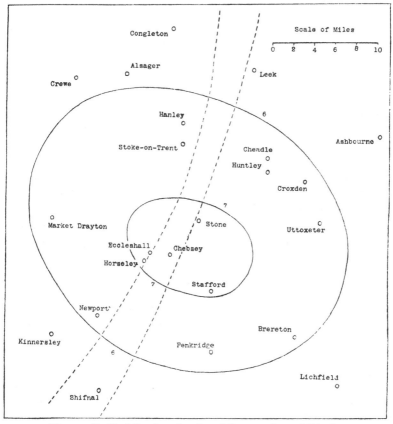

Fig. 55. Stafford Earthquake of 1916 Jan. 14: Central District.

6 seconds, increasing in strength from the beginning and dying away towards the end; in the middle there were two maxima separated by half a second. At Brereton, the shock was in two distinct parts, the first being the stronger, and the interval between the two series being 2 or 3 seconds. At Kinnersley, the shock was again in two parts, separated by an interval of 2 seconds, the second part being the stronger.

Over the greater part of the disturbed area, and in some directions even as far as the isoseismal 3, the shock consisted of two distinct parts. The two parts, however, coalesced along the synkinetic band (represented by broken-lines in Fig. 55), which crosses at right angles the axis of the isoseismal 7 and at a short distance to the west of the centre. The band is very slightly curved, the concavity facing the west, showing that the impulse in the eastern focus occurred before, but very slightly before, that in the other. That the two impulses were very nearly equal in strength is clear from the wide observation of the double shock. On the east side of the synkinetic band, however, 56 per cent. of the observers regarded the first part as the stronger; on the west side, 56 per cent. regarded the second part as the stronger. Thus, the impulse in the eastern focus must have been slightly the stronger.

The mean duration of the shock was 3·1 seconds over the whole disturbed area, 4·0 seconds within the isoseismal 7, and 4·0, 3·2, 2·4 and 2·5 seconds in the successive isoseismal zones. The mean duration of the interval between the two parts was 2·2 seconds over the whole disturbed area, 2·0 seconds within the isoseismal 7, and 2·1, 2·4, 2·4 and 1·1 seconds in the successive isoseismal zones.

Sound-Phenomena. The boundary of the sound-area lies between the isoseismals 4 and 3, and is 177 miles long, 157 miles wide, and contains about 21,000 square miles. The sound was heard by 77 per cent. of the observers throughout the whole area, by 98 per cent. within the isoseismal 7, by 96, 83 and 57 per cent. in the successive isoseismal zones and by 31 per cent. outside the isoseismal 4.

The sound was compared to passing waggons, etc., in 53 per cent. of the records, thunder in 7, wind in 9, loads of stones falling in 8, the fall of a heavy body in 9, explosions in 12, and miscellaneous sounds in 2, per cent.

2. **1916** JAN. 15, 10.45 *a.m.* Cat. No. 1180; Intensity 4; No. of records 7 from 7 places; Fig. 55.

A slight shock was felt at Alsager, Cheadle, Coton Hill (near Stone), Croxden, Horseley, Huntley and Stone. With one exception, these places lie within the isoseismal 6 of the preceding earthquake. Alsager lies 1 mile to the north. The epicentre is probably not far from the centre of the isoseismal referred to, and therefore between the two epicentres of the principal earthquake. The shock was a slight one and consisted of a single series of vibrations.

ORIGIN OF THE STAFFORD EARTHQUAKES

From the seismic evidence, we obtain the following elements of the originating fault: (i) its mean direction lies between N. 68° W. and N. 62° W., or about N. 65° W.; (ii) its hade is to the N.; and (iii) the fault-line must pass a short distance to the south of the centre of the isoseismal 7. Close to Eccleshall, there is a small post-Triassic fault which is parallel to the longer axis of the isoseismal 7 and hades to the N. If this fault were continued to the east, it would occupy the position assigned to the fault-line by the seismic evidence.

Judging from the form of the isoseismal 7 and the course of the synkinetic band (which must pass between the epicentres), it follows that the eastern epicentre must lie about 2 miles N.E. of Stafford, and the western about $1\frac{1}{2}$ miles N.W. of Eccleshall, the distance between them being thus about 8 or 9 miles.

In 1678, and again in 1863, there may have been small slips within one or other focus. With these possible exceptions, there has been no perceptible movement along the earthquake-fault for several centuries. Then suddenly came the twin-earthquake of 1916, the movements in the two foci being practically simultaneous, though that in the eastern focus was the earlier by a second or less than a second. In little more than 15 hours later, the twin movement was followed by a simple displacement within the interfocal region of the fault.

DERBY EARTHQUAKES

The most important seismic centres in Derbyshire are those with which the twin earthquakes of 1903, 1904 and 1906 were connected. One of these centres lies about $1\frac{1}{2}$ miles E. of Ashbourne, the other about 3 miles W. of Wirksworth. Of the three earthquakes recorded in the county before 1903 (see last chapter), none can be attributed with confidence to these centres.

1. **1903** MAR. 24, 1.30 *p.m.* Cat. No. 963; Intensity 7 (nearly 8); Centre of isoseismal 7 in lat. 53° 3·1′ N., long. 1° 41·5′ W.; No. of records 1136 from 528 places, 63 negative records from 56 places, and 48 records from 32 mines; Figs. 56, 57 (*Quart. Journ. Geol. Soc.* vol. 60, 1904, pp. 215–232).

Isoseismal Lines and Disturbed Area. On the map of the earthquake (Fig. 56) are shown five isoseismal lines. The innermost, corresponding to an intensity of nearly 8, includes places at which slight damage is known to have occurred to buildings. In none was the injury more serious than the cracking of a poorly-built wall, or the overthrow of a few chimney-pots. The curve is 16½ miles long, 8½ miles wide, and contains 112 square miles. The centre of the curve coincides with the village of Kniveton, about 3 miles N.E. of Ashbourne, and its longer axis is directed N. 32½° E. The isoseismal 7 is 23½ miles long, 17½ miles wide, and 272 square miles in area, with its longer axis directed N. 33° E. The isoseismal 6 is 48 miles long, 36 miles wide, and 1348

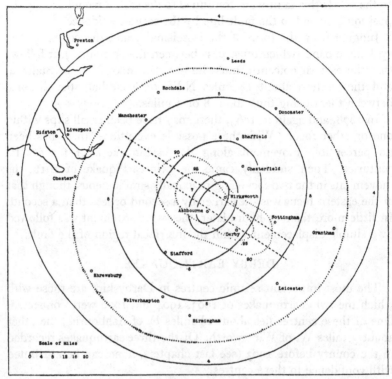

Fig. 56. Derby Earthquake of 1903 Mar. 24.

square miles in area. Its longer axis is parallel to that of the isoseismal 7, the distance between the curves being 12 miles on the north-west side and 8 miles on the south-east side. The isoseismal 5 is 76 miles long, 69 miles wide, and contains 4060 square miles, its distance from the isoseismal 6 being 17 miles towards the north-west and 16 miles towards the south-east. Still more nearly circular is the isoseismal 4, which is 129 miles from N.W. to S.E., 126 miles wide, and 12,000 square miles in area.

The shock was also felt at a few places outside this isoseismal—at Settle, Aysgarth, Richmond and Easby, which are respectively $7\frac{1}{2}$, 21, 27 and 27 miles N. of the isoseismal, and at Boston, 12 miles to the east. If the disturbed area were bounded by a circle concentric with the isoseismal 4 and passing through Richmond and Easby, it would contain 25,000 square miles.

Nature of the Shock. The following accounts are given to illustrate the twin character of the earthquake and its variation throughout the disturbed area. Ashbourne and Darley Dale are close to the longer axis of the inner isoseismals, the former being very near the epicentre; Duffield lies on the continuation of the shorter axis of the same curves; while Quarndon and Derby are respectively $1\frac{1}{2}$ and 3 miles from it.

At Ashbourne, two distinct shocks were felt, the first being slightly the stronger, the impression produced by both shock and sound being that a heavy piece of furniture was rapidly rolled in the room upstairs from E. to W., and, then, after a pause of a second or two, rolled a short way back again. At Darley Dale, there were also two parts, each of which began with a low distant rumbling like the rushing of a strong wind, and culminated in a violent shock as it passed underneath the house, the second part being the stronger. At Derby, there were again two shocks, each lasting 3 seconds; separated by an interval of $\frac{1}{2}$ a second. At Quarndon, a rumbling sound was first heard; then came a violent shock; the rumbling continued for about 2 seconds, and, before it ceased, a second shock was felt, but not so strong as the first, the rumbling gradually dying away. At Duffield, only one shock was observed, and that during the loudest part of the sound.

With some exceptions, such as that last mentioned, the double shock was observed in every part of the disturbed area. Towards the north, it was clearly perceptible at Preston, Lytham, Aysgarth, Settle, Richmond and Doncaster; towards the east at Grantham, Eagle and Boston; towards the south at Barnt Green, Mere Hall and Hagley; and towards the west at Shrewsbury and Vicar's Cross (near Chester). Throughout the whole disturbed area, the double shock is distinctly recorded by 68 per cent. of the observers. In some parts, this percentage rises to 80; in no large area does it fall below 48.

The single shock was felt chiefly within a narrow rectilinear band (the synkinetic band), about 5 miles wide, running centrally across the inner isoseismals in the direction N. 56° W., that is, at right angles to the longer axes of the isoseismals. In the map (Fig. 56), the boundaries of this band are represented by broken-lines. Outside the band, the interval between the two parts of the shock was one of rest and quiet, its average duration over the whole disturbed area being 3 seconds. Close to the synkinetic band, as at Derby, the interval was much shorter, though still distinct; within the band, the shock appeared continuous, the ends of the two parts overlapping, with two maxima of intensity at places (like Quarndon) near one of the boundaries, and only one maximum at places (like Duffield) near the central axis of the band.

From the fact that the double shock was noticed at places near the boundary of the disturbed area, it is evident that the two impulses were of nearly equal strength. Dividing the whole area into two portions by the synkinetic band, 60 per cent. of the observers on the north-east side, and 63 per cent. of those on the south-west side, regarded the first part as the stronger.

The following conclusions may be drawn from the above account: (i) the double shock owed its origin to two distinct impulses of nearly equal strength; (ii) the impulses originated in two detached foci arranged along a line parallel, or nearly parallel, to the axes of the iso-seismal curves; and (iii) as is shown by the rectilinearity of the synkinetic band, the two impulses occurred at exactly the same instant.

From the form of the isoseismal 7, it is probable that one epicentre was situated near Ashbourne and the other 3 miles W. of Wirksworth, their centres being therefore about 8 or 9 miles apart.

Seismographic Records and Velocity of the Earth-Waves. Records of the earthquake were given by an Omori horizontal pendulum at Birmingham, by a Milne seismograph at Bidston (near Birkenhead), and by a Wiechert pendulum at Göttingen.

The first of these is reproduced in Fig. 57. Birmingham lies 41 miles S. 11° W. of the centre of the isoseismal 7. The instrument is erected in the N.–S. plane, and the movements of the ground are magnified by it 13·7 times. The first movement (on the right), an abrupt displacement to the east, occurred at 1h. 30m. 19s. p.m., G.M.T. The diagram is chiefly remarkable for the two prominent displacements to the west[1], which occurred at 1h. 30m. 23s. and 1h. 30m. 28s., and which no doubt correspond to the two parts of the shock so widely observed. In each case, the period of the vibration was 0·8 second. At 1h. 30m. 31s., another oscillation

---

[1] The larger curve, on which the vibrations are superposed, was due to the oscillation of the undamped pendulum.

of some importance took place, followed by a series of 13 ripples with an average period of 0·84 second. These are all that are shown in Fig. 57, but the original record continues with a series of 79 still smaller ripples, with an average period of 1·03 seconds, the last visible (under the microscope) occurring at 1h. 32m. 3s. The total duration of the disturbance, as registered at Birmingham, was 1m. 44s. The range of motion of the ground from E. to W. was ·078 mm. during the first prominent displacement and ·075 mm. during the second. Birmingham, however, lies S. 11° W. from the epicentre, and thus the total displacements must have been 0·41 and 0·39 mm., respectively. These, with a period of 0·8 second, correspond to maximum

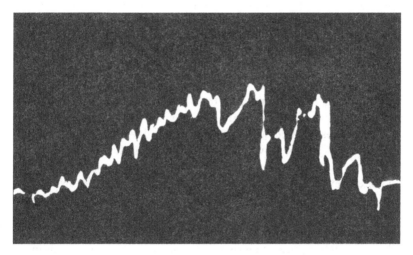

Fig. 57. Seismographic Record at Birmingham of the Derby Earthquake
of 1903 Mar. 24.

accelerations of 12·6 and 12·0 mm. per second per second, showing how nearly equal in strength the two parts of the shock were.

Bidston lies 65 miles W.N.W. of the epicentre and 8½ miles S.W. of the synkinetic band. The time of the first displacement was 1h. 30m. 44s. The record gives evidence of but one impulse, which moved the pendulum towards the west. The maximum amplitude of oscillation was about 0·7 mm. Bidston being close to the synkinetic band, the interval between the two prominent vibrations was too short to allow of their separate photographic registration.

At the time of the earthquake, rather strong pulsations were being registered by the Wiechert pendulum at Göttingen; and, on this account, the measurement of the epochs may err by as much as 5 seconds on either side of the time given. The preliminary tremors began at 1h. 33m. 32s. p.m., G.M.T.; their period was about 1 second and their amplitude about ·0001 mm. They were succeeded by a series of larger waves, beginning at 1h. 34m. 20s., with a period of between 2 and 3 seconds and an amplitude of ·0007 mm. The total duration of the movement was about 1¾ minutes.

The distances of Birmingham and Göttingen from the epicentre are 66 kms. and 808 kms. respectively. The interval between the arrival of the first vibrations at these places was 193 seconds. Thus, the preliminary tremors travelled with a velocity of 3·8 kms. per second.

Sound-Phenomena. On the map (Fig. 56), two isacoustic lines (corresponding to percentages of 95 and 90) are indicated by dotted lines. The inner line is 33 miles long and 16 miles wide, and the outer 49 miles long and 19 miles wide. The main feature of both curves is their expansion in the direction of the synkinetic band due to the coalescence of sound-vibrations from the two foci. The boundary of the sound-area is represented by the outer dotted line in Fig. 56, the area being 101 miles long in the direction of the longer axes of the isoseismal lines, 98 miles wide, and containing about 7800 square miles, or nearly two-thirds of the whole disturbed area. Within the isoseismal 7, 97 per cent. of the observers heard the earthquake-sound; and, in the zones between successive isoseismals, the percentages are 89, 80 and 65 (the last zone being limited by the boundary of the sound-area).

The sound was compared to passing waggons, etc., in 53 per cent. of the records, thunder in 21, wind in 5, loads of stones falling in 8, the fall of a heavy body in 4, explosions in 7, and miscellaneous sounds in 3, per cent.

Observations in Mines. Most of the mines are situated between two lines running east and north-east from the centre of the isoseismal 7. Towards the south-west, the earthquake passed unnoticed in the pits of Cannock Chase. Towards the west, it was perceived in mines as far as Bucknall near Stoke-on-Trent (19 miles from the centre); towards the north-west at Monsal Wade near Buxton (18 miles, 117 yards deep); towards the north-east at Eckington (22 miles); towards the east at Hucknall Torkard (20 miles, 500 yards deep) and Bulwell (20 miles, 300 yards deep); and towards the south at Swadlincote (20 miles, 470 yards deep).

The general impression produced by the earthquake was that an explosion or fall of rock had occurred in some distant part of the mine

In the pits at Clay Cross and Morton (between Alfreton and Chesterfield) both parts of the shock were felt, the first part being the stronger and, at Clay Cross, accompanied by the louder noise. In three pits, at Glapwell, Pilsley and Swancote (all in the Alfreton district), the shock was strong enough to detach small pieces of shale from the roof. The sound differed slightly from that observed on the surface, in being less intermittent and more monotonous. The observations lead to the following conclusions: (i) the shock, as a rule, was not felt in the more distant mines; (ii) in several cases, the sound appeared to be overhead rather than below; for instance, at Clay Cross, Manners colliery (near Ilkeston), Pilsley and Swadlincote, the sound was like that of a train passing overhead; (iii) at Swadlincote (near Burton-on-Trent), the sound was not heard at all at a depth of 220 yards; in a seam 350 yards from the surface, it was heard by some men but attracted little notice; at a depth of 400 yards, the men were alarmed by a rumbling noise passing overhead, like a railway-train going over a wooden bridge; and, lastly, in a seam 470 yards deep, a heavy rumbling noise was heard; thus, the intensity of the sound seems to have increased with the depth.

2. **1903** MAR. 24, *about* 1.45 *p.m.* Cat. No. 964.

A slight tremor was felt at Abbotshulme (near Rocester), Bakewell and Tissington, and at Abbotshulme was accompanied by a rumbling sound.

3. **1903** MAR. 24, *about* 5 *p.m.* Cat. No. 965.

A slight shock was felt at Brailsford, Fenny Bentley, and Onecote (near Leek).

4. **1903** MAY 3, 9.22 *p.m.* Cat. No. 966; Intensity 5; Centre of isoseismal 4 in lat. 53° 2·4′ N., long. 1° 39·9′ W.; No. of records 73 from 52 places and 35 negative records from 30 places; Fig. 58.

On the map of this earthquake, the isoseismals of intensities 5, 4 and 3 are represented by continuous lines. The isoseismal of intensity nearly 8 of the principal earthquake of Mar. 24 (no. 1) is shown by the broken-line. The isosesimal 5 is 10¾ miles long, 5¼ miles wide, and about 44 square miles in area. The isoseismal 4 is 19 miles long, 12 miles wide, and contains about 179 square miles. The centre of this curve is about 1 mile S.W. of Hognaston, and the longer axis is directed N. 25° E. The isoseismal 3, which bounds the disturbed area, is 31 miles long, 24 miles wide, and about 585 square miles in area. The distances between the isoseismals 5 and 4, and 4 and 3, are respectively 3·7 and 6·4 miles on the north-west side, and 3·0 and 5·0 miles on the south-east side.

The shock is usually described as a sudden shiver or brief tremor, and in all places it consisted of only one part.

The boundary of the sound-area is indicated by the dotted line. It is 24 miles long, 17 miles wide, and contains about 320 square miles.

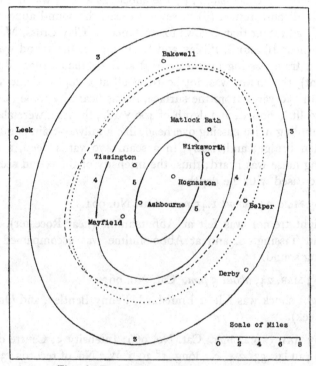

Fig. 58. Derby Earthquake of 1903 May 3.

The sound was heard by 92 per cent. of all the observers. It was compared to passing waggons, etc., in 45 per cent. of the records, thunder in 39, wind in 6, loads of stones falling in 3, explosions in 3, and miscellaneous sounds in 3, per cent.

5. 1904 JULY 3, 2.28 p.m. Cat. No. 979; Intensity 3; No. of records 5 from 5 places (*Quart. Journ. Geol. Soc.* vol. 61, 1905, pp. 8–17).

A slight quiver was felt at Ambergate, Cromford, Matlock Bath, Mayfield and Wirksworth (see Fig. 60). At Ambergate, a noise like a loud peal of thunder accompanied the tremor. As four of the five places mentioned are close to the north-east of the Wirksworth epicentre of the principal earthquakes, it is probable that the shock originated in the corresponding focus.

6. **1904 JULY 3, 3.21 p.m.** Cat. No. 980; Intensity 7; Centre of isoseismal 7 in lat. 53° 0·4′ N., long. 1° 41·6′ W.; No. of records 1467 from 653 places and 46 negative records from 44 places; Fig. 59.

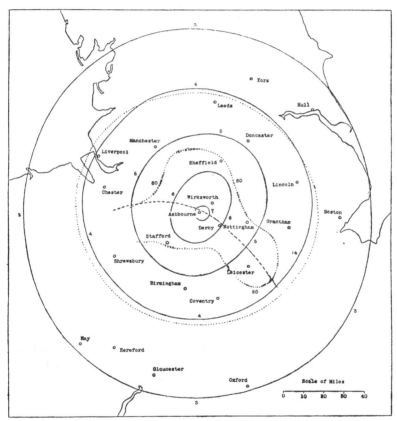

Fig. 59. Derby Earthquake of 1904 July 3, 3.21 p.m.

Isoseismal Lines and Disturbed Area. The five isoseismal lines, represented by continuous lines in Fig. 59, correspond to intensities 7 to 3. The isoseismal 7 is approximately circular in form, 7¼ miles in diameter, and 41 square miles in area. Its centre lies about 1½ miles E. of Ashbourne. The next isoseismal, of intensity 6, is 37 miles long, 27 miles wide, and 804 square miles in area. The direction of its longer axis is N. 31° E., or very nearly parallel to that of the isoseismal 7 of the principal earthquake of 1903 (no. 1). The distance between the isoseismals 7 and 6 is 12 miles on the north-west side and 8 miles on the south-east side. The most important feature of the two curves is, however, their excentricity, the distance between their centres in the direction of the longer axis being about 2 miles. In the neighbourhood of Wirksworth and Matlock Bath, the intensity was also 7 or nearly 7. The observations are insufficient to draw a second isoseismal 7 in that district; but it must clearly be smaller than the other and their centres must be separated by a distance of 6 or 7 miles. Thus, the two epicentres are approximately coincident with those of the earthquake of 1903.

The isoseismal 5 is 72 miles long, 65 miles wide, and contains 3600 square miles; its distance from the isoseismal 6 is 21 miles on the north-west side and 17 miles on the south-east side. The next two isoseismals are very nearly circular, the isoseismal 4 being 114 miles long, 113 miles wide, and about 10,120 square miles in area; the isoseismal 3, which forms the boundary of the disturbed area, is 181 miles long, 179 miles wide, and contains about 25,000 square miles.

Comparing the dimensions above given with those for the earthquake of 1903 (no. 1), we see that the isoseismals 7 to 4 of the latter are all larger than those of the earthquake of 1904. In 1903, a few buildings were slightly injured over an area of 112 square miles, while, in 1904, there was practically no damage. On the other hand, the earthquake of 1904, owing to its occurrence on a Sunday afternoon, was traced to a much greater distance, its disturbed area being nearly double that of the earthquake of 1903.

Nature of the Shock. As in 1903, this earthquake was undoubtedly a twin earthquake, though only one out of every five observers records the existence of two parts or two maxima. This small proportion is due, partly to the inequality in strength of the two parts, partly to their incomplete separation, an intermediate tremor being observed within the central area. Thus, at Ashbourne (1½ miles from the centre of the isoseismal 7), the shock consisted of a single series of vibrations, which increased in strength to a maximum and then died away; at Birmingham (37 miles), two series of vibrations were felt, the first being distinctly the stronger. The intermediate tremor was sensible as far as Farnsfield (27 miles), while the twin-shock was felt to the north at Bradford (54 miles), to the west at Ellesmere (50 miles), to the south at Stourport (52 miles), and to the east at Hough (44 miles), or over a district of roughly 8000 square miles in area.

This district is, however, traversed by a synkinetic band, the boundaries of which cannot be traced with accuracy; but the course of its median line is shown by the broken-line in Fig. 59. The synkinetic band is thus hyperbolic in form, the concavity facing the southwest. The width of the band increases towards its extremities, being 15 or 16 miles at a distance of 30 miles from the centre.

Outside the synkinetic band, the duration of the interval between the two parts varied from 1 to 4 seconds. The average over the whole area is 2·1 seconds, and this average is practically uniform throughout, being 2·2 seconds within the isoseismals 6, 2·0 seconds between the isoseismals 6 and 5, and 2·1 seconds outside the latter curve.

The positions of the two epicentres can be determined with greater accuracy than in the earthquake of 1903. The south-western epicentre must be near the centre of the isoseismal 7, and the other near Wirksworth or Matlock Bath. These positions are supported by numerous observations on the direction of the shock in Derby and Nottingham. These places lie on the south-west and north-east sides respectively of the synkinetic band, and the observations of the direction at each must refer as a rule to the movement from the nearer focus. At Derby, the mean direction was N. 53° W. or exactly in a line with the centre of the isoseismal 7; at Nottingham, it was N. 51° W., or along a line passing through a point about 3 miles W. of Wirksworth.

The impulse at the south-western focus was clearly the stronger. Moreover, although the two impulses were nearly simultaneous, the curvature of the synkinetic band shows that the north-eastern focus was in action a short time, perhaps a second or less, before the other.

Sound-Phenomena. The boundary of the sound-area, which differs but slightly from the isoseismal 4, is represented by the outer dotted line in Fig. 59. It is 121 miles long from N.W. to S.E., 113 miles wide, and includes about 10,700 square miles. The inner dotted line represents an isacoustic line of audibility-percentage 80. It is incomplete towards the west, and evidently extends some distance farther in that direction. The isacoustic line is thus distorted in both directions along the synkinetic band. Within the isoseismal 7, 94 per cent. of the observers heard the sound, and the percentages were 93,

79 and 56 within the successive isoseismal zones, and 38 between the isoseismal 4 and the boundary of the sound-area.

The sound was compared to passing waggons, etc., in 45 per cent. of the records, thunder in 26, wind in 15, loads of stones falling in 5, the fall of a heavy body in 4, explosions in 2, and miscellaneous sounds in 2, per cent.

7. **1904** JULY 3, 11.8 *p.m.* Cat. No. 981; Intensity 4; Centre of isoseismal 4 in lat. 53° 2·8′ N., long. 1° 39·5′ W.; No. of records 76 from 42 places and 2 negative records from 2 places; Fig. 60.

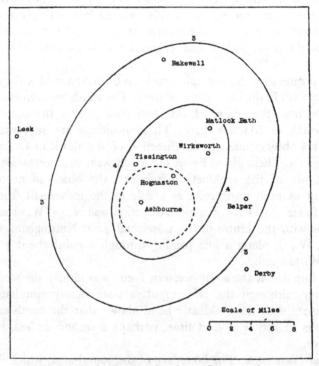

Fig. 60. Derby Earthquake of 1904 July 3, 11.8 p.m.

In the map of this after-shock, the continuous lines represent the isoseismals 4 and 3, and the broken-line the isoseismal 7 of the preceding earthquake. The isoseismal 4 is 16 miles long, 10 miles wide and contains 125 square miles. Its centre is about ½ mile S.W. of Hognaston, and the direction of its longer axis N. 27° E. The isoseismal 3 is 27 miles long, 20 miles wide, and contains 427 square miles. The distance between the two isoseismals is 6 miles on the north-west side and 4½ miles on the south-east side. The shock and sound were also observed, though very faintly, at Derby (1½ miles S.E. of the isoseismal 3) and at Leek (4 miles W.).

The shock is uniformly described as a continuous series of rapid vibrations, of mean duration 3·5 seconds. The sound was heard by 96 per cent. of the observers, and was compared to passing waggons, etc., in 28 per cent. of the records, thunder in 39, wind in 11, loads of stones falling in 6, and explosions in 17, per cent.

8. **1906** AUG. 27, 5.56 *a.m.* Cat. No. 1004; Intensity 5; Centre of isoseismal 5 in lat. 53° 0·8′ N., long. 1° 42·3′ W.; No. of records 131 from 79 places and 85 negative records from 79 places; Fig. 61 (*Geol. Mag.* vol. 5, 1908, pp. 301–303).

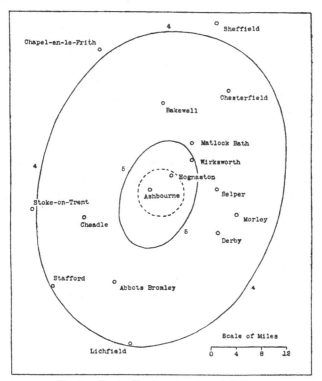

Fig. 61. Derby Earthquake of 1906 Aug. 27.

The continuous lines on the map represent isoseismals of intensities 5 and 4, and the broken-line the isoseismal 7 of the principal earthquake of 1904 (no. 6). The isoseismal 5 is 17 miles long, 11 miles

18–2

wide, and 147 square miles in area. Its centre lies 1 mile S.E. of
Ashbourne, and the longer axis is directed N. 25° E. The isoseismal 4
is 48 miles long, 36 miles wide and contains about 1360 square miles.
The distance between the isoseismals 5 and 4 both to the north-west
and south-east is 13 miles. The shock was also felt at eight places
outside the isoseismal 4, at distances from it between 1½ and 5½ miles.
The whole disturbed area must therefore contain about 2100 square
miles.

To the great majority of the observers, the shock appeared to consist of a single series of vibrations, with a mean duration of 2·7
seconds. According to eleven observers, however, the shock consisted of two distinct parts, separated by an interval of mean duration
2·6 seconds, one part being so much weaker than the other that it
usually escaped notice. The first part was the stronger at Breaston,
Clifton, Edlaston and Morley, and the second at Bradbourne and Kirk
Ireton. Now, Morley lies on the minor axis of the isoseismals;
Breaston, Clifton and Edlaston to the south of that axis; and Bradbourne and Kirk Ireton on the north side. Thus, there must have
been a second focus to the north of that near Ashbourne, and the synkinetic band must have been curved with its concavity to the north-
east, that is, the stronger impulse in the south-western focus was also
the first in action. The evidence is not sufficient to determine the
position of the north-eastern epicentre. The mean interval between
the two parts was, however, nearly the same as in 1904, and it is
therefore probable that the north-eastern focus coincided with that
of 1903 and 1904, lying about 3 miles W. of Wirksworth.

Towards the east, the sound-area extends as far as the isoseismal 4,
but in other directions the boundary is unknown. The sound was
heard by 88 per cent. of the observers within the isoseismal 5, by
64 per cent. between the isoseismals 5 and 4, and by 71 per cent. of
all the observers. It was compared to passing waggons, etc., in 61 per
cent. of the records, thunder in 13, wind in 4, loads of stones falling
in 10, the fall of a heavy body in 9, and explosions in 4, per cent.

ORIGIN OF THE DERBY EARTHQUAKES

Of the eight Derby earthquakes of the years 1903–1906, the three
stronger were twin earthquakes. In all probability, they originated
in the same twin-foci, one epicentre being 1 mile E. of Ashbourne,
the other 3 miles W. of Wirksworth. The distance between the epicentres is 7 or 8 miles.

The three twin earthquakes showed a continual decrease in strength, the intensity of the first being 7 (nearly 8), of the second 7, and of the third 5, and the areas included within the isoseismal 4 respectively 12,000, about 10,120, and about 1360, square miles. Moreover, the twin earthquake of 1903 was attended by 3 minor shocks, that of 1904 by 2, and that of 1906 by none. Each of the earlier twin earthquakes was followed by a simple earthquake originating in the interfocal region of the fault, the first after 40 days, the second after 8 hours. The earlier after-shock was of intensity 5 and the later of intensity 4, the areas within the isoseismal 4 being 179 square miles in 1903 and 125 square miles in 1904.

The longer axes of the inner isoseismals of the principal earthquakes were directed N. 33° E. in 1903 and N. 31° E. in 1904; those for the interfocal after-shocks were directed N. 25° E. in 1903 and N. 27° E. in 1904. In each of the three twin earthquakes, the hade of the fault within the south-western focus is to the N.W. Now, if the hade within the north-eastern focus be towards the S.E., the greater expansion of the isoseismals for each focus on the side towards which the fault hades would cause a displacement of the north end of the longer axis of the compound isoseismals towards the east. Thus, the divergence between the isoseismal axes of the principal earthquakes of 1903 and 1904 and of the other three earthquakes points to a change of hade in the interfocal region of the fault, the true mean direction of the originating fault being about N. 26° E. This agrees almost exactly with that deduced from the earthquake of 1906, namely, N. 25° E., the course of the isoseismals of this earthquake being unaffected by the weak impulse within the north-eastern focus.

In each of the three twin earthquakes, the impulse in the Ashbourne focus was the stronger. In 1903, the impulses occurred simultaneously; in 1904 the Wirksworth focus, and in 1906 the Ashbourne focus, was the first in action. The twin movement in 1903 was followed by two extremely slight slips, the first probably in the Wirksworth focus, the second in the Ashbourne focus, and these by a more important slip in the interfocal region. The twin movement in 1904 was preceded by a very slight slip in the Wirksworth focus, and followed by a more important slip in the interfocal region. The twin movement in 1906 was too small to affect materially the stresses in the neighbouring regions of the fault.

LEICESTER EARTHQUAKES

The Leicester earthquakes are associated with a pair of connected foci. One epicentre lies about 2 miles S.S.W. of Loughborough or about 1 mile N.W. of Woodhouse Eaves; the other is close to the village of Tugby, which lies 17 miles S. 56° E. of the former epicentre.

1. **1839** DEC. 23, *about* 4 *p.m.* Cat. No. 410 (*Gent. Mag.* vol. 11, 1839, p. 198).

A smart shock at Woodhouse Eaves, preceded by a loud rumbling noise as of a waggon passing. The shock was strong enough to loosen pieces of mortar from the roof of the church.

2. **1893** AUG. 4, 6.41 *p.m.* Cat. No. 892; Intensity 5; Centre of isoseismal 5 in lat. 52° 44·6' N., long. 1° 13·8' W.; No. of records 391 from 298 places and 103 negative records from 97 places; Fig. 62 (*Roy. Soc. Proc.* vol. 57, 1894, pp. 87–95; *Quart. Journ. Geol. Soc.* vol. 61, 1905, pp. 1–3).

The continuous lines on the map represent isoseismals of intensities 5, 4 and 3. The isoseismal 5 is 18 miles long, 11¼ miles wide, and 161 square miles in area. Its longer axis is directed N. 60° W., and its centre lies 2 miles S. 20° W. of Loughborough. The isoseismal 4 is 46 miles long, 32 miles wide, and contains 1170 square miles. Its longer axis is directed N. 50° W., and its distances from the isoseismal 5 are 8 miles on the north-east side and 5 miles on the south-west. The isoseismal 3, which forms the boundary of the disturbed area, is 59 miles long, 47 miles wide, and contains 2200 square miles. While, however, the two latter curves are normal as regards their relative position, the isoseismal 5 is excentric with regard to both, the distance of its centre from that of the isoseismal 4 being 8½ miles. This excentricity and the intensity of the shock at various places in the south-eastern portion of the isoseismal 4 imply the existence of a second focus in the neighbourhood of Tugby, about 17 miles S. 56° E. of the former.

Over a large part of the disturbed area—ranging from Burton-on-Trent to Ketton and from Nottingham to Burbage—the shock consisted of two distinct parts separated by an interval of mean duration 2·5 seconds. The first part was the stronger at Borrowash, Burton-on-Trent and Uppingham, while the two parts were equally strong at Ketton. The evidence derived from the nature of the shock is incomplete, but it would seem that the impulse within the north-western focus was the stronger and preceded that at the other by not more than 1 or 2 seconds.

The sound was heard by 98 per cent. of the observers within the isoseismal 5 and by 91 per cent. of all the observers within the disturbed area. It was compared to passing waggons, etc., in 51 per

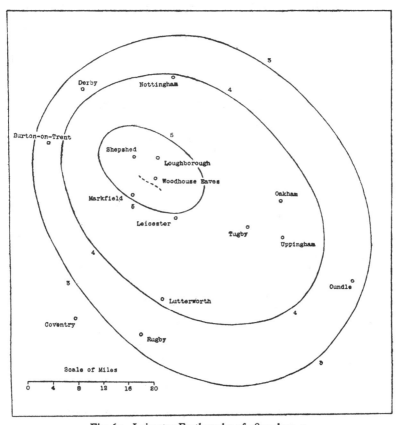

Fig. 62. Leicester Earthquake of 1893 Aug. 4.

cent. of the records, thunder in 34, loads of stones falling in 2, the fall of a heavy body in 2, explosions in 9, and miscellaneous sounds in 2, per cent.

3. 1904 JUNE 21, *about* 3.30 *a.m.* Cat. No. 977; Intensity 3; No. of records 2 from 2 places.

A slight quiver was felt at Groby and Markfield, both of which are close to the north-western epicentre of the earthquake of 1893.

4. **1904 JUNE 21, 5.28 *a.m.*** Cat. No. 978; Intensity 5; Centre of isoseismal 5 in lat. 52° 35·2′ N., long. 0° 59·5′ W.; No. of records 249 from 130 places and 56 negative records from 44 places; Fig. 63 (*Quart. Journ. Geol. Soc.* vol. 61, 1905, pp. 3–7).

The isoseismals of intensities 5 and 4 are represented by the continuous lines in Fig. 63. The isoseismal 5 is 23 miles long, 17 miles

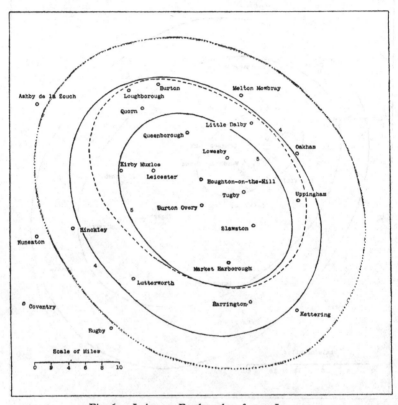

Fig. 63. Leicester Earthquake of 1904 June 21.

wide, and 314 square miles in area. Its centre is about ½ mile N.E. of Burton Overy, and its longer axis is directed N. 42° W. The isoseismal 4 is 33½ miles long, 26 miles wide, and 681 square miles in area. Its distance from the isoseismal 5 is 3 miles towards the northeast and 5¾ miles towards the south-west. The shock was also felt at 15 places outside this isoseismal and within the boundary of the sound-area (represented by the dotted line in Fig. 63). The disturbed area is thus 43 miles long, 36 miles wide, and contains about 1200 square miles.

In the central part of the disturbed area, the shock consisted of two distinct parts, separated by a brief interval of rest and quiet. At Slawston, for example, the first and stronger part lasted 4 seconds, the interval 2 seconds, and the second part about 2 seconds; the sound was also in two parts, the first and louder being compared to thunder, the second to a rushing wind. Nearer the boundary, as, for instance, at Harrington, the shock consisted of one continuous series of vibrations 3 seconds in duration. At Queeniborough, the two parts of the shock were about equal in strength; but at seven other places (Burton Hall, Kirby Muxloe, Leicester, Little Dalby, Lowesby, Quorn and Slawston) the first part was the stronger.

The boundary of the area within which the double shock was felt is represented by the broken-line in Fig. 63. It is 29 miles long, 20½ miles wide, and contains an area of 464 square miles. Its centre lies about ½ mile S.W. of Houghton-on-the-Hill or 2½ miles N.N.W. of the centre of the isoseismal 5; and its axis is nearly parallel to those of the isoseismal lines, being directed N. 40° W. Thus, though the two foci must have occupied portions of the same fault, they were not coincident, but, at the same time, were not completely detached. The earthquake was therefore a double earthquake, rather than a twin.

The boundary of the sound-area (represented by the dotted line in Fig. 63) coincides approximately with that of the disturbed area. The sound was unusually loud within a central area bounded approximately by the isoseismal 5, which is very nearly concentric with the boundary of the sound-area. It was, however, heard by nearly all observers close up to the boundary; for the percentage of audibility was 97 over the whole area, 98 within the isoseismal 5, 96 between the isoseismals 5 and 4, and 93 between the isoseismal 4 and the boundary of the sound-area. The sound was compared to passing waggons, etc., in 62 per cent. of the records, thunder in 17, wind in 4, loads of stones falling in 7, the fall of a heavy body in 5, explosions in 4, and miscellaneous sounds in 1, per cent.

ORIGIN OF THE LEICESTER EARTHQUAKES

For the earthquake of 1893, the seismic evidence gives the following elements of the originating fault. In the north-western focus, (i) the mean direction of the fault is N. 60° W.; (ii) the hade is to the N.E.; and (iii) the fault-line passes a short distance to the south-west of the centre of the isoseismal 5, or about 1 or 2 miles to the south-west of Woodhouse Eaves. In the south-eastern focus, it is probable that the hade is to the S.W. and that the fault-line passes a short distance to the north-east of Tugby in a direction between N. 60° W. and N.W. For the principal earthquake of 1904, the elements are: (i) the mean direction of the fault is N. 48° W.; (ii) the hade is to the S.W.; and (iii) the fault-line passes a short distance to the north-east of the centre of the isoseismal 5, and therefore not far from Tugby. It is thus probable that the two earthquakes originated in a single fault which changes hade in the interfocal region.

It may be remarked that this change of hade furnishes an explanation of the divergence between the axes of the isoseismals 5 and 4 in the earthquake of 1893. For, at the north-western focus, the isoseismal 4 diverges farther to the north-east than to the south-west; while, at the south-eastern focus, it would diverge farther to the south-west than to the north-east. Thus, the axis of the compound isoseismal 4 (shown in the map, Fig. 62) should be tilted more to the north-west than the axis of the isoseismal 5; and the same should hold for the compound isoseismal 3.

In Fig. 62, an important strike-fault on the north-east side of the Charnwood anticlinal axis is represented by the broken-line[1]. Its mean direction near Woodhouse Eaves is N. 63° W., it probably hades to the N.E., and it passes about 2 miles to the south-west of the centre of the isoseismal 5. Owing to the covering of Triassic rocks, it cannot be traced farther to the south-east than the village of Cropston; but there is no reason for supposing that it dies out there. Assuming, then, that the fault referred to extends to the neighbourhood of Tugby, trending there more nearly in a south-easterly direction, the recent history of the district appears to be as follows:

The first movement, that of 1839, evidently occurred in the north-western focus. In 1893, the displacements in the two foci took place nearly simultaneously, that in the north-western focus being the earlier by 1 or 2 seconds. In 1904, the first movement, a very slight one, occurred in the north-western focus, and this was followed two hours later by a double (not a twin) slip in the south-eastern focus, the second part of the double slip showing a focal migration of about 2 or 3 miles to the north-west.

---

[1] W. W. Watts, *Proc. Geol. Assoc.* vol. 17, 1902, Plate XIX.

NORTHAMPTON EARTHQUAKES

The earthquakes of the Northampton district, two or three in number, appear to be connected with a pair of associated foci. One epicentre lies about 2 miles E. of Market Harborough, the other some, but an unknown, distance to the north-east.

1650 Dec. 10, a shock felt at Northampton (Short, vol. 2, p. 170).

1. **1750** OCT. **11**, *about* 0.30 *p.m.* Cat. No. 112; Intensity 7 (nearly 8); Centre of isoseismal 7 in lat. 52° 29' N., long. 0° 51' W.; No. of records 73 from 43 places; Fig. 64 (*Phil. Trans.* vol. 46, 1752, pp. 701–750; 1792, pp. 284–285; *Gent. Mag.* vol. 20, 1750, p. 473; *London Mag.* 1750, p. 474; *Notes and Queries*, 1792, pp. 284–285; etc.).

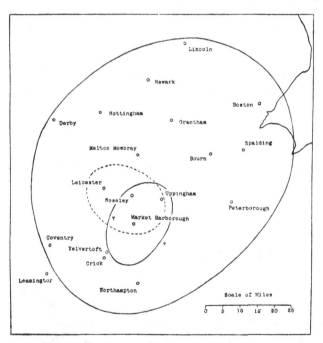

Fig. 64. Northampton Earthquake of 1750 Oct. 11.

This was the last and strongest shock of the year 1750, the "year of earthquakes" as Stukeley called it. The continuous lines represent, though only approximately, the isoseismal 7 and the boundary of the disturbed area; and the broken-line the isoseismal 5 of the Leicester earthquake of 1904. At several places within and near the isoseismal 7, buildings were slightly injured; for instance, a few chimneys were in part thrown down or an old wall slightly cracked at Northampton, Leicester, Uppingham, etc. The isoseismal 7 is 25 miles long from N.N.E. to S.S.W., 20 miles wide, and about 400 square miles in area. Its centre lies about 2 miles E. of Market Harborough. The boundary of the disturbed area is 85 miles long from N.E. to S.W., 68 miles wide, and contains about 4500 square miles. Though neither curve can be regarded as accurately drawn, the excentricity of the inner curve with regard to the other is distinct. This excentricity is a marked characteristic of a twin earthquake in which the impulse at one focus is much stronger than at the other. The shock consisted of two distinct parts, separated by an interval of a few seconds, at Northampton, Noseley (7 miles N. of the south-western epicentre), Yelvertoft (13 miles W.S.W.) and Hill Morton (2 miles S.E. of Rugby). The earthquake was evidently a twin, of which the stronger impulse originated in the Market Harborough focus. The position of the second epicentre is unknown, but it cannot be much less than 20 miles from the other and in the north-east direction from it. It will be seen from the map that the line joining the epicentres is nearly at right angles to the longer axis of the isoseismal 5 of the Leicester earthquake of 1904.

2. **1768** JAN. 3, 1 *a.m.* Cat. No. 140 (*Ann. Reg.* 1768, p. 59).

A slight shock felt at, and in the neighbourhood of, Crick, a village which lies close to the isoseismal 7 of the preceding earthquake and about 14 miles S.W. from its centre.

DONCASTER EARTHQUAKES

In the neighbourhood of Doncaster, there are two epicentres, one about 8 miles S.E. of Doncaster and ½ mile N. of Bawtry, the other about 4 miles E. of Crowle. That the corresponding foci are connected is shown by the occurrence of the twin earthquake of 1905, but, except on this occasion, they have rarely been in action during the period—a little more than two centuries—over which our record extends.

1682 Feb. 13, an earthquake at Doncaster is recorded by Short (vol. 2, p. 170).

1703 Nov. —, an earthquake in the north of England, felt at Lincoln and at Beckingham (about 5 miles E. of Newark) (Cat. No. 77; Milne, vol. 31, p. 96; *Gent. Mag.* vol. 20, 1750, p. 56; vol. 21, 1751, p. 357; *London Mag.* 1750, p. 103).

1818 Apr. —, a smart shock felt from one side of Lincolnshire to the other and across Holderness in Yorkshire (Milne, vol. 31, p. 118).

1. **1873** APR. 29, *about* 2.40 *p.m.* Cat. No. 768; Intensity 4 (*Nature,* vol. 8, 1873, p. 13; *Times,* Apr. 30).

A shock, accompanied by a sound like that of a heavy body falling, at Doncaster; felt also at Bawtry (8 miles S.E. of Doncaster) and Conisbro' (5 miles S.W.).

1874 Apr. 27, Roper records an earthquake at Doncaster (p. 39).

2. **1902** APR. 13, *about* 11.50 *a.m.* Cat. No. 961; Intensity 4; Centre of isoseismal 4 in lat. 53° 33·4′ N., long. 0° 41·5′ W.; No. of records 34 from 28 places and 77 negative records from 76 places; Fig. 66 (*Geol. Mag.* vol. 1, 1904, pp. 536–537).

The isoseismal 4 (represented by the broken-line in Fig. 66) is 34 miles long, $22\frac{1}{2}$ miles wide, and contains about 600 square miles. Its centre lies $3\frac{1}{2}$ miles S.S.E. of Crowle, the longer axis being directed N. 65° E. Outside this isoseismal, a tremulous motion was felt at Clarborough ($3\frac{1}{4}$ miles to the S.), Hull, Beverley, and Bexford ($2\frac{1}{2}$, 7 and 17 miles to the N.E.), and Leeds, Horsforth and Rawdon (24, 28 and 30 miles to the N.W.). The sound was also heard at Clarborough, Hull and Horsforth. The total extent of the disturbed area may thus amount to about 3000 square miles. For an earthquake of such low intensity, this points to a great depth of focus.

The shock consisted of a single series of vibrations, increasing in intensity to a maximum and then dying away. Its mean duration was about 4 seconds.

The sound was heard by 98 per cent. of the observers, and was compared to passing waggons, etc., in 48 per cent. of the records, thunder in 38, and wind in 14, per cent.

3. **1905** APR. 23, *about* 1.30 *a.m.* Cat. No. 986; No. of records 2 from 2 places.

A slight shock was felt at Epworth and a rumbling sound was heard at Norton. Epworth lies between the two epicentres, and Norton about 8 miles N. of Doncaster.

4. **1905** APR. 23, 1.37 *a.m.* Cat. No. 987; Intensity 7; Centre of S.W. portion of isoseismal 7 in lat. 53° 26·3′ N., long. 1° 1·2′ W.; Centre of N.E. portion of isoseismal 7 in lat. 53° 36·5′ N., long. 0° 43·7′ W.; No. of records 1428 from 662 places and 68 negative records from 66 places; Figs. 65, 66 (*Quart. Journ. Geol. Soc.* vol. 62, 1906, pp. 5–12).

Isoseismal Lines and Disturbed Area. The continuous lines in Fig. 65 represent isoseismal lines of intensities 7 to 3. The isoseismal 7 consists of two portions (Figs. 65, 66), which are approximately circular in form. The south-western and larger portion is 18¼ miles long from N.E. to S.W., 17¾ miles wide, and 244 square miles in area. Its centre lies ½ mile N. of Bawtry. The north-eastern portion is 9¼ miles long from N.E. to S.W., 8¾ miles wide, and 63 square miles in area. Its centre is about 4 miles E. of Crowle, and 17 miles N.E. of the centre of the larger portion.

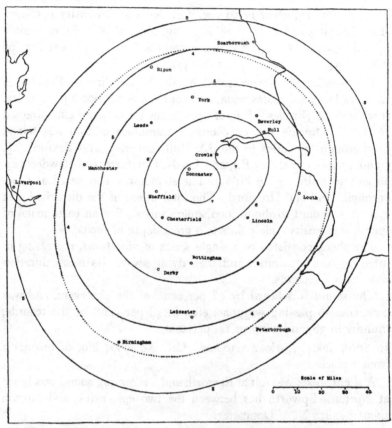

Fig. 65. Doncaster Earthquake of 1905 Apr. 23.

The isoseismal 6 (Figs. 65, 66) is 58 miles long, 44 miles wide, and contains 2050 square miles. The direction of its longer axis is N. 52° E. Its distances from the south-western portion of the isoseismal 7 are 14 miles on the north-west side and 12½ miles on the south-east side, and, from the north-eastern portion of the same curve, 14 and 16 miles respectively. The isoseismal 5 is 90 miles long from N.E. to S.W., 76 miles wide, and contains 5300 square miles; its distance from the isoseismal 6 is 14 miles on the north-west side, and 18 miles on the south-east side. The isoseismal 4 is 126 miles long from N.E. to S.W., 108 miles wide, and about 10,700 square miles in area, and is distant 16 miles in both directions from the isoseismal 5. The isoseismal 3, which bounds the disturbed area, is 166 miles long from N.E. to S.W., 130 miles wide, and about 17,000 square miles in area, its distance from the isoseismal 4 being 14 miles towards the north-west and 21 miles towards the south-east.

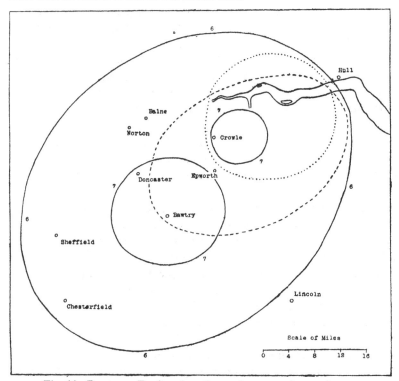

Fig. 66. Doncaster Earthquakes of 1902 Apr. 13 and 1905 Apr. 23.

Nature of the Shock. The twin character of the shock was clearly recognised throughout a district overlapping the isoseismal 5 by a few miles in every direction, and was sensible to some observers as far as the isoseismal 4. Over the whole disturbed area, 32 per cent. of the observers noticed either two maxima in a continuous series of vibrations, or two detached series of vibrations separated by an interval of a few seconds. For instance, at Balne (near Doncaster), the movement was continuous and contained two maxima, the first of which was the stronger; at Sheffield, the shock consisted of two detached parts separated by an interval of 1 second; at Humberstone (near Leicester), the shock was a single series of vibrations, lasting from 3 to 4 seconds, the intensity of which increased to a maximum and then died away. The mean duration of the interval between the two parts of the shock was 3·5 seconds, and there is no evidence whatever of the existence of a synkinetic band. Thus, the interval between the epochs of the two impulses must have been greater than the time required for the vibrations to travel from one focus to the other. The two parts of the shock differed but little in strength, the first being slightly the stronger over the greater part of the disturbed area; while the second was the stronger within a small and nearly circular area, the boundary of which is represented by the dotted line in Fig. 66. This area, which is about 20 miles in diameter, includes the centre of the north-eastern portion of the isoseismal 7, its own centre lying about 6 miles from it to the N.E.

It is evident that the two parts of the shock originated in two detached foci; though, in the intervening region of the fault, there must have been a displacement sufficient to account for the tremor connecting the two series of vibrations. The two epicentres must be close to the centres of the two portions of the isoseismal 7, and are therefore separated by a distance of about 17 miles. It follows, from the relative intensity of the two parts of the shock, that the impulse within the south-western focus took place a few seconds before the other, and that it was slightly the stronger, the latter inference being confirmed by the larger size of the south-western portion of the isoseismal 7; also, that every point of the disturbed area was reached first by the vibrations from the south-western focus. The second part of the shock was the more intense near the north-eastern epicentre owing to the proximity of the corresponding focus.

Seismographic Record. The earthquake was registered by an Omori horizontal pendulum at Birmingham, which lies 75 miles S. 28° W. of the south-western epicentre. The record is very minute, the range (or double amplitude) of the largest waves (the second, third and fourth) being less than ·01 inch, and corresponding to a movement of the ground of about ·0005 inch. The remaining vibrations are represented by mere notches on the trace, and in parts these cannot be separated. Altogether, there appear to be about 30 vibrations in about 15 seconds. The exact time of the first movement cannot be determined.

Sound-Phenomena. The boundary of the sound-area is indicated by the dotted curve in Fig. 65. It is 135 miles long, 115 miles wide, and contains about 12,000 square miles. The sound was heard by 93 per cent. of the observers in the whole disturbed area, and this high percentage is maintained to a considerable distance from the epicentres, being 94 within the isoseismal 7 and between the isoseismals 7 and 6, and 93, 86 and 67 in successive zones bounded by the isoseismals. There is no sensible variation in the audibility with the direction from the epicentres, a result which accords with the absence of a synkinetic band.

The sound was compared to passing waggons, etc., in 43 per cent. of the records, thunder in 29, wind in 12, loads of stones falling in 4, the fall of a heavy body in 4, explosions in 6, and miscellaneous sounds in 2, per cent.

ORIGIN OF THE DONCASTER EARTHQUAKES

The only known element of the fault in which the earthquake of 1902 (no. 2) originated is its mean direction, which is N. 65° E. That of the corresponding fault for the principal earthquake of 1905 (no. 4) is about N. 52° E. From the relative positions of the isoseismals of the later earthquake, it may be inferred that the fault hades to the N.W. in the south-western focus and to the S.E. in the north-eastern focus. That the two earthquakes were connected with the same fault is probable for the following reasons: (i) both earthquakes occurred in a district which is remarkably free from seismic action; (ii) the north-eastern epicentre of the principal earthquake is close to the minor axis of the isoseismal of the earthquake of 1902; and (iii) the directions assigned to the originating fault are nearly the same.

The list of seismic events in this district is a brief one. The earthquake of 1873 probably originated in the south-western focus. In 1902, the north-eastern focus came into action, and, in 1905, a much more important and double movement occurred in both foci, the first slip in the south-western focus and rapidly extending across the interfocal region until another large slip took place in the north-eastern focus.

CHAPTER XIV

EARTHQUAKES OF THE EAST OF ENGLAND

FOR convenience, the counties of Lincoln, Cambridge, Norfolk and Suffolk are grouped in one district. Except in the first county, earthquakes are slight and infrequent. In Lincolnshire, there are several well-defined centres. The eastern epicentre of the twin Doncaster earthquake of 1905 lies 4 miles E. of Crowle, this centre being also responsible for the Doncaster earthquake of 1902 Apr. 13. The western epicentre lies ½ mile N. of Bawtry and therefore just outside the county. The other centres are (i) about 11 miles N.E. of Gainsborough, (ii) in the neighbourhood of Lincoln, (iii) about 5 miles N.E. of Grantham, and (iv) near Coningsby (7 miles S. of Horncastle). Besides the earthquakes of these centres, there are the following recorded shocks of which the seismic origin is not established.

1114, a violent shock at Crowland, by which Crowland Church was much damaged (E. J. Lowe, p. 12).

1186 Apr. 25, an earthquake in Lincolnshire (Short, vol. 1, p. 129).

1563 Sept., an earthquake in different parts of the country and especially in Lincolnshire and Northamptonshire (Stow, p. 1112; Short, vol. 1, p. 227).

1564 Sept. and Nov., earthquakes in Northamptonshire and Lincolnshire (Short, vol. 2, p. 169).

1885 June 17, about 11 a.m., an earthquake slightly felt and heard at Killingholme, about 10 miles W.N.W. of Grimsby (*Mon. Met. Mag.* vol. 20, 1885, p. 95).

The epicentre of the following earthquake is unknown:

1818 Apr. 30, a smart shock felt from one side of Lincolnshire to the other and across the Humber into Holderness (Cat. No. 285; Milne, vol. 31, p. 118).

GAINSBOROUGH EARTHQUAKE

1703 DEC. 28, 5.3 *p.m.* Cat. No. 78; Intensity at least 5; Centre of disturbed area in lat. 53° 28′ N., long. 0° 33′ W.; Fig. 67 (W. Thoresby, *Phil. Trans.* vol. 23, 1704, pp. 1555–1558; R. Pickering, *Phil. Trans.* vol. 46, 1752, p. 624; Milne, vol. 31, p. 96; *Notes and Queries*, vol. 3, 1863, p. 405).

The boundary of the disturbed area is represented by the line in Fig. 67. It is 65 miles long, 60 miles wide, and contains about 3000 square miles. The centre is 17 miles N. of Lincoln.

LINCOLN EARTHQUAKES

1123, an earthquake at Lincoln (Short, vol. 2, p. 167).

1. 1142 DEC. Cat. No. 19 (Symeon of Durham, vol. 2, p. 307; John of Hexham, p. 17).

"An Earthquake was heard three times in the City of Lincoln infra natale Domini."

1180 about Sept. 29, two or three shocks in England; according to Short, "an earthquake fatal to many great buildings in England, especially to Lincoln Church" (Mallet, 1852, p. 28; Short, vol. 1, p. 128).

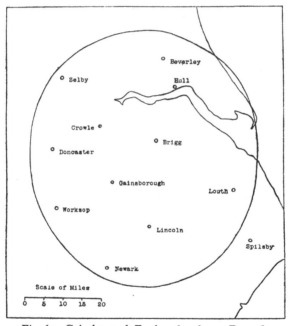

Fig. 67. Gainsborough Earthquake of 1703 Dec. 28.

2. 1185 APR. 15. Cat. No. 24; Intensity 8 (Matthew Paris, vol. 1, p. 434; vol. 3, p. 206; *Flor. Hist.* vol. 2, p. 97; Roger de Hoveden, vol. 2, pp. 303–304; R. de Diceto, col. 628; *Annal. Monast.* vol. 3, pp. 23, 446; vol. 4, p. 385; Holinshed, vol. 2, pp. 188–189; etc.).

"On the mondaie in the weeke before Easter, chanced a sore earthquake through all the parts of this land, such a one as the like had not beene heard of in England sithens the beginning of the world

For stones that laie couched fast in the earth, were remooued out of their places, stone houses were overthrowne, and the great church of Lincolne was rent from the top downwards" (Holinshed).

GRANTHAM EARTHQUAKES

1. **1750** SEPT. 3, *about* 6.45 *a.m.* Cat. No. 110; Centre of disturbed area in lat. 52° 58′ N., long. 0° 35′ W.; Records from 11 places; Fig. 68 (*Phil. Trans.* vol. 46, 1752, pp. 725–726, 731–750; *Gent. Mag.* vol. 20, 1750, p. 378; Milne, vol. 31, p. 99).

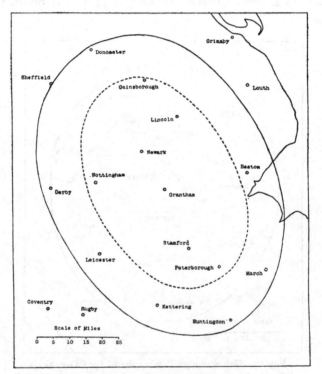

Fig. 68. Grantham Earthquakes of 1750 Sept. 3 and 1792 Feb. 25.

The boundary of the disturbed area is represented by the broken-line in Fig. 68. It is 68 miles long, 46 miles wide, and contains about 2500 square miles. Its centre lies about 5 miles N.E. of Grantham, and the longer axis is directed about N.N.W.

2. **1792** FEB. 25, *about* 8.30 *p.m.* Cat. No. 173; Intensity 5; Centre of disturbed area in lat. 52° 58′ N., long. 0° 35′ W.; Records from

17 places and negative records from 4 places; Fig. 68 (E. Turnor, *Phil. Trans.* 1792, pp. 283–288; Milne, vol. 31, p. 109).

The boundary of the disturbed area is represented by the continuous line in Fig. 68. It is 93 miles long, 72 miles wide, and contains about 5200 square miles. Its centre coincides with that of the preceding earthquake and its longer axis is parallel to that of the same earthquake. The sound was usually compared to that of a passing vehicle or to thunder.

CONINGSBY EARTHQUAKES

1. **1818** FEB. 6. Cat. No. 281; Intensity 4 (*Ann. de Chim. et de Phys.* vol. 9, 1818, p. 434; *Journ. of Sci.* vol. 5, 1818, p. 135; *Gent. Mag.* vol. 88, 1818, p. 171).

A slight shock, lasting several seconds and accompanied by a noise like the firing of guns, felt at Coningsby (7 miles S. of Horncastle), and also, it is said, at Kirton-in-Lindsey and the east end of Holderness.

2. **1818** FEB. 20, *about* 3 *p.m.* Cat. No. 284 (*Gent. Mag.* vol. 88, 1818, p. 364).

A similar slight shock felt at the same places.

CAMBRIDGESHIRE EARTHQUAKES

1105, an earthquake at Ely (*Gent. Mag.* vol. 20, 1750, p. 56).

1165 JAN. 26. Cat. No. 21 (Matthew Paris, vol. 1, p. 338; *Annal. Monast.* vol. 3, p. 442; vol. 4, p. 381; *Flor. Hist.* vol. 2, p. 79; B. de Cotton, p. 73; *Materials for the Hist. of Thomas Becket*, vol. 5, p. 164; Gervase of Cant. vol. 1, p. 196; Stow, p. 221; F. Blomefield, *Hist. of Norwich*, 1741, p. 21; etc.).

"There was an earthquake took place throughout Ely, and Norfolk, and Suffolk,...which threw down persons who were standing up, and made bells ring" (*Flor. Hist.*).

1166 Jan. 20, an earthquake in Ely, Norfolk and Suffolk (Short, vol. 1, p. 123), possibly the same as the preceding.

NORFOLK EARTHQUAKES

1199 May 22, an earthquake in Norfolk (*London Mag.* 1750, p. 102).

1. **1480** DEC. 28. Cat. No. 44; Intensity 8 (F. Blomefield, *Hist. of Norwich*, 1741, p. 121).

An earthquake felt at Norwich and almost the whole of England, by which buildings were thrown down and much damage was done.

2. **1487** DEC. 21. Cat. No. 45 (F. Blomefield, p. 126).

An earthquake at Norwich.

1601 Dec. 25, noon, an earthquake in Norfolk (Blomefield, vol. 3, p. 358).

1884 July 16, an earthquake at Norwich (Fuchs, p. 139).

BUNGAY EARTHQUAKE

1757 JAN. 10. Cat. No. 129; Centre of disturbed area in lat. 52° 29′ N., long. 1° 26′ E.; Fig. 69 (Roper, p. 23, quoting from the *Norwich Mercury* for Jan. 15).

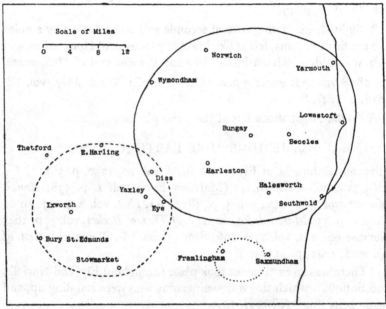

Fig. 69. Bungay Earthquake of 1757 Jan. 10, Ixworth Earthquake of 1869 Jan. 9 and Framlingham Earthquake of 1901 Oct. 19

The boundary of the disturbed area is represented approximately by the continuous line in Fig. 69. It is 32 miles long, 27 miles wide, and contains about 680 square miles. Its centre lies 1 mile N. of Bungay, and the longer axis is directed a few degrees N. of E.

SUFFOLK EARTHQUAKES

1869 Nov. 13–14, near midnight, a slight shock at Haverhill (Perrey, *Mém. Cour.* vol. 24, 1875, p. 57).

IXWORTH EARTHQUAKE

1869 JAN. 9, 11.17 *a.m.* Cat. No. 746; Intensity 4; Centre of disturbed area in lat. 52° 18·1′ N., long. 0° 57·4′ E.; No. of records 14 from 11 places; Fig. 69 (*Times*, Jan. 12, 15, 16).

The disturbed area, the boundary of which is represented by the broken-line in Fig. 69, is 20 miles long from E. to W., 18 miles wide, and contains about 280 square miles. Its centre is at the village of Walsham-le-Willows or 9 miles S. of East Harling.

The shock consisted of a quiver, accompanied by a rumbling sound like thunder or the rapid passing of a heavy carriage. At Yaxley and West Harling, pheasants crowed and flew from the ground in alarm.

FRAMLINGHAM EARTHQUAKES

1. **1901** OCT. 19, *about* 7.25 *p.m.* Cat. No. 957; Intensity 4; Centre of isoseismal 4 in lat. 52° 12·6′ N., long. 1° 15·0′ E.; No. of records 18 from 11 places and 27 negative records from 23 places; Fig. 69 (*Geol. Mag.* vol. 1, 1904, pp. 539–540).

The isoseismal 4 is represented by the dotted line in Fig. 69. It is nearly circular, 6½ miles long from E. to W., 6 miles wide, and contains 31 square miles. Both shock and sound were observed at Rendlesham (2½ miles S. of the isoseismal); and the sound was also heard, but no shock felt, at Dennington (1 mile N.W.) and Earl Soham (3 miles W.). The centre of the isoseismal lies 1 mile S.W. of Swefling and 15 miles N.E. of Ipswich. There was a sudden shock followed by a brief tremulous movement lasting for a few seconds. The sound, usually of brief duration, was heard by all the observers.

2. **1901** OCT. 22, *about* 9.15 *a.m.* Cat. No. 958; Intensity 4; Centre of isoseismal 4 in lat. 52° 12·7′ N., long. 1° 15·2′ E.; No. of records 22 from 13 places and 27 negative records from 23 places; Fig. 69.

The isoseismal 4 coincides very nearly with that of the preceding earthquake. The shock and sound were observed at Earl Soham, the shock was also felt at Snape (1 mile S.E. of the isoseismal), and the sound was heard at Cretingham (3½ miles W.). The shock was similar to that of the first earthquake, consisting of one prominent vibration followed by tremulous motion and lasting altogether about 3 seconds. In 15 out of 17 records, the sound was compared to types of short duration.

CHAPTER XV

EARTHQUAKES OF THE SOUTH-WEST
OF ENGLAND

THE earthquakes of the south-west of England—the counties of Cornwall, Devon and Somerset—are usually of slight intensity. One shock in 1248 is credited with damage to Wells cathedral, and another in 1275 with the destruction of St Michael's Church, Glastonbury. Of the rest, two (those of Penzance in 1757 and Cheddar in 1852) may have attained the degree 6; six others were of intensity 5, and twelve of intensity 4; the remaining earthquakes, except a few of unknown intensity, were mere tremors.

The seismic centres of Cornwall are chiefly grouped in two portions of the county. In the south-west, there are five centres, namely, those of Penzance, Helston, Falmouth, Truro and St Agnes. Mid-Cornwall possesses only one centre, that of St Austell. In north-east Cornwall, there are four centres, namely, those of Camelford, Altarnon, Liskeard and Launceston. The earthquakes of Devonshire belong to four centres, those of Okehampton, Dartmouth, Exmouth and Barnstaple. In Somerset, half the earthquakes belong to unknown centres. The three Taunton earthquakes appear to be conected with two epicentres, one near Wiveliscombe and the other near Taunton. Five other earthquakes originated in a centre close to Wells.

PENZANCE EARTHQUAKES

1. **1757 JULY 15,** *about* 6.30 *p.m.* Cat. No. 131; Intensity about 6; No. of records 19 from 18 places (Rev. W. Borlase, *Phil. Trans.* vol. 50, 1759, pp. 499–506; *Gent. Mag.* vol. 27, 1757, p. 335; J. Wesley's *Journal*, vol. 2, 1827, p. 370).

Dr Borlase's paper is notable as being the first detailed study of a British earthquake.

The disturbed area occupies nearly all Cornwall, the boundary towards the east passing just beyond Camelford and Plymouth. How far to the west it extended is unknown, for the shock was strong in the Scilly Isles. The epicentre was probably not far from Penzance, for, at this place, the intensity was about 6, while it was 5 within most of the district to the west of St Agnes and Falmouth.

At St Agnes, according to John Wesley, there was first a rumbling noise underground, hoarser and deeper than common thunder; then followed a trembling of the earth, which afterwards waved violently to and fro once or twice. The shock was felt in mines near St Just, Lannant and Godolphin; in others near Gwinear and Chacewater, the noise only was observed, as if the loose rubbish of the mines were set in motion.

2. **1852** NOV. 17, *morning.* Cat. No. 710 (Perrey, *Mém. Cour.* vol. 16, 1864, p. 18; Edmonds, p. 84).

A shock, lasting 15 or 20 seconds, at Sancreed (3 miles W. of Penzance).

3. **1855** MAY 30, 3 *p.m.* Cat. No. 714; Intensity 4 (Perrey, *Mém. Cour.* vol. 12, 1861, p. 15; Edmonds, p. 84 and *Edin. New Phil. Journ.* vol. 3, 1856, pp. 280–285).

A shock at Penzance and in the neighbourhood.

4. **1904** MAR. 3, *about* 1.5 *p.m.* Cat. No. 976; Intensity 5; Centre of isoseismal 5 in about lat. 50° 4·2′ N., long. 5° 27·6′ W.; No. of records 76 from 46 places and 13 negative records from 12 places; Fig. 70 (*Geol. Mag.* vol. 1, 1904, pp. 487–490).

Isoseismal lines of intensities 5 and 4 are represented by the continuous lines in Fig. 70, and these show that the epicentre is submarine. Little more than half of each curve traverses the land, and the form of the remaining portions over the sea-area can only be conjectured from their trend near the coast. If, however, the isoseismal 5 be completed, the centre of the curve must be close to a point about 3½ miles S. of Marazion. This curve is 13½ miles long, probably 10 miles wide, and 110 square miles in area. The isoseismal 4 is 19 miles long, about 15½ miles wide, and contains 230 square miles, its distance from the preceding isoseismal towards the north being 2½ miles. The longer axes are directed a few degrees N. of E. The earthquake was also observed at a few places outside the isoseismal 4, the sound being heard at Clowance, Sennen, St Just-in-Penwith and Zennor, which are respectively ½, ¾, 1 and 1½ miles from the curve, and a slight shock was also felt at Clowance, Sennen and Zennor.

The shock consisted of a single series of vibrations, which increased gradually in strength, and then faded away, the mean duration being about 4 seconds.

The sound was heard by all the observers. It was compared to passing waggons, etc., in 18 per cent. of the records, thunder in 22, wind in 2, loads of stones falling in 3, the fall of a heavy body in 2, explosions in 52, and miscellaneous sounds in 2, per cent. Thus, in 58 per cent. of the records, the comparison is to a type of short duration.

This brevity of the sound is suggestive of heavy gun-firing from a vessel about 3 or 4 miles S. of Marazion, but there were no battleships in the bay on Mar. 3; and, as the Secretary of the Admiralty informs me, there is no record of any firing in the district on that day. Nor can the disturbance be due to a fault-slip precipitated by mining operations; for, though some old workings run out beneath the sea near Marazion, there is none in the required position.

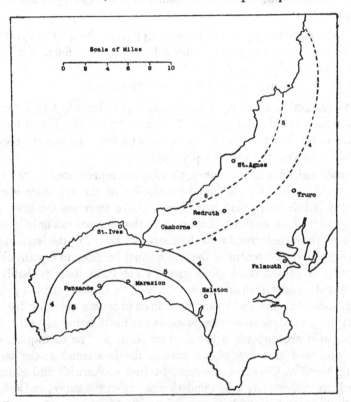

Fig. 70. Penzance Earthquake of 1904 Mar. 3 and St Agnes Earthquake of 1905 Jan. 20.

The earthquake was therefore probably due to a slip, 3 or 4 miles in length, along a submarine fault about $3\frac{1}{2}$ miles S. of Marazion, and occurring at a comparatively small depth below the surface. The direction of the earthquake-fault must be parallel, or nearly parallel, to that of the lodes in the neighbourhood of Marazion.

Mr Clement Reid has suggested[1] that an Eocene basin may lie

[1] Quart. Journ. Geol. Soc. vol. 60, 1904, pp. 113–117.

under the sea in Mount's Bay and the western part of the English Channel; and it is possible that the last of the series of movements resulting in the formation of this basin was that which caused the Penzance earthquake of 1904.

HELSTON EARTHQUAKES

1. **1842** FEB. 17, *about* 8.30 *a.m.* Cat. No. 569; Intensity 4; Centre of isoseismal 4 in lat. 50° 8·3′ N., long. 5° 11·2′ W.; No. of records 19 from 16 places, negative records from 5 places, and 8 records from mines; Fig. 71 (W. J. Henwood, *Phil. Mag.* vol. 21, 1842, pp. 153–156 and *Cornwall Geol. Soc. Trans.* vol. 5, 1843, p. 459 n.).

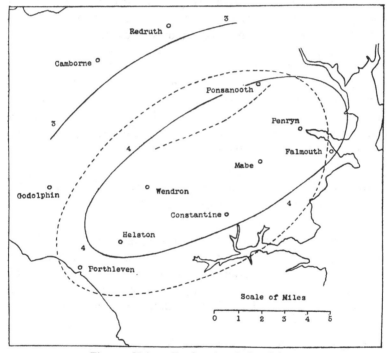

Fig. 71. Helston Earthquake of 1842 Feb. 17.

The isoseismal 4 is represented by the continuous line in Fig. 71 (the broken-line being the corresponding curve for the earthquake of 1898 Apr. 1). It is 12½ miles long, 5 miles wide, and contains about 49 square miles. Its centre is 2¾ miles N. 77° E. of Wendron, its longer axis being directed N. 62° E. Outside this curve, the shock was felt towards the north at Wheal Basset, Lanner and Comfort, and towards the west at Porthleven and Godolphin. The total disturbed area was thus about 160 square miles. Still farther to the north, the sound was heard (without any accompanying tremor) at Camborne, Tuckingmill and Pool. This overlapping of the disturbed area by the sound-area may indicate (as in the earthquake of 1898 Apr. 1) that the originating fault hades to the S.; but the inference is rendered somewhat uncertain by some vague reports of the sound being heard on the south side of the disturbed area. The shock consisted of a brief tremor of intensity 4 at most of the places where it was observed.

The shock was felt in many of the mines—in those of Poldory (130 fathoms deep), Tresavean (near Redruth) at all the depths worked, Wheal Vyvyan (near Constantine, 20 or 30 fathoms), Wheal Vor (near Helston, 80 and 175 fathoms), Penhale (near Helston, 50 fathoms), Great Work (near Godolphin), Wheal Penrose (near Porthleven), and West Wheal Virgin (near St Hilary, 100 fathoms). The sound was also heard in the Wheal Vor, Great Work, Wheal Penrose, and West Wheal Virgin mines, in the last named resembling the noise made by a shaft crushing in. The area disturbed underground was thus almost the same as that at the surface, if it did not extend beyond it towards the west.

2. 1892 MAY 16, *about* 10.30 *p.m.* Cat. No. 876; Intensity 3; Centre of disturbed area in lat. 50° 8·5′ N., long. 5° 13·1′ W.; No. of records 9 from 8 places; Fig. 72 (*Geol. Mag.* vol. 10, 1893, pp. 296–297).

The boundary of the disturbed area is represented by the broken-line in Fig. 72. In form, it is nearly circular, about 12½ miles in diameter, and containing about 120 square miles. Its centre lies about 1¼ miles N.E. of Wendron. The shock was slight, and was accompanied by a loud rumbling sound, which lasted for 2 or 3 seconds.

3. 1892 MAY 17, *about* 1.30 *a.m.* Cat. No. 877; Intensity 4 (nearly 5); Centre of isoseismal 4 in lat. 50° 8·5′ N., long. 5° 18·0′ W.; No. of records 70 from 40 places; Fig. 72 (*Geol. Mag.* vol. 10, 1893, pp. 297–302).

The isoseismals 4 and 3 are represented by the continuous lines in Fig. 72. The isoseismal 4 is 20 miles long, 15 miles wide and contains about 224 square miles. Its centre lies 3½ miles N. 15° W. of Helston, and the direction of its longer axis is approximately E. The isoseismal

3, which bounds the disturbed area, is 29 miles long, 22 miles wide, and contains about 480 square miles. Towards the north the distance between the isoseismals is 5 miles, and towards the south 2 miles. The intensity of the shock was nearly 5 at Helston, Nancegollen, Penzance, Porthleven and Wendron.

At Penryn, which is near the east end of the longer axis, the shock consisted of two distinct vibrations of equal intensity, each accompanied and followed, but not preceded, by tremulous motion. The

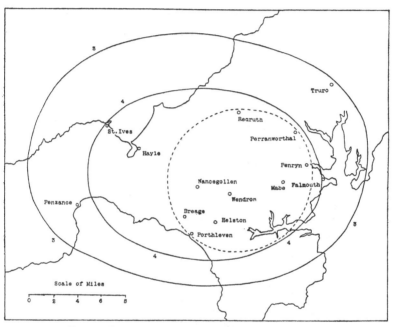

Fig. 72. Helston Earthquakes of 1892 May 16 and 17.

double shock was also observed at Breage, Perranarworthal, Redruth and Hayle. At the last two places, the first part was the stronger.

The sound was heard by all the observers, and was compared to passing waggons, etc., in 23 per cent. of the records, thunder in 23, loads of stones falling in 7, the fall of a heavy body in 2, explosions in 43, and miscellaneous sounds in 2, per cent. At Penryn, Redruth, Breage and Perranarworthal, each part of the shock was accompanied by an explosive sound.

4. 1898 MAR. 29, *about* 10.25 *p.m.* Cat. No. 927; Intensity 4; Centre of disturbed area in lat. 50° 7·3′ N., long. 5° 2·9′ W.; No. of records

9 from 7 places and negative records from 8 places; Fig. 73 (*Quart. Journ. Geol. Soc.* vol. 56, 1900, pp. 1–7).

The boundary of the disturbed area, which is an isoseismal of intensity less than 4, is indicated by the nearly circular broken-line in Fig. 73. It is 10½ miles long, 9 miles wide, and contains 74 square miles. Its centre is 1¼ miles S. 82° E. of Wendron and is close to that of the earthquake of 1892 May 16 (no. 2). The direction of its longer axis is roughly parallel to the longer axes of the isoseismal lines of the next earthquake.

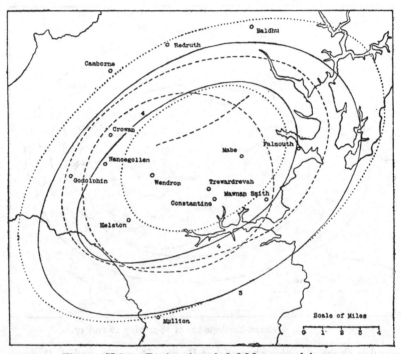

Fig. 73. Helston Earthquakes of 1898 Mar. 29 and Apr. 1.

There were two distinct shocks, the first of which occurred at about 10.25 p.m. and the second about five minutes later. The sound-area coincided with the disturbed area, and the sound was usually compared to the low rumble of distant thunder.

5. 1898 APR. 1, 9.55 *p.m.* Cat. No. 928; Intensity 4; Centre of isoseismal 4 in lat. 50° 7·8′ N., long. 5° 2·3′ W.; No. of records 89 from 56 places and negative records from 14 places; Fig. 73.

Two isoseismal lines of intensities 4 and 3 are represented by continuous lines in Fig. 73. The isoseismal 4 is 13 miles long, 7½ miles wide, and contains 76 square miles. Its centre is 2 miles E. of Wendron and about 1 mile from that of the preceding earthquake. Its longer axis passes approximately through the centre of that earthquake and is directed N. 57° E. The isoseismal 3, which forms the boundary of the disturbed area, is 19 miles long, 12 miles wide, and contains 175 square miles. The distance between the isoseismals is 1·7 miles on the N.W. side and 2·7 miles on the S.E. side.

Over the greater part of the disturbed area, the nature of the shock was nearly uniform, and the following account from Constantine may be regarded as typical: There were two shocks, each lasting about 5 seconds, with an interval of about 15 seconds between them, the second being the stronger; in both, the intensity increased gradually and then died away, being greatest near the middle. Two distinct shocks, as in this account, were recorded at 25 places, and at 6 other places two sounds were heard but no shock felt. The second shock was at all places the stronger and was accompanied by the louder noise. The estimates of the length of the interval between the two parts differ rather widely, the average of 20 estimates being a little more than a minute. Even supposing this to be in excess of the true value, it is evident that the first part must have been felt over its disturbed area before the occurrence of the second part at its focus. The boundary of the area within which the double shock was felt is represented in Fig. 73 by the oval broken-line. It is 15 miles long, 10 miles wide, and includes an area of 116 square miles. Its longer axis is directed N. 65° E. and is therefore nearly parallel to the axes of the isoseismal lines. The centre of the curve is 0·8 mile N. 35° E. of that of the isoseismal 4, and the distance between these two curves is 1·5 miles towards the N.W. and 0·8 mile towards the S.E. It is thus evident that the boundary of the double-shock area does not coincide with an isoseismal line corresponding to any value between 4 and 3; in other words that the foci of the two parts of the shock were not coincident. As the area in question is that over which the weaker of the two parts of the shock was felt, it corresponds therefore to the disturbed area of the first part of the shock, while the isoseismal 3 bounds the disturbed area of the second part.

The sound was heard by 94 per cent. of the observers, and the sound-area not only included the whole of the disturbed area, but, as shown by the outer dotted line, overlapped it towards the south-west, north and north-east. It is 22 miles long, 13 miles wide and contains 225 square miles. Near the central part of the sound-area, the sound was very loud and was a far more striking feature than the shock. The inner dotted line in Fig. 73 represents a line of equal sound-intensity, separating the places at which the sound was loud from those at which it was distinctly fainter. It is $9\frac{1}{2}$ miles long, $6\frac{1}{2}$ miles wide and 48 square miles in area, and is approximately concentric with the boundary of the double-shock area and also with that of the sound-area. The sound was compared to passing waggons, etc., in 25 per cent. of the records, thunder in 37, wind in 6, loads of stones falling in 2, the fall of a heavy body in 2, explosions in 24, and miscellaneous sounds in 4, per cent.

6. 1898 APR. 2, *about* 3 *p.m.* Cat. No. 929; Intensity 3; No. of records 4 from 3 places.

This shock was felt at Mabe, Mawnan Smith and Trewardrevah, and consisted of a weak tremor accompanied by a noise like distant thunder. The epicentre appears to be near, perhaps slightly to the east of, the centre of the double-shock area.

ORIGIN OF THE HELSTON EARTHQUAKES

The Helston earthquakes originated in two centres, one near Nancegollen and the other about 2 miles E. of Wendron, the distance between them being about 3 miles. The earthquake of 1842 (no. 1) occurred in the Wendron centre. That of 1892 May 16 (no. 2) also originated in this centre, but in a region about $1\frac{1}{2}$ miles farther west; its focus was of small dimensions, probably less than a mile, in the horizontal direction. The next earthquake, that of 1892 May 17 (no. 3) took place chiefly in the Nancegollen centre, but a second and weaker impulse may have occurred in the Wendron centre. The earthquake of 1898 Mar. 29 (no. 4) originated in the Wendron centre, its focus being in the same region as that of 1892 May 16 (no. 2) and between 1 and 2 miles in length. Three days later, on Apr. 1, a slip (no. 5) 5 or 6 miles in length occurred in the Wendron centre, its epicentre lying about $\frac{1}{2}$ mile to the N.E. of the preceding and about a mile W.S.W. of that of the earthquake of 1842. The focus of the first and weaker part of the shock lay in the same region but a short distance farther north, that is, as will be seen, higher up the fault-surface. The slip on Apr. 2 (no. 6) also took place in the Wendron centre, but probably a little farther east and not far from the focus of the earthquake of 1842.

From the seismic evidence, we should infer that the originating fault of the earthquakes of 1842 and 1898 is directed about N. 57° E. and hades to the S.E., while that of the earthquakes of 1892 is directed nearly E. and probably hades to the N. The south-easterly hade within the Wendron centre is confirmed by the overlapping of the disturbed area by the sound-area towards the north in 1898 and perhaps also in 1842.

On the Geological Survey map of the disturbed area, very few faults are marked, and there is none that will agree even approximately with the seismic conditions. There is, however, no contradiction between these conditions and the known geological structure of the district. To the south, in the neighbourhood of Mullion, the strike of the beds is parallel to the isoseismal axes of 1898, and there is a thrust-plane in the same direction which hades to the S.E. Near the epicentre, the general trend of the lodes is E.N.E. and several elvan-courses are parallel or nearly parallel to the isoseismal axes. One of these, represented by the short broken-line in Figs. 71 and 73, is such that, if its surface be faulted and hade to the S.E., it would satisfy the seismic conditions implied by the earthquakes of 1842 and 1898. At its western end, it trends in a more westerly direction, and, if continued (with possibly a change of hade), it would occupy the position required by the earthquakes of 1892.

FALMOUTH EARTHQUAKE

The following earthquake is apparently connected with a centre in the neighbourhood of Falmouth, but no details are known:

1696. Cat. No. 76 (*Gent. Mag.* vol. 20, 1750, p. 56; *London Mag.* 1750, p. 103).

An earthquake felt at Falmouth.

TRURO EARTHQUAKES

1. **1859** OCT. 21, *about* 6.45 *p.m.* Cat. No. 720; Intensity 4 (Perrey, *Mém. Cour.* vol. 16, 1864, p. 65; Edmonds, pp. 93–95; etc.).

An earthquake felt over the greater part of Cornwall, from Penzance to St Austell, and as far north as St Tudy. It was felt and heard in the Wheal Ellen mine (St Agnes) and also in mines near St Austell.

2. **1859** OCT. 21, *about* 7.45 *p.m.* Cat. No. 721 (Edmonds, p. 95). A shock felt at Truro.

3. **1860** JAN. 13, 10.30 *p.m.* Cat. No. 723; Intensity 5; Records from 22 places, negative record from 1 place, and 7 records from mines (Edmonds, pp. 95–100).

The disturbed area covers the greater part of Cornwall, from St Just on the west to Callington on the east, the most northerly places of observation being Bodmin, Liskeard and Callington. The shock was not felt in the Scilly Isles. It was felt in mines at Redruth (10 to 190 fathoms), Gwennap (208 fathoms), and near Par (4 miles E. of St Austell). The sound was heard in mines as far west as Lelant and St Ives (a slight trembling being felt here at a depth of 130 fathoms). Towards the east, neither shock nor sound was observed in mines at Menheniot (near Liskeard). The epicentre was probably near Truro.

4. **1860** JAN. 13, 11.30 *p.m.* Cat. No. 724 (Perrey, *Mém. Cour.* vol. 16, 1864, p. 73). A shock felt at Truro, Liskeard and Callington.

ST AGNES EARTHQUAKE

1905 JAN. 20, 1.50 *a.m.* Cat. No. 984; Intensity 5; Centre of isoseismal 4 in lat. 50° 22′ N., long. 5° 18′ W.; No. of records 44 from 24 places and negative records from 9 places; Fig. 70 (*Geol. Mag.* vol. 5, 1908, pp. 298–300).

Portions of the isoseismals 5 and 4 are represented by the broken-lines in Fig. 70, and, from their form, it is evident that the epicentre is submarine. Continuing the isoseismal 4 in what appears to be its probable course, it includes an area 31 miles long, 22 miles wide, and containing about 530 square miles. The centre of the curve is about 6 miles N.W. of St Agnes, while the direction of its longer axis is nearly N.E. or parallel to the adjoining coast-line. On the south-east side, the distance between the two isoseismals is 3 miles.

The shock consisted of a single prominent vibration, resembling a thud or blow, accompanied by a brief series of rapid tremors. The mean duration of the shock was 3·5 seconds.

The sound was heard by all the observers. It was compared to passing waggons, etc., in 27 per cent. of the records, thunder in 42, wind in 3, loads of stones falling in 3, the fall of a heavy body in 9, explosions in 12, and miscellaneous sounds in 3, per cent.

From the seismic evidence, it appears that the mean direction of the originating fault is about N.E., its distance from land being about 5 or 6 miles. The focus must have been several miles in length, and, as shown by the closeness of the isoseismals 5 and 4, it must have been situated at a small depth.

ST AUSTELL EARTHQUAKE

1886 JAN. 20, *about* 7 *a.m.* Cat. No. 802; Intensity 4 or 5 (*Nature*, vol. 33, 1886, p. 301; etc.).

A shock, lasting 4 or 5 seconds, felt at St Austell, St Blazey and Mevagissey, and accompanied by a loud noise. The disturbed area was not less than 8 miles in diameter, and the epicentre was probably submarine and about 2 miles S.E. of St Austell.

CAMELFORD EARTHQUAKES

1. **1837** OCT. 27. Cat. No. 343 (Parfitt, I, p. 651).

Several shocks, accompanied by a dull rumbling noise, at Camelford.

2. **1837** NOV. 24. Cat. No. 344 (Milne, vol. 31, p. 121).

A shock at Camelford.

3. **1891** MAR. 26, *about* 11.30 *a.m.* Cat. No. 859; Intensity 4; Centre of disturbed area in lat. 50° 40' N., long. 4° 36·7' W.; No. of records 17 from 16 places and negative records from 8 places; Fig. 74 (*Geol. Mag.* vol. 9, 1892, pp. 299–304).

The boundary of the disturbed area, an isoseismal of intensity less than 4, is 17 miles long, 13 miles wide, and contains about 170 square miles. Its centre is 4 miles N. 35° E. of Camelford, and its longer axis is directed nearly N.

At most places, two distinct shocks were felt, separated by an interval of 2 or 3 seconds. The first part was the stronger at Boscastle, Poundstock and St Juliot, all lying to the north of the epicentre; the second at Michaelstow to the south. With one exception, the double shock was recorded at all the places which determine the boundary of the disturbed area. The sound-area was co-extensive with the disturbed area and overlapped it slightly towards the east.

ALTARNON EARTHQUAKES

1. **1852** AUG. 12, *about* 8 *a.m.* Cat. No. 708 (Perrey, *Bull.* vol. 20, 1853, pp. 54–55).

A shock felt at Wetherham (near St Tudy), Liskeard and Calstock (4 miles S.W. of Tavistock), and accompanied by a noise like thunder.

2. **1889** OCT. 7, *about* 1.45 *p.m.* Cat. No. 826; Intensity 4; Centre of disturbed area in lat. 50° 34′ N., long. 4° 32′ W.; Records from 24 places; Fig. 74 (*Geol. Mag.* vol. 8, 1891, pp. 368–371).

The boundary of the disturbed area, an isoseismal of intensity 4, is represented by the continuous line in Fig. 74. It is 25 miles long,

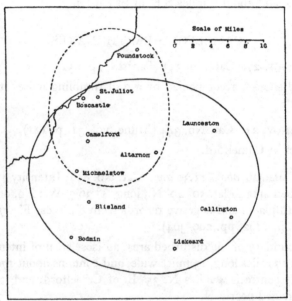

Fig. 74. Altarnon Earthquake of 1889 Oct. 7 and
Camelford Earthquake of 1891 Mar. 26.

20 miles wide, and contains about 400 square miles. Its centre is about 2¾ miles S.W. of Altarnon and the direction of its longer axis nearly E.

The shock was a mere tremor, with a mean duration of 3 seconds. The sound was usually compared to the passing of heavy waggons, thunder or the roaring of a chimney on fire.

LISKEARD EARTHQUAKES

1. **1759** FEB. 24, 10 *p.m.* Cat. No. 134 (*Gent. Mag.* vol. 29, 1759, p. 142).

A slight shock felt at and near Liskeard, the disturbed area being about 12 miles long from E. to W. and 6 miles wide.

2. **1837** OCT. 20, *about* 2 *p.m.* Cat. No. 342; Intensity 4 (W. J. Henwood, *Cornwall Geol. Soc. Trans.* vol. 5, 1843, p. 141 n.; Parfitt, I, p. 651).

A tremulous motion felt at Camelford, Northhill, Liskeard (in Cornwall), and Mary Tavy, Peter Tavy, Buckland Monachorum and North Bovey (in Devon). The sound was also heard at St Tudy (5 miles S.S.W. of Camelford). The position of the epicentre is unknown, but it probably lies a few miles N. of Liskeard.

LAUNCESTON EARTHQUAKES

1. **1783** AUG. 9. Cat. No. 158 (*Gent. Mag.* vol. 53, 1783, p. 708).

An earthquake felt at Launceston.

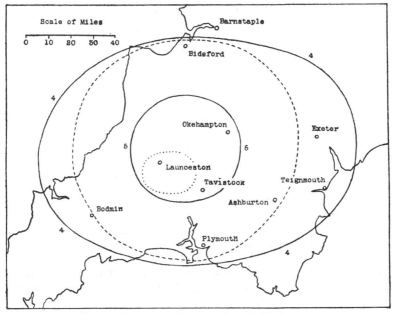

Fig. 75. Launceston Earthquakes of 1883 June 25.

2. **1883** JUNE 25, 1.39 *p.m.* Cat. No. 792; Intensity 5; Centre of isoseismal 5 in lat. 50° 41′ N., long. 4° 14′ W.; No. of records 56 from 40 places; Fig. 75 (Parfitt, I, pp. 655–657; III, p. 548; *Nature*,

vol. 28, 1883, pp. 199, 206; *Mon. Met. Mag.* vol. 18, 1883, pp. 89–90; etc.).

The continuous lines on the map (Fig. 75) are isoseismal lines of intensities 5 and 4. The isoseismal 5 is very nearly a circle, 24 miles in diameter and 490 square miles in area, its centre being about 5 miles E. 27° N. of Launceston. The isoseismal 4 is 74 miles long from E. to W., 52 miles wide, and contains about 3020 square miles.

The shock was a strong tremor, of mean duration about 3 seconds. The sound was heard by all the observers, and was compared to passing waggons, etc., in 79 per cent. of the records, thunder in 14, and explosions in 7, per cent.

3. 1883 JUNE 25, 2.7 *p.m.* Cat. No. 793; Intensity 4; Centre of isoseismal 4 in lat. 50° 41′ N., long. 4° 14′ W.; No. of records 23 from 16 places; Fig. 75.

The broken-line in Fig. 75 represents the boundary of the disturbed area, which is an isoseismal of intensity 4. It is nearly a circle, 53

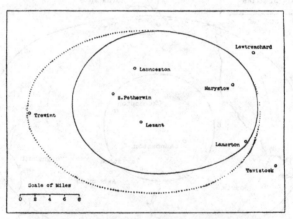

Fig. 76. Launceston Earthquake of 1896 Jan. 26.

miles long from E. to W., 50 miles wide, and about 2080 square miles in area. Its centre is practically coincident with that of the isoseismal 5 of the preceding earthquake.

The shock was slighter and of less duration than in the preceding earthquake. The sound was heard by all the observers, and was compared to passing waggons, etc., in 67 per cent. of the records and thunder in 33 per cent.

4. **1896** JAN. 26, 6.50 *a.m.* Cat. No. 910; Intensity 3 (nearly 4); Centre of disturbed area in lat. 50° 36′ N., long. 4° 19′ W.; No. of records 36 from 30 places and negative records from 20 places; Fig. 76 (*Geol. Mag.* vol. 7, 1900, pp. 166–167).

The continuous line represents the boundary of the disturbed area, which is 12 miles long, 8¾ miles wide, and contains 86 square miles. The direction of its longer axis is E., and the centre of the area lies 1½ miles S. 42° E. of Launceston.

The shock was a slight tremor. The sound was heard by all the observers. The boundary of the sound-area, indicated by the dotted line, is 15 miles long, 10 miles wide, and contains 124 square miles. It overlaps the disturbed area by 3 miles towards the west and by 1¼ miles towards the south.

EARTHQUAKES OF UNKNOWN EPICENTRES IN CORNWALL

1. **1752** JULY 15. Cat. No. 116 (Parfitt, I, p. 648).

An earthquake felt from St Mary's (Scilly Isles) to Plymouth.

2. **1756** DEC. 26. Cat. No. 128 (Perrey, p. 143).

Several shocks in Cornwall.

3. **1771** JULY —. Cat. No. 148 (Perrey, p. 147).

An earthquake in the Scilly Isles.

1792 Jan. 22, a reported earthquake in Cornwall (Roper, p. 27).

4. **1826** AUG. 30, 1 *a.m.* Cat. No. 319 (Letter to author).

A slight and momentary shock, accompanied by a deep rumbling sound, at St Tudy; felt also in most of the surrounding parishes.

1839 Jan. 21, a reported earthquake in the Scilly Isles (W. J. Henwood, *Cornwall Geol. Soc. Trans.* vol. 5, 1843, p. 141 n.).

1847 May 22, at night, a slight tremor felt on the shore of Mount's Bay, between Newlyn and Mousehole, by two coastguardsmen (Edmonds, p. 83).

5. **1858** APR. 2. Cat. No. 717 (Perrey, *Mém. Cour.* vol. 16, 1864, p. 41; Edmonds, p. 116).

An earthquake at Plymouth and Liskeard.

1858 Nov. 11, 0.20 p.m., a shock at Tolvaddon (1 mile E. of St Michael's Mount); the ground trembled for 2 or 3 seconds, and, after an interval of 3 or 4 seconds, the tremor was repeated. Both tremors were accompanied by a sound like those of distant guns at sea (Edmonds, pp. 84–86).

OKEHAMPTON EARTHQUAKES

1. **1752** FEB. 23. Cat. No. 113 (Mrs Bray, *Borders of the Tamar and Tavy*, vol. 1, p. 240).

A smart shock felt at many places on Dartmoor and in its immediate neighbourhood (Manaton, Moreton Hampstead, Widdecombe, etc.). In Widdecombe, houses were injured and one of the pinnacles of the church tower was thrown down.

2. **1858** SEPT. *early in.* Cat. No. 718 (G. W. Ormerod, *Quart. Journ. Geol. Soc.* vol. 15, 1859, p. 190).

A shock, accompanied by sound, at Cheriton Bishop and Fingle Bridge (1 mile S. of Drewsteignton).

3. **1858** SEPT. 28, *about 7.45 p.m.* Cat. No. 719; Centre of disturbed area in lat. 50° 42′ N., long. 3° 56′ W.; Records from 17 places and negative records from 16 places; Fig. 77 (G. W. Ormerod, *Quart. Journ. Soc.* vol. 15, 1859, pp. 188–191; Parfitt, I, p. 652).

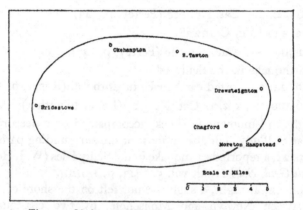

Fig. 77. Okehampton Earthquake of 1858 Sept. 28.

The boundary of the disturbed area is represented by the continuous line in Fig. 77. It is 15 miles long, 8 miles wide, and contains 95 square miles. Its centre is 4 miles S.E. of Okehampton and about 8 miles W. of that of the preceding earthquake, and the direction of its longer axis is about E. The sound (without any accompanying shock) was also heard at Crediton and Druids (1 mile N.W. of Ashburton). The sound-area must thus contain about 600 square miles.

1858 Sept. 30, in the evening, a noise was heard by a single observer at Trusham (2 miles N. of Chudleigh). This village lies 15 miles E.S.E. of the centre of the disturbed area of the preceding earthquake and close to the boundary of its sound-area (G. W. Ormerod, p. 189). 1866 Oct. 5, 8.55 p.m., a shock felt by a single observer at Chagford (Parfitt, I, p. 653).

DARTMOUTH EARTHQUAKE

1886 JAN. 4, 10.19 *a.m.* Cat. No. 801; Intensity 5; Centre of disturbed area in lat. 50° 20′ N., long. 3° 37′ W.; No. of records 30 from 14 places; Fig. 78 (*Nature*, vol. 33, 1886, p. 234; Parfitt, III, pp. 548–552; etc.).

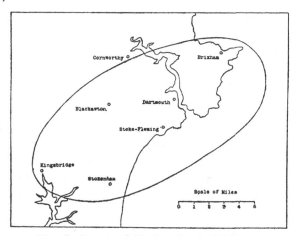

Fig. 78. Dartmouth Earthquake of 1886 Jan. 4.

The boundary of the disturbed area is represented by the continuous line in Fig. 78. It is 16 miles long, 8 miles wide, and contains about 100 square miles. Its centre lies 2 miles S.W. of Dartmouth, and the direction of its longer axis is about E.N.E. The duration of the shock was about 2 seconds.

A slight shock occurred on the southern border of Dartmoor on 1923 Dec. 25 at about 7.40 p.m. The disturbed area is an elongated ellipse, 14 miles long, 7½ miles wide, and containing about 82 square miles. Its centre is in lat. 50° 25·9′ N., long. 3° 49·4′ W. or about 1 mile N.E. of South Brent. The longer axis is directed about N. 74° E. The shock was a slight tremor and was accompanied by a rumbling sound. One or two of the accounts seem to imply that the disturbance was transmitted through the air and are suggestive of a meteoritic origin, but the evidence on the whole seems in favour of its seismic nature. In size and form of disturbed area, it resembles the Okehampton earthquake of 1858 Sept. 28 (on the northern border of Dartmoor) and the Dartmouth earthquake of 1886 Jan. 4.

EXMOUTH EARTHQUAKES

1. **1813** MAR. 21, 6 *a.m.* Cat. No. 261 (Howard, vol. 1, 1818, after Table 79).

A shock felt at Exmouth, Kenton, Sidmouth, Budleigh Salterton, Starcross, etc.

2. **1865** JUNE 14, 1.20 *a.m.* Cat. No. 739 (Perrey, *Mém. Cour.* vol. 19, 1867, p. 73).

A shock, accompanied by a rather loud detonation, at Teignmouth, Dawlish, Starcross, etc.

1871 Aug. 26, 4.25 a.m., a shock felt by a single observer at Teignmouth (Parfitt, I, p. 654).

1873 Apr. 14, 9.45 p.m., a disturbance, which may have been due to gun-firing, windows and shutters being shaken at Budleigh Salterton, Kenton and Starcross, while a noise like thunder was heard at Mamhead and Powderham Castle (Parfitt, I, p. 654).

1883 May 6, 5.55 —, a shock felt by two persons at Teignmouth, and a sound like the discharge of a gun heard by two others (Parfitt, I, p. 655).

BARNSTAPLE EARTHQUAKES

1. **1756** FEB. 27, 6 *p.m.* Cat. No. 124 (Parfitt, I, p. 650).

A rumbling noise at Ilfracombe.

2. **1789** MAY 5, 3.15 *a.m.* Cat. No. 165 (*Gent. Mag.* vol. 59, 1789, p. 437).

A shock, preceded by a rumbling noise, at Barnstaple.

3. **1809** MAY 3. Cat. No. 253 (Parfitt, I, p. 651).

A shock, accompanied by a rumbling noise, at Barnstaple.

1863 Oct. 5, about 10.30 p.m., a reported earthquake at Barnstaple (*Morning Herald*, Oct. 8).

1894 Jan. 23, about 10 p.m., a slight shock, with a noise like underground thunder, felt by a single observer at Arlington.

1894 Jan. 23, about 8.40 a.m., a series of small vibrations (intensity 3), preceded and accompanied by a noise like the rumbling of a heavy cart, felt by a single observer at Berrynarbor.

4. **1894 JAN. 23**, *about 9 a.m.* Cat. No. 902; Intensity 4; Centre of isoseismal 4 in lat. 51° 8·0′ N., long. 3° 46·0′ W.; No. of records 56 from 48 places and negative records from 13 places; Fig. 79 (*Geol. Mag.* vol. 3, 1896, pp. 553–556).

The continuous lines in Fig. 79 represent the isoseismals 4 and 3 of this earthquake, and the dotted line part of the boundary of the disturbed area of the Taunton earthquake of 1885 Jan. 22. The

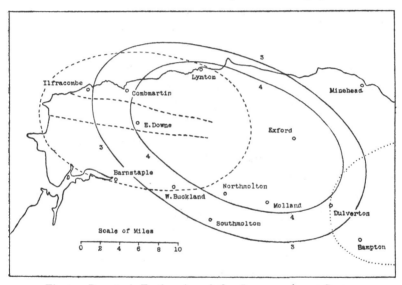

Fig. 79. Barnstaple Earthquakes of 1894 Jan. 23 and 1920 Sept. 9.

isoseismal 4 is 22¾ miles long, 12¼ miles wide, and contains 228 square miles. Its centre lies 5¾ miles W. of Exford, and its longer axis is directed N. 68° W. The isoseismal 3 is 29¾ miles long, 16½ miles wide, and contains 389 square miles. The distance between the two isoseismals is 1¼ miles on the north side and 3 miles on the south.

As a rule, the shock consisted of a tremulous motion, which increased gradually in strength and then died away.

The sound was heard by all the observers and was compared to passing waggons, etc., in 40 per cent. of the records, thunder in 43, wind in 5, the fall of a heavy body in 3, explosions in 5, and miscellaneous sounds in 3, per cent. At East Downe, two prominent vibrations were felt in the midst of the tremulous motion; and a sound was heard, as of a heavy cartload of iron crossing a wooden floor, which grew and died away with the trembling, encountering, as it were, two large stones at the instants when the prominent vibrations were felt.

5. **1920** SEPT. 9, *about* 11.15 *p.m.* Cat. No. 1187; Intensity 4; Centre of disturbed area in lat. 51° 9·9′ N., long. 4° 0·0′ W.; Records from 15 places; Fig. 79 (*Nature*, vol. 106, 1920, p. 132).

The broken-line in Fig. 79 represents the boundary of the disturbed area. It is 22 miles long, 13 miles wide, and contains 230 square miles. Its centre lies ½ mile N. of East Downe, or about 10 miles W. of that of the preceding earthquake, and its longer axis is directed N. 82° W.

6. **1920** SEPT. 10, *about* 0.15 *a.m.* Cat. No. 1188.

Two slighter shocks, separated by an interval of ten minutes, were felt in the same district about an hour later.

ORIGIN OF THE BARNSTAPLE EARTHQUAKES

From the evidence of the earthquake of 1894 (no. 4), it follows that the originating fault must be directed N. 68° W. in the neighbourhood of Exford, and must hade to the south; also that the focus was at a rather small depth below the surface, so that the fault-line must pass close to the centre of the isoseismal 4. The earthquake of 1920 shows that, farther west, near East Downe, the direction must be about N. 82° W. According to the late Dr Hicks[1], "a great thrust fault extends continuously along the northern boundary of the Morte Slates, from the coast near Ilfracombe to the Exe Valley. On the south side there is evidence generally also of a well-marked fault." Both faults, which are represented by broken-lines in Fig. 79, hade towards the S. The mean direction of the northern fault is N. 73° W. near East Downe and N. 63° W. near Exford. It is therefore probable that the Barnstaple earthquakes are connected with this northern boundary fault. As the distances from it of the epicentres of the earthquakes of 1894 and 1920 are respectively ½ and 1 mile S. of the fault-line, it follows that the foci of both are at a small depth below the surface. This is also evident from the small and elongated disturbed areas and the closeness of the isoseismal lines in the earthquake of 1894.

---

[1] *Quart. Journ. Geol. Soc.* vol. 52, 1896, pp. 254–272.

The first movements with which we are acquainted were very small slips, almost creeps, near the western ends of the faults as traced. These occurred in 1756, 1789 and 1809, and possibly also in 1863 and a few hours before the chief movement of the series in 1894. The focus in this earthquake must have been 9 or 10 miles in horizontal length. Then, after a lapse of 26 years, followed the slip near East Downe in 1920, the focus of which was about 8 miles long. As the distance between their epicentres was about 10 miles, it follows that the nearer margins of the foci almost coincided, if they did not slightly overlap. An hour later, the series of movements closed with two small movements in the western or East Downe focus (see also p. 319).

EARTHQUAKES OF UNKNOWN EPICENTRES
IN DEVON

1536, a slight earthquake is said to have been felt at Plymouth (Ll. Jewitt, *Hist. of Plymouth*, 1873, p. 88).

1. **1727** JULY 19, *between 4 and 5 a.m.* Cat. No. 82 (Parish Register of St Pancras, Exeter, quoted in J. C. Cox's *Parish Registers of England*, p. 212, and *Notes and Queries*, vol. 1, 1862, p. 177).

"All the houses in Exeter did shake with an Earthquake that people was shakt in their beds from one side to the other, and was al over England, and in some places beyound sea, but doed little damage; tis of a certain truth."

2. **1750** FEB. 20, *about 1 a.m.* Cat. No. 101 (Rev. W. Barlow, *Phil. Trans.* vol. 46, 1752, pp. 692–695).

A shock felt at and near Plymouth.

1774 Oct. 25, about 10 p.m., a slight shock, lasting 2 seconds, felt in several house in Exeter (Parfitt, I, p. 651).

1866 Sept. 13, about 9.45 p.m., two distinct shocks, about five minutes apart, at East Budleigh, the second and more violent being also felt at Branscomb and Budleigh Salterton (*Brit. Rainfall*, 1866, p. 48; *Times*, Sept. 19).

1869 Aug. 29, 1 p.m., a shock, accompanied by a rumbling sound, felt to the S.E. of Exeter (Parfitt, I, p. 654).

1869 Aug. 29, 1.15 p.m., a noise, without any accompanying shock, heard to the S.E. of Exeter (Parfitt, I, p. 654).

1871 Mar. 23, about 8.20 p.m., a slight shock at Plymouth and Devonport (*Manchester Guardian*, Mar. 25).

1881 Apr. 29, 8.15 a.m., a shock at Ashburton (*Brit. Rainfall*, 1881, p. 10).

TAUNTON EARTHQUAKES

1. **1747** JULY 1, *between* 10 *and* 11 *p.m.* Cat. No. 96; Intensity 5 (Rev. J. Forster, *Phil. Trans.* vol. 45, 1750, pp. 398–400; *London Mag.* 1750, p. 103).

A shock, accompanied by a rumbling noise like distant thunder, felt at Taunton and as far as Exeter (28 miles from Taunton).

2. **1868** JAN. 4, 5.10 *a.m.* Cat. No. 743; Intensity 4; Centre of disturbed area in lat. 50° 59′ N., long. 2° 48′ W.; Records from 13 places; Fig. 80 (*Times*, Jan. 6 and 10).

The boundary of the disturbed area is represented by the continuous line in Fig. 80. It is 23 miles long, 11 miles wide, and contains

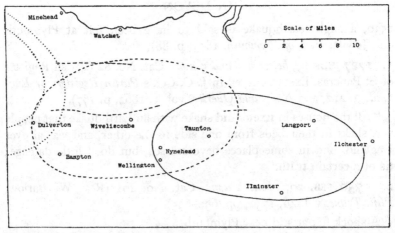

Fig. 80. Taunton Earthquakes of 1868 Jan. 4 and 1885 Jan. 22.

about 200 square miles. Its centre is 6 miles S. 78° E. of Taunton, and its longer axis is directed N. 83° W. The shock was accompanied by a low rumbling noise.

1868 Jan. 8, 4.25 *a.m.*, an undulation, followed by a trembling and low rumbling noise, felt by a single observer at Nynehead Court (near the west end of the disturbed area of the preceding earthquake).

3. **1885** JAN. 22, *between* 8.30 *and* 9 *p.m.* Cat. No. 797; Intensity 4; Centre of disturbed area in lat. 51° 3′ N., long. 3° 18′ W.; No. of records 27 from 23 places; Fig. 80 (*Nature*, vol. 31, 1885, pp. 289, 339; *Times*, Jan. 26).

The boundary of the disturbed area is represented by the broken-line in Fig. 80. It is 22 miles long, 12 miles wide, and contains about 210 square miles. Its centre lies $\frac{1}{2}$ mile N.E. of Wiveliscombe, and its longer axis is directed E. The shock was a mere trembling, accompanied by a rumbling noise.

ORIGIN OF THE TAUNTON EARTHQUAKES

The dotted lines in Fig. 80 represent portions of the isoseismal lines of the Barnstaple earthquake of 1894. The Taunton earthquakes of 1868 and 1885 appear from this map and from Fig. 79 to be closely connected with the Barnstaple earthquakes of 1894 and 1920. The broken-lines in Fig. 80 between Watchet and Wiveliscombe represent portions of the northern and southern boundary faults of the Morte Slates as traced by the late Dr Hicks[1]. The Taunton earthquakes thus seem to be connected with the southern, rather than with the northern, boundary fault; and it is interesting to notice the continual westerly migration of the epicentre from 1868 to 1920. The four earthquakes were closely alike in the forms of their isoseismal lines, the dimensions of the disturbed areas of the earthquakes of 1868, 1885 and 1920 and of the isoseismal 4 of the earthquake of 1894 being almost identical. The relations of these earthquakes are considered in Chapter XXIII.

WELLS EARTHQUAKES

1. **1248** DEC. 21. Cat. No. 32; Intensity 8 (Matthew Paris, vol. 3, pp. 42, 305; *Annal. Monast.* vol. 4, p. 439; Stow, p. 285; etc.).

"In the same year, on the day of our Lord's Advent, which was the fourth day before Christmas, an earthquake occurred in England, by which (as was told to the writer of this work by the Bishop of Bath, in whose diocese it occurred) the walls of buildings were burst asunder, the stones were torn from their places, and gaps appeared in the ruined walls. The vaulted roof which had been placed on the top of the church of Wells by the great efforts of the builder, a mass of great size and weight, was hurled from its place, doing much damage, and fell on the church, making a dreadful noise in its fall from such a height, so as to strike great terror into all who heard it... This earthquake was the third which had occurred within three years on this side

[1] *Quart. Journ. Geol. Soc.* vol. 53, 1897, pp. 438–458. The broken-line to the north-east of Taunton represents an E.–W. fault, the course of which is given in the Geological Survey map, new series, sheet 295.

the Alps: one in Savoy, and two in England; a circumstance unheard of since the beginning of the world, and therefore the more terrible" (Matthew Paris, trans. by Rev. J. A. Giles, vol. 2, 1853, p. 286). 1681 Jan. 3, according to Roper (p. 16), an earthquake at Wells.

2. **1852** APR. 1, 5.30 *a.m.* Cat. No. 704 (*Times*, Apr. 12).

A shock felt at Chaddar and Winscombe (1 mile N.W. of Axbridge), both places being near the western end of the disturbed areas of the earthquakes of 1893.

3. **1893** DEC. 30, 11.20 *p.m.* Cat. No. 898; Intensity 4 (nearly 5); Centre of isoseismal 4 in lat. 51° 13′ N., long. 2° 40′ W.; No. of records 45 from 27 places and negative records from 20 places; Fig. 81 (*Geol. Mag.* vol. 7, 1900, pp. 108–112).

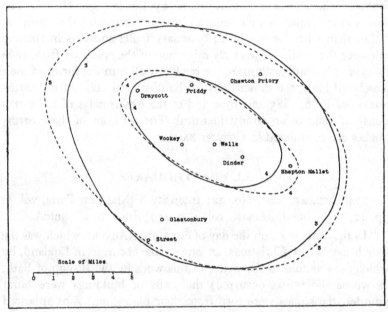

Fig. 81. Wells Earthquakes of 1893 Dec. 30 and 31.

The continuous lines in Fig. 81 represent the isoseismals 4 and 3. The isoseismal 4 is 9½ miles long, 5 miles wide, and 38 square miles in area. The direction of its longer axis is N. 60° W., and its centre is 1 mile N. 35° W. of Wells. The isoseismal 3, which bounds the disturbed area, is 15½ miles long, 13¼ miles wide, and contains 159 square miles. The distance between the isoseismals is 3 miles on the north-east side and 5 miles on the south-west.

In all parts, the shock was a continuous series of vibrations, like that felt in an upper room when a heavy waggon passes. Its mean duration was 4 seconds. The sound was heard by 91 per cent. of the observers.

4. **1893** DEC. 31, 0.28 *a.m.* Cat. No. 899; Intensity 4 (nearly 5); Centre of isoseismal 4 in lat. 51° 12′ N., long. 2° 39′ W.; No. of records 41 from 25 places and negative records from 20 places; Fig. 81.

The broken-lines in Fig. 81 represent the isoseismals 4 and 3. The isoseismal 4 is 11½ miles long, 4¾ miles wide, and includes 43 square miles. The direction of its longer axis is N. 60° W. Its centre is ¾ mile N. 63° W. of Wells, and ¼ mile from the corresponding centre for the preceding shock, the line joining the two centres being parallel to the longer axes of both curves. The disturbed area, which is bounded by an isoseismal of intensity 3, is 19¾ miles long, 12½ miles wide, and contains 180 square miles. The distance between the two isoseismals is about 3 miles on the north-east side and about 5 miles on the south-west.

In nature and duration, the second shock resembled the first. The sound was heard by all the observers. Taking the two earthquakes together, the sound was compared to passing waggons, etc., in 26 per cent. of the records, thunder in 11, wind in 5, loads of stones falling in 13, the fall of a heavy body in 11, explosions in 26, and miscellaneous sounds in 8, per cent.

5. **1893** DEC. 31, *about* 4 *a.m.* Cat. No. 900; Intensity about 3; No. of records 4 from 4 places.

A slight shock was felt at Priddy, Street, Wells and Wookey. At Street and Wells, it was accompanied by sound. The epicentre probably coincides with those of the two preceding earthquakes or lies slightly to the west of them.

EARTHQUAKES OF UNKNOWN EPICENTRES IN SOMERSET

1. **1122** JULY 25. Cat. No. 15 (*Anglo-Saxon Chron.* vol. 2, p. 217; *Annal. Monast.* vol. 1, p. 10).

"And the eighth night before the calends of August was a very violent earthquake all over Somersetshire and in Glocestershire" (*Anglo-Saxon Chron.*).

2. **1201 MAY 22.** Cat. No. 25 (*Annal. Monast.* vol. 2, p. 74; vol. 4, p. 390; B. de Cotton, p. 92).

A great earthquake in Somerset.

3. **1275 SEPT. 11.** Cat. No. 36; Intensity 8 (*Annal. Monast.* vol. 2, pp. 119, 384, 386; vol. 3, p. 266; vol. 4, pp. 264, 265, 467–469; *Flor. Hist.* vol. 3, p. 46; *Eulog. Hist.* vol. 3, pp. 138, 142; *N. Triveti Annales*, p. 293; John de Oxenedes, p. 247; Henry of Knyghton, col. 2461; W. Rishanger, p. 86; etc.).

An earthquake felt throughout all England, and especially at Glastonbury, where the church of St Michael's was thrown down.

There is some doubt as to the date of this earthquake. The great majority of the chroniclers give the year as 1275, and most of these the date as Sept. 11. Other earthquakes on 1274 Dec. 5 and 1276 Sept. 11, which are also said to have destroyed the church of St Michael's, are no doubt the same (1274, John Capgrave, pp. 163–164; 1276, Stow, p. 305).

4. **1680 JAN. 4.** Cat. No. 68 (*Gent. Mag.* vol. 20, 1750, p. 56; *London Mag.* 1750, p. 103).

An earthquake in Somerset.

1726 Nov. 6, about 6 a.m., an earthquake at Ilchester (Mallet, 1852, p. 124).

1745 June 15, Roper (p. 18) records an earthquake in Somersetshire on this date, without giving any authority.

5. **1752 MAR. 31.** Cat. No. 114 (Perrey, p. 135).

An earthquake felt at Bristol and at various places in Somerset.

6. **1752 APR. 16.** Cat. No. 115 (Perrey, p. 174).

An earthquake in Somerset.

7. **1852 APR. 3, 3 *a.m.*** Cat. No. 705 (T. Godwin-Austin, *Quart. Journ. Geol. Soc.* vol. 8, 1852, pp. 233–234).

A shock lasting about 2 seconds.

8. **1852 APR. 3, 5.35 *a.m.*** Cat. No. 706; Intensity 4.

A shock felt from the Mendip Hills to Bristol, at Wells, Cheddar, Pensford, Dundry, Bristol, Clifton, Cotham, Kingsdown, Westbury-upon-Trym, and Henbury; lasting a few seconds and accompanied by a low rumbling noise.

1852 Dec., Perrey (*Bull.* vol. 23, 1856, p. 46) records, without details, the occurrence of an earthquake in Somerset.

CHAPTER XVI

EARTHQUAKES OF THE SOUTH
OF ENGLAND

IN the south of England—Dorsetshire, Wiltshire and Hampshire (except near Portsmouth)—earthquakes are few in number, of slight intensity, and of little interest.

DORSETSHIRE EARTHQUAKES

1583 Jan. 13, a disturbance evidently due to a landslip (Stow, p. 1173).

1583 July 1, an earthquake in Dorsetshire (Short, vol. 2, p. 169).

1588, an earthquake in Dorsetshire (Sir R. Baker, p. 398).

1689, Lyme Regis said to have been destroyed by an earthquake (Roper, p. 16, quoting from *Tablet of Memory*, p. 53).

1726 Oct. 25, an earthquake in London, Dorchester and other places (Lowe, p. 44).

1. 1750 MAY 4, *about 10 a.m.* Cat. No. 109; Intensity 4 (*Phil. Trans.* vol. 46, 1752, pp. 689–691).

A shock with sound at Winborne (Wimborne or Winterborne?), Cashmoor, Shapeele and Eastbrook.

1761 Feb. 6, between 11 and 12 p.m., a shock, accompanied by a rumbling noise, at Sturminster (*Ann. Reg.* 1761, p. 69).

2. 1761 JUNE 9, 11.45 *a.m.* Cat. No. 135 (Milne, vol. 31, p. 104; *Gent. Mag.* vol. 31, 1761, p. 282).

A strong shock at Sherborne, Shaftesbury, and (according to Roper, p. 23) for 13 miles round.

1762 Mar. 20, an earthquake at Shaftesbury (Perrey, p. 146).

1839 Dec. 24, an earthquake on the coast of Dorsetshire (Perrey, p. 176).

1877 Nov. 2, two reported earthquakes (apparently felt by only one observer) at Bullpits (7 miles W.N.W. of Shaftesbury). At about 8.10 a.m., a very slight vibration was felt, accompanied by a rumbling noise. At 11.20 a.m., a stronger vibration (intensity 4) lasting about 6 seconds, accompanied by a heavier rumbling noise (*Nature*, vol. 17, 1878, p. 38).

WILTSHIRE EARTHQUAKES

1684, an earthquake at Salisbury (Short, vol. 2, p. 170).

1692 June 18, an earthquake at Clarendon (3 miles E. of Salisbury) (Perrey, p. 128).

1793 SEPT. 28, *immediately after* 4 *p.m.* Cat. No. 182 (*Gent. Mag.* vol. 63, 1793, pp. 950–951).

A rather strong earthquake felt at Salisbury and Shaftesbury (18 miles W. of Salisbury). At Fovant (9 miles W.), the casement of a window is said to have been thrown out, at Swallowcliffe (11 miles W.) a chimney thrown down, and at Hindon (14 miles W.) the bells in most houses rang.

1793 Oct. 5, an earthquake at Shaftesbury, which may be the same as the preceding (*Mon. Met. Mag.* vol. 3, 1868, p. 4).

1884 Dec. 25, about 10.20 p.m., a slight shock at Ramsbury apparently felt by a single observer (*Nature*, vol. 31, 1885, p. 200).

HAMPSHIRE EARTHQUAKES

1081 MAR. 27. Cat. No. 6 (Matthew Paris, vol. 1, p. 26; vol. 2, p. 18; *Flor. Hist.* vol. 2, p. 11; B. de Cotton, p. 51; Holinshed, vol. 2, p. 23; Speed, p. 431; Roger of Wendover, vol. 2, p. 21).

"A most fearefull Earthquake, with a warring noise, did shake the ground" (Speed), "at the first hour of the night"(Roger of Wendover).

1864 Sept. 7, morning, and Sept. 17, 9.50 p.m., slight shocks, accompanied by a noise as of the fall of a heavy body, at Headley, 7 miles W. of Hazlemere (*Times*, Sept. 20).

1883 Oct. 10, early morning, two distinct tremors felt (*Nature*, vol. 28, 1883, p. 623).

CHAPTER XVII

EARTHQUAKES OF THE SOUTH-EAST OF ENGLAND

THE counties included in this district are the east of Hampshire, Sussex, Surrey, Kent, Middlesex, Hertford and Essex, in which the surface-rocks for the most part belong to somewhat recent formations. The earthquakes are infrequent, occur usually at long intervals, and are sometimes of a strength much above the average. To one centre, that of Colchester, belongs the most destructive earthquake known in this country.

CHICHESTER EARTHQUAKES

1. **1553.** Cat. No. 48 (R. Fabyan, p. 711).

An earthquake felt at different places, but especially at Southsea, between Easter and Whitsuntide.

2. **1638** *end of the year.* Cat. No. 61 (Mallet, 1852, p. 75).

Several shocks, causing great damage, at Chichester.

3. **1707** OCT. 25, 3.30 *p.m.* Cat. No. 79; Intensity 5 (Milne, vol. 31, p. 96).

A shock felt at Chichester and along the coast from Havant to Shoreham.

4. **1734** OCT. 25, 1 *a.m.* Cat. No. 87 (*Gent. Mag.* vol. 4, 1734, pp. 571, 625; Milne, vol. 31, p. 96).

A shock felt at Portsmouth, Milton and other places.

5. **1734** OCT. 25, *about* 3.50 *a.m.* Cat. No. 88; Intensity 5 (*Phil. Trans.* vol. 39, 1738, pp. 361–366; *London Mag.* 1734, pp. 551, 605).

A strong earthquake, felt especially at Portsmouth, Chichester and Lewes, and at intermediate places along and near to the coast, but apparently at no great distance inland. At Lewes, two shocks were felt. At Havant, there was a quick tremulous motion lasting about 2 or 3 seconds, repeated after a short interval.

6. **1750** MAR. 29, *about* 5.45 *p.m.* Cat. No. 105; Intensity 7; Centre of disturbed area in about lat. 50° 48′ N., long. 1° 4′ W.; Fig. 82 (*Phil. Trans.* vol. 46, 1752, pp. 646–647, 649–656, 688; *Gent. Mag.* vol. 20, 1750, p. 137; *London Mag.* 1750, p. 139).

This was one of the more important of a remarkable series of earthquakes. The boundary of the disturbed area is approximately a semicircle about 150 miles in diameter with its centre near Portsmouth. It was felt as far as Bridport, Bath, Hertford, Loughton, Bromley, etc., and was strongest in the neighbourhood of Portsmouth and in the Isle of Wight. The disturbed area probably contained about 18,000 square miles. At Portsmouth, a jarring was felt, succeeded immediately by three or four slow vibrations, the whole movement being accompanied by a noise like distant thunder.

7. **1750** MAR. 30, 3.30 *a.m.* Cat. No. 106 (*Phil. Trans.* vol. 46, 1752, p. 651).

A shock felt in the Isle of Wight at Newport, St Helens, etc.

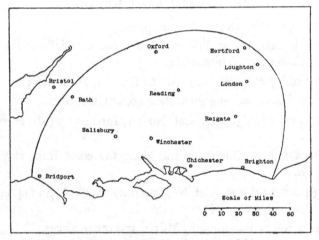

Fig. 82. Chichester Earthquake of 1750 Mar. 29.

8. **1750** APR. 6, 3 *or* 4 *a.m.* Cat. No. 107 (Milne, vol. 31, p. 99). A shock felt in the Isle of Wight.

9. **1750** SEPT. 18, 6 *p.m.* Cat. No. 111 (Milne, vol. 31, p. 99). A shock felt at Portsmouth and in the Isle of Wight.

10. **1811** NOV. 30, *about* 2.40 *a.m.* Cat. No. 256; Intensity 5; Centre of disturbed area in lat. 50° 50′ N., long. 0° 50′ W.; Records from 11 places; Fig. 83 (Milne, vol. 31, p. 115; *Edin. Adver.*).

The boundary of the disturbed area is represented by the broken-line in Fig. 83. It is 30 miles long, 20 miles wide, and contains about 470 square miles. Its centre is about 2 miles W.S.W. of Chichester, and the direction of its longer axis is about N.E. At Chichester, the shock was accompanied by a noise like the rolling of thunder ending with a crash like that of an explosion. 1812, according to Perrey (p. 151), an earthquake was felt at Portsmouth.

11. **1814** DEC. 6, *a few minutes before 2 a.m.* Cat. No. 264; Intensity 6 (*Ann. Reg.* 1814, p. 166).

A shock felt very generally in and near Portsmouth and as far as Arundel.

12. **1821** FEB. 1, *about 2.30* —. Cat. No. 304 (*Ann. Reg.* 1821, p. 20).

A tremulous motion, accompanied by a noise like the rumbling of heavy carriages over stones, at Alfriston.

13. **1824** DEC. 6, *1.45 p.m.* Cat. No. 317; Intensity 5; Centre of disturbed area in about lat. 50° 50′ N., long. 0° 50′ W.; Records from 8 places; Fig. 83 (*Phil. Mag.* vol. 65, 1825, pp. 70–71; *Ann. de Chim. et de Phys.* vol. 27, 1824, p. 380).

The boundary of the disturbed area lies inside that of the earthquake of 1811 (no. 10) and is approximately concentric with it. It is about 25 miles long from N.E. to S.W., 17 miles wide, and contains about 330 square miles.

The remaining earthquakes belonging to this centre form an interesting series. With the exception of the last (no. 22), they were studied by a small Committee, the report of which was written by Mr. J. P. Gruggen and read before the Royal Society (p. 3).

14. **1833** SEPT. 18, *about 10 a.m.* Cat. No. 330.

The shock was felt from Liphook (16 miles N. of Chichester) to Birdham (3 miles S.W.), and possibly at Longfleat (near Poole, about 52 miles from Chichester). It resembled the thud caused by the fall of a heavy body, followed by a vibration. It was felt by a boat's crew in Chichester harbour, the impression conveyed being as if the boat had suddenly struck on a rock. In Stanstead Forest and elsewhere, pheasants crowed as they usually do when the evening gun at Portsmouth is fired.

15. **1833** NOV. 13, *2.40 a.m.* Cat. No. 331.

The area disturbed was nearly the same as in the preceding earthquake but extended a little farther to the north. The shock consisted of a series of rapid undulations and was preceded by a low sound.

328 EARTHQUAKES OF THE SOUTH-EAST OF ENGLAND

16. 1833 NOV. 13, 6 *a.m.* Cat. No. 332.

A much slighter shock than the preceding, felt in Chichester and its immediate vicinity.

17. 1834 JAN. 23, 2.45 *a.m.* Cat. No. 333; Intensity 5; Centre of disturbed area in lat. 50° 52′ N., long. 0° 52′ W.; Records from 16 places; Fig. 83.

The boundary of the disturbed area is represented by a continuous line in Fig. 83. It is 38 miles long, 26 miles wide, and contains about 780 square miles. The centre is 4 miles W.N.W. of Chichester, and the longer axis is directed N. 60° E. The shock is said to have been

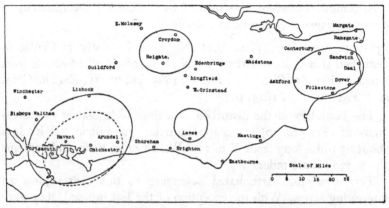

Fig. 83. Chichester Earthquakes of 1811 Nov. 30 and 1834 Jan. 23, Lewes Earthquake of 1864 Apr. 30, Canterbury Earthquake of 1776 Nov. 27, and Reigate Earthquake of 1551 May 25.

felt at Longfleat. The shock usually consisted of several undulations, preceded and accompanied by a sound as of the continuous falling of heavy pieces of furniture.

18. 1834 FEB. 20, 2 *a.m.* Cat. No. 334.

A slight shock at Chichester.

19. 1834 AUG. 27, 10.25 *p.m.* Cat. No. 336; Records from 8 places.

Though this shock was apparently slighter than that of Jan. 23 (no. 17), it was felt farther to the west, by about 1 mile at Bishop Waltham and 7 miles at Southampton. The disturbed area must therefore contain about 1100 square miles. At Chichester, some loose chimney-tops were partially thrown down. The shock was a tremulous motion and was preceded by a low rumbling sound.

20. **1834** SEPT. 21, 11.20 *a.m.* Cat. No. 337.

A slight shock at Chichester.

1834 Oct. 5, according to Perrey (p. 155) a slight shock was felt at Chichester, but no reference is made to it in Mr Gruggen's report.

21. **1835** JAN. 12, 8 *a.m.* Cat. No. 338.

A slight shock, slighter than any of the recent series, at Chichester.

22. **1835** AUG. 3. Cat. No. 339 (*Times*, Aug. 24).

A slight shock at Chichester.

1888 Dec. 28, 11 a.m., a shock and rumbling sound in the neighbourhood of Emsworth Common (7 miles W. of Chichester), with a violent rustling of trees. The whole account is suggestive of distant gunfiring (*Nature*, vol. 39, 1889, p. 231).

LEWES EARTHQUAKES

1. **1864** APR. 30, 11.6 *p.m.* Cat. No. 731; Intensity 5; Centre of disturbed area in lat. 50° 56′ N., long. 0° 2′ E.; Records from 7 places; Fig. 83 (*Times*, May 3 and 4).

The boundary of the disturbed area is represented by the continuous line near Lewes in Fig. 83. It is approximately circular, 10 miles in diameter and about 80 square miles in area. Its centre lies 4 miles N. of Lewes.

A rushing noise, like that of hail driven against a window, was also heard after midnight at Lewes and Maresfield.

1864 June 1, an earthquake in Sussex (Perrey, *Mém. Cour.* vol. 21, 1870, p. 25).

2. **1864** AUG. 21, 1.27 *a.m.* Cat. No. 733; Intensity 5 (Rev. E. B. Ellman, *Brit. Ass. Rep.* 1864, pt. ii, pp. 16–17; *Times*, Aug. 22).

A shock, accompanied by a rumbling sound like the passing of a heavily-laden waggon over a rough stone surface, felt at Lewes and to a distance of 15 miles round the town.

In addition to the above, there are several records of earthquakes of doubtful origin in various parts of Sussex:

1833 Apr., about the 2nd, a shock at Horsham (Mallet, 1854, p. 237).

1871 May 21, noon, a slight quivering of the ground felt at Hastings, apparently by a single observer (*Nature*, vol. 20, 1879, p. 161).
1883 Jan. 16, 9.9½ a.m., a shock lasting about 4 seconds felt at Hastings (*Nature*, vol. 27, 1883, p. 293).
1887 Aug. 17, about 7.18 p.m., a slight vibration accompanied by a heavy rumbling noise at Forest Row (3½ miles S.E. of E. Grinstead) (*Times*, Aug. 19).

CANTERBURY EARTHQUAKES

For nearly four and a half centuries, the county of Kent has been free from strong earthquakes, but, before the end of the fourteenth century, it was visited by two or three shocks that caused injury to property.

1246 June 1, "there happen'd so great an Earthquake in England, that the like had been seldom seen or heard. In Kent it was more violent than in any other Part of the Kingdom, where it overturn'd several Churches" (Cat. No. 29; Perrey, p. 120).

1. **1382 MAY 21.** Cat. No. 41; Intensity 8 (W. Thorn, cols. 2157–2158; Henry of Knyghton, col. 2644; T. Walsingham, vol. 2, p. 67; *Fasc. Zizan.* pp. 272–273; *Eulog. Hist.* vol. 3, p. 356; *Hist. Monast. S. August. Cant.* p. 68; *Annal. Monast.* vol. 3, p. 480; *Chron. Angliae*, p. 351; R. Fabyan, p. 531; *John Hardyng's Chron.* pp. 339–340; Holinshed, vol. 2, pp. 754–755; etc.).

"An erthquake in Englande, that the lyke thereof was never seen in Englande before that daye nor sen" (R. Fabyan). "A great earthquake in England at nine of the clocke, fearing the hearts of many, but in Kent it was most vehement, where it suncke some Churches, and threwe them down to the earth" (Stow). Holinshed gives the time as about 1 p.m. The date is given as 1380 May 21 by Prestwich (p. 544) and 1381 June 12 by John Capgrave (p. 238).

This earthquake was followed on May 24 by another, less violent. The area disturbed by it is unknown, and it may have been of submarine origin.

2. **1580 MAY 1,** *after midnight.* Cat. No. 54 (Stow, pp. 1163–1164; R. Furley, *Hist. of the Weald of Kent*, vol. 2, pt. 2, pp. 513–514).

"An Earthquake felt in divers Places in Kent, namely, at Ashford, Great Chart, etc., which made the People rise out of their Beds, and run to their Churches, where they called upon God by earnest Prayer to be merciful unto them."

3. **1756** JUNE 1. Cat. No. 125 (Milne, vol. 31, p. 103).

A shock at Ashford, accompanied by a noise like the report of a cannon or waggon passing.

4. **1776** NOV. 27, 8.15 *a.m.* Cat. No. 152; Intensity 6; Centre of disturbed area in about lat. 51° 11′ N., long. 1° 8′ E.; Fig. 83 (*Gent. Mag.* vol. 46, 1776, p. 574).

The continuous curve near Canterbury in Fig. 83 bounds the places (Ashford, Canterbury, Dover, Folkestone, Sandwich, etc.) at which the shock was felt most strongly. It is 24 miles long, 16 miles wide, and contains about 300 square miles. Its centre lies about 7 miles S.S.E. of Canterbury, and its longer axis is directed about E.N.E. The shock was also felt at Calais (24 miles S.E. of the curve and 34 miles from its centre). The shock lasted 8 seconds and was attended by a distant rumbling noise.

5. **1831** MAR. 2. Cat. No. 327 (Mallet, 1854, p. 224).

A strong shock felt at Dover, Ramsgate, Margate and Deal.

6. **1848** OCT. 19, 7 *a.m.* Cat. No. 696 (Edmonds, p. 116; Perrey, *Mém. Cour.* vol. 23, 1873, p. 12).

A slight trembling, lasting 4 or 5 seconds and accompanied by a subterranean noise, at Sandwich.

REIGATE EARTHQUAKES

1. **1551** MAY 25, *about noon.* Cat. No. 47; Centre of disturbed area in about lat. 51° 16′ N., long. 0° 10′ W.; Records from 8 places; Fig. 83 (Strype, vol. 2, 1721, p. 272; Stow, p. 1021).

The continuous curve near Reigate in Fig. 83 bounds the places at which the shock was distinctly felt. It is 17 miles long from E. to W., 14 miles wide, and contains about 190 square miles. Its centre lies about 6 miles S.W. of Croydon.

2. **1758** JAN. 24, 2 *a.m.* Cat. No. 132; Intensity 4 (J. Burrow, *Phil. Trans.* vol. 50, 1759, pp. 614–617).

The disturbed area lies to the south of the above and partly overlaps it. It includes the following places: Edenbridge, E. Grinstead, Lingfield, Starborough Castle and Worth. The centre of the area was probably near Lingfield.

1880 June 29, 9 p.m., a low rumbling sound, lasting 2 or 3 seconds, was heard by two persons in the same room at Capel (about 8 miles S.W. of Reigate); it was repeated five or six times in the course of four or five minutes; the last three or four times, the rumbling was accompanied by a slight vibration sensible only to the feet (*Nature*, vol. 22, 1880, p. 230).

Though belonging to another part of Surrey, the following reported earthquake may be entered here:

1750 Mar. 14, early morning, a slight shock felt by a single observer at East Molesey, about 2 miles W. of Kingston (Rev. S. Hales, *Phil. Trans.* vol. 46, 1752, pp. 684–687).

MAIDSTONE EARTHQUAKE

1860 SEPT. 3, 3.30 *p.m.* Cat. No. 725; Centre of disturbed area in about lat. 51° 16′ N., long. 0° 22′ E. (Edmonds, p. 124).

A slight shock felt at Maidstone, Sevenoaks, Tonbridge, Wrotham and other places in Kent. The disturbed area is about 15 miles long from E. to W., 10 miles wide, and contains about 120 square miles. Its centre lies about 7 miles W. of Maidstone.

The following disturbances in Kent have been recorded as earthquakes, but their seismic origin is doubtful. That of 1727 was probably a landslip:

1581 Apr. and May 1, a trembling of the earth in Kent (Sir R. Baker, p. 398).

1727 Aug. 10, an earthquake felt at Skeat Hill near Dartford. On the same morning, ground sank in a meadow near Farningham (Rev. E. Barrel, *Phil. Trans.* vol. 35, 1729, pp. 305–306).

1749, a shock of intensity 4 felt at Blackheath; a noise like the firing of guns was also heard (Hart, *Hist. of Lee*, 1882, p. 91).

1828 Feb. 23, a few minutes past 8 a.m., an earthquake reported at Boughton-under-Blean near Faversham (*Gent. Mag.* vol. 98, 1828, p. 194).

LONDON EARTHQUAKES

Many of the earthquakes felt in London are of external origin. There is, however, a centre situated within the present city, which is responsible for the remarkable shocks of 1750 and possibly for others. Four earthquakes (those of 1247, 1580, 1634–35 and 1726) were felt over wide areas, though perhaps more strongly in London than elsewhere.

1. **1247** FEB. 13. Cat. No. 31 (Matthew Paris, vol. 3, pp. 20, 299; *Annal. Monast.* vol. 1, pp. 136, 285; vol. 2, p. 338; vol. 4, p. 438; *Flor. Hist.* vol. 2, p. 329; vol. 3, pp. 240–241; *Brut y Tywysogion*, p. 333; B. de Cotton, p. 126; *Eulog. Hist.* vol. 3, p. 138; etc.).

"At various places in England, especially at London, and there mostly on the banks of the Thames, an earthquake was felt, which shook buildings, and was very injurious and terrible in its effects" (Matthew Paris).

2. **1580** APR. 6, *about 6 p.m.* Cat. No. 53; Intensity 8 (W. Stukeley, *Phil. Trans.* vol. 46, 1752, p. 660; Stow, p. 1163; Camden, p. 244; R. Furley, *Hist. of the Weald of Kent*, vol. 2, pt. 2, p. 513; *London Mag.* 1750, p. 103; *Notes and Queries*, vol. 12, 1891, p. 294; etc.).

A very great earthquake in London and almost generally throughout England. Stukeley says that no houses were thrown down. A piece of the Temple Church, however, fell, also some stones from St Paul's Church, a stone fell from the top of Christ Church, which killed an apprentice, and various chimneys were thrown down by the shock. At Dover, part of the cliff fell into the sea with a portion of the castle wall. A small part of Saltwood Castle and of Sutton Church (both in Kent) fell down. At these places, and also in East Kent, three shocks were felt—at 6 p.m., 9 p.m., and 11 p.m. The shock of 6 p.m. was also felt in France and Belgium.

3. **1601** DEC. 24. Cat. No. 58 (Perrey, p. 125).

A shock in London without much damage.

1634–35, an earthquake in England, London being shaken (Short, vol. 1, p. 313).

4. **1726** OCT. 25. Cat. No. 81 (Lowe, p. 44).

An earthquake in London, Dorchester and other places.

1750 Feb. 17, about 7 p.m., a reported shock in London (*Phil. Trans.* vol. 46, 1752, p. 604).

5. **1750** FEB. 19, *about 0.40 p.m.* Cat. No. 99; Intensity 7; Fig. 84 (*Phil. Trans.* vol. 46, 1752, pp. 601–614, 616, 624; *Gent. Mag.* vol. 20, 1750, p. 89; *London Mag.* 1750, p. 91; J. Wesley's *Journal*, vol. 2, 1827, p. 131; *Horace Walpole's Letters*, ed. by P. Cunningham, vol. 2, pp. 198–203).

This was the first undoubted earthquake of the remarkable series felt during the year 1750. The boundary of the disturbed area, represented by the broken-line in Fig. 84, is about 20 miles long, 15 miles wide and contains about 240 square miles, its centre coinciding with London Bridge. The shock was unusually strong for so small a disturbed area. Some damage was caused by it, but hardly sufficient to entitle the shock to the intensity 8. A slaughter-house with a hay-loft over it was thrown down in Southwark, part of a chimney in Leadenhall Street, a chimney in Billiter Square, several chimneys

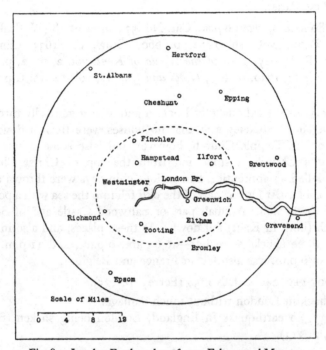

Fig. 84. London Earthquakes of 1750 Feb. 19 and Mar. 19.

and part of a house near Horsleydown, and some chimneys at Lime-house and Poplar. The intensity was apparently greatest in the districts bordering the river. "There were three distinct shakes, or wavings to and fro, attended with a hoarse rumbling noise, like thunder" (Rev. J. Wesley).

6. 1750 FEB. 19, *about* 1.10 *p.m.* Cat. No. 100 (Milne, vol. 31, p. 97; *Gent. Mag.* vol. 20, 1750, p. 89).

A second shock, about half an hour after the first, felt at Kings-bridge and Paynesbridge.

7. **1750** MAR. 19, *between* 1 *and* 2 *a.m.* Cat. No. 102 (*Gent. Mag.* vol. 20, 1750, p. 137).

A slight shock in London.

8. **1750** MAR. 19, *about* 5.30 *a.m.* Cat. No. 103; Intensity 7 or nearly 8; Fig. 84 (*Phil. Trans.* vol. 46, 1752, pp. 613–641, 647–649, 669–683; *Gent. Mag.* vol. 20, 1750, p. 137; *London Mag.* 1750, pp. 138–140; *Notes and Queries*, vol. 3, 1899, p. 330; *Encyc. Londinensis*, vol. 13, 1815, p. 100; J. Wesley's *Journal*, vol. 2, 1827, pp. 133–134; C. Wesley's *Journal*, vol. 2, 1849, p. 70; *Horace Walpole's Letters*, ed. by P. Cunningham, vol. 2, pp. 198–203).

This shock was even stronger, and lasted a longer time, than that of Feb. 19 (no. 5). The boundary of the disturbed area, represented by the continuous line in Fig. 84, is nearly circular, about 39 miles in diameter, and contains about 1200 square miles. Its centre is about 2 or 3 miles N. of London Bridge. Several chimneys were thrown down in various parts of London, part of a house in Old Street and two uninhabited houses in Whitechapel fell. Stones also fell from the new spire of Westminster Abbey. "It was a strong and jarring motion attended with a rumbling noise, like that of distant thunder" (J. Wesley). The mean duration of the shock was 5·4 seconds. The sound was compared to passing waggons, etc., in 6 per cent. of the records, thunder in 35, wind in 18, loads of stones falling in 6, and explosions in 35 per cent.

Some miscellaneous effects of the shock may be noted: (i) At Bloomsbury, Mr Michael Russell, F.R.S., had two china figures placed on a cabinet in his dining-room, with their faces towards the west; after the earthquake, they were found to be facing north-east —an early, if not the earliest, record of the rotation of bodies by an earthquake. (ii) Many animals in and near London were much disturbed by the shock—cats started up, dogs howled, sheep ran about, a horse refused to drink, the water being so much agitated, in several ponds fish leaped out of the water and were seen to dart away in all directions.

9. **1750** MAR. 20, 2 *a.m.* Cat. No. 104 (J. Parsons, *Phil. Trans.* vol. 46, 1752, pp. 635–636).

A trembling motion in London.

1750 Mar. 20, 4 a.m., a trembling motion in London, apparently felt by only one observer (J. Parsons, *Phil. Trans.* vol. 46, 1752, pp. 635–636).

1750 June 18, a loud report like that of a cannon (without any tremor) at London and Norwich (Roper, p. 20).

1755 Apr., a reported shock at Stepney (Mallet, 1852, p. 161).

10. **1758** DEC. 20. Cat. No. 133 (Mallet, 1853, p. 134).

A slight shock in London and neighbourhood.

1864 Apr. 30, 6.20 a.m., a single observer awakened by the oscillation of his bed in London (*Times*, May 6).

1865 Jan. 26, about 11.45 p.m., a loud rumbling sound, with violent shaking of windows, lasting 6 or 8 seconds, in London (*Times*, Feb. 4).

1875 Jan. 8, about 7.50 or 7.55 a.m., the crockery on the washing-table jingled for a few seconds, and, after a few seconds, it was resumed (extract from a letter to the author).

1886 Apr. 8, 5.35 a.m., a door was heard to vibrate regularly for 3 or 4 seconds without any motion being felt, in London (*Nature*, vol. 33, 1886, p. 559).

The following reported earthquakes may be noted here:

1865 Jan. 29, 10.30 p.m., the railway-station at Sunbury (Middlesex) was shaken as if by a heavy gust of wind; the noise lasted 4 or 5 seconds (*Times*, Feb. 9).

1865 Jan. 30, 4 a.m., the same.

ST ALBANS EARTHQUAKE

1250 DEC. 13. Cat. No. 34 (Matthew Paris, vol. 3, pp. 87–88, 97, 314; *Flor. Hist.* vol. 2, p. 367; Holinshed, vol. 2, p. 420).

"On the day of St Lucia, about the third hour of the day, an earthquake occurred at St Alban's and the adjacent districts..., where from time immemorial no such event had ever been seen or heard of; for the land there is solid and chalky, not hollow or watery, nor near the sea; wherefore such an occurrence was unusual and unnatural, and more to be wondered at. This earthquake, if it had been as destructive in its effects as it was unusual and wonderful, would have shaken all buildings to pieces; it came on with a trembling motion, and attended by a sound as if it were dreadful subterranean thunder. A remarkable circumstance took place during the earthquake, which was this: the pigeons, jackdaws, sparrows and other birds which were perched on the houses and on the branches of trees, were seized with fright, as though a hawk were hovering over them, and suddenly expanding their wings, took to flight, as if they were mad, and flew backwards and forwards in confusion...; but, after the trembling motion of the earth and the rumbling noise had ceased, they returned to their usual nests" (Matthew Paris, trans. by Rev. J. A. Giles, vol. 2, 1853, pp. 401–402).

1755 Mar., Hertfordshire "the hills were shaken and masses thrown down," though no distinct shock is mentioned (Mallet, 1852, p. 151). 1768 Dec. 27, an earthquake in Hertfordshire (Perrey, p. 146).

COLCHESTER EARTHQUAKES

The Colchester earthquake of 1884 is remarkable as the most destructive earthquake known in this country and as visiting a district hitherto free from disturbance. It has been described in the following papers: G. H. Kinahan, *Roy. Dublin Soc. Proc.* vol. 4, pp. 318–325; R. Meldola, *Geol. Assoc. Proc.* vol. 9, pp. 20–42; R. Meldola, *Hertfordshire Nat. Hist. Soc. Trans.* vol. 4, 1886, pp. 23–32 (with notes by J. Hopkinson); R. Meldola and W. White, "Report on the East Anglian Earthquake of April 22nd, 1884," *Essex Field Club Special Memoirs*, vol. 1, 1885, 224 pp.; G. J. Symons, *Mon. Met. Mag.* vol. 19, 1884, pp. 49–62; W. Topley, *Nature*, vol. 30, 1884, pp. 17–18, 60–62; H. B. Woodward, *Norfolk and Norwich Nat. Hist. Soc. Trans.* vol. 4, pp. 31–35; *Nature*, vol. 29, 1884, pp. 602–603; vol. 30, 1884, pp. 18–19, 31–32, 57, 77, 101–102, 124–125, 145, 170.

The report presented to the Essex Field Club by Messrs Meldola and White includes all that is important in the remaining papers. It is a valuable account of the damage wrought by the earthquake and many of the details in the following pages are drawn from it. In order to obtain more ample materials for the construction of the isoseismal lines and for the study of the twin character of the shock, I made special inquiries towards the close of 1900 and again in 1908. These inquiries and the examination of many local newspapers have nearly doubled the number of records of the earthquake.

1. **1884** FEB. 18, *about* 1.20 *a.m.* Cat. No. 794; No. of records 3 from 2 places.

A slight shock was felt at Peldon and West Mersea, and, it is said, by many others throughout the island of Mersea. At West Mersea, a sound like a clap of thunder was heard. The shock apparently originated in the south-western of the two epicentres of the principal earthquake.

2. 1884 APR. 22, 9.18 *a.m.* Cat. No. 795; Intensity 9; South-western epicentre in lat. 51° 48·5′ N., long. 0° 53·0′ E.; North-eastern epicentre in lat. 51° 51·4′ N., long. 0° 55·9′ E.; No. of records 446 from 293 places; Figs. 85–87.

Isoseismal Lines and Disturbed Area. The destructive energy of the earthquake was chiefly manifested in the villages lying a few miles

Fig. 85. Colchester Earthquake of 1884 Apr. 22.

to the east and south-east of Colchester (Fig. 86). In Colchester itself, more than 400 buildings (including 4 churches and 6 chapels) were injured. At Abberton, less than half-a-dozen chimneys were left standing. Every house and cottage at Peldon was damaged, and about 70 per cent. of the chimney-stacks were thrown down. About half the chimney-stacks at Rowhedge fell, and about three-quarters of those along the river-front. At Wivenhoe, the damage was widespread, being greatest on the lower ground facing the river, where 60 or 70 per cent. of the chimney-stacks were thrown down. East Donyland, Fingringhoe, Langenhoe and West Mersea also suffered seriously, though to a less extent than the above places. Altogether, within an area of about 150 square miles, more than 1200 buildings (including 20 churches and 11 chapels) had to be repaired. The number of places at which houses were damaged is 32 within the isoseismal 8, and, to a very slight extent, 21 others outside that isoseismal.

Shortly after the earthquake, a Mansion House Fund was raised for the benefit of those owners of property who were unable to bear the whole cost of the necessary repairs; and the following table, showing the distribution and amount of the damage, was prepared by the Committee entrusted with the disposal of the fund.

	POPULATION IN 1881	NO. OF BUILD-INGS REPAIRED	CHURCHES AND CHAPELS
Abberton	244	26	2
Aldham	433	1	0
Alresford	259	4	0
Bergholt, E.	1191	1	0
Bergholt, W.	1067	1	0
Bradwell-by-Sea	987	28	1
Coggeshall	2998	2	1
Colchester	28395	414	10
Donyland, E.	1264	207	1
Fingringhoe	659	49	2
Fordham	699	1	0
Langenhoe	234	19	1
Layer Breton	293	3	1
Layer-de-la-Haye	687	27	1
Mersea, E.	280	7	1
Mersea, W.	1082	58	3
Messing	641	0	1
Peldon	458	72	1
Salcot	233	1	0
Tollesbury	1434	1	0
Tolleshunt d'Arcy	828	0	1
Wigborough, Gt	421	20	1
Wigborough, Lit.	91	12	1
Wivenhoe and Elmstead	3147	259	2
Total		1213	31

The continuous lines in Fig. 85 represent isoseismal lines of intensities 8 to 4 and about 3. The isoseismal 8 is drawn on a larger scale in Fig. 86, in which are also represented the isoseismal 9 and two other isoseismals indicated by broken-lines.

Assuming that there are five persons on an average in every house, the percentage of houses needing repair at different places may be calculated from the above table, and from these percentages may be drawn curves of equal percentages of houses injured. In Fig. 86, the

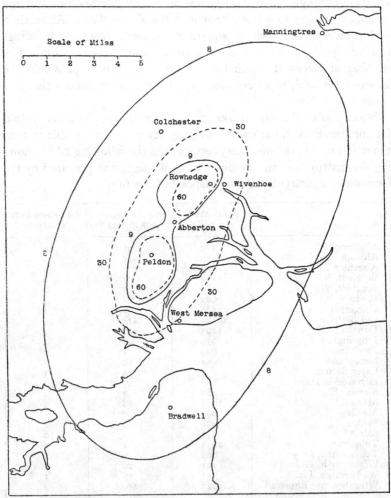

Fig. 86. Colchester Earthquake of 1884 Apr. 22: Central District.

broken-lines are the curves corresponding to percentages of 60 and 30 respectively, the smaller continuous curve to a percentage of 50 and therefore representing an isoseismal of intensity 9. The curve of percentage 60 consists of two detached portions, with centres 3¾ miles apart, the south-west centre being ¼ mile S. of Peldon, and the north-east centre 1¼ miles W. of Wivenhoe. The line joining the centres is directed N. 28° E.

The isoseismal 9 is continuous and dumb-bell shaped, 7 miles long and 2 miles in greatest width, the area included within it being 10 square miles. The isoseismal 8 is 18 miles long, 10½ miles wide, and contains 151 square miles. Its centre is 1 mile S.E. of Abberton, in lat. 51° 49·6′ N., long. 0° 54·9′ E., and its longer axis, which is directed N. 28° E., is parallel to the line joining the centres of the two detached portions of the curve of percentage 60 of damaged houses. The isoseismal 7 is 52 miles long, 32 miles wide, and about 1300 square miles in area. For the next two isoseismals, the evidence is somewhat incomplete. The isoseismal 6 is 120 miles long, about 82 miles wide, and contains about 7700 square miles; and the isoseismal 5 is about 190 miles long. The isoseismal 4 is roughly the arc of a circle of radius 130 miles, with its centre near Colchester. If the isoseismal were completed, its area would be about 53,000 square miles. The outermost isoseismal, of intensity about 3, is also approximately circular and with a mean radius of 180 miles, the complete circle containing about 100,000 square miles. The shock is also said to have been felt at Exeter, 42 miles S.W. of the isoseismal 3.

The Colchester earthquake is one of the few British earthquakes that have been felt on the continent. There can be little doubt that it was observed at Boulogne and Ostend, which are respectively 42 and 95 miles from the epicentre.

Nature of the Shock. According to 42 observers at 29 different places, the shock consisted of two distinct parts, separated by an interval the average length of which was 2·4 seconds. The places are widely distributed over the disturbed area, Cromer being 79 miles to the north of the epicentres, Leamington and Cheltenham 110 and 133 miles to the west, and Brighton 83 miles to the south-west. The first part was the stronger according to six observers and the second according to four, but there is no law to be seen in the distribution of the places of observation. Probably, the two parts were very nearly equal in strength; and this inference is supported by the wide extent of country over which the double shock was observed and by the approximate equality of the two curves in which 60 per cent. of the houses were injured, the north-eastern curve containing 2½ square miles and the south-western curve 3¼ square miles. The mean duration of the whole shock was 5·3 seconds.

342 EARTHQUAKES OF THE SOUTH-EAST OF ENGLAND

Instrumental Records. The shock was recorded by magnetographs at Greenwich and Kew observatories. At Greenwich, small movements of the light needles of the earth-current registers occurred at 9.20 a.m. The needles were not under the influence of any earth-current at the time, as the circuits were interrupted in order to determine the zero-positions of the needles. The declination and horizontal force magnets were also put into very slight vibration. At Kew, the same instruments were deflected, the horizontal force magnet considerably and the declination magnet slightly. At Cambridge observatory, observations were being made for determining the level of the transit-instrument, and the mercury was disturbed as if by a waggon passing along the adjoining road. Between 9.15 and 9.30 a.m., the ink-line of Messrs Reynolds and Branson's recording barometer at Leeds thickened suddenly, while a general movement downwards was shown. The line continued thick until 1.30 p.m., but this can hardly have been due to the Colchester earthquake.

Sound-Phenomena. The boundary of the sound-area is represented approximately by the dotted line in Fig. 85, a curve which is nearly a portion of a circle 108 miles in mean radius, and 37,000 square miles in area. Within the isoseismal 8, the sound was heard by 96 per cent. of the observers, and, outside this isoseismal, by 91 per cent. In the whole sound-area, the percentage of audibility was 92. The sound was compared to passing waggons, etc., in 36 per cent. of the records, thunder in 18, wind in 10, loads of stones falling in 7, the fall of a heavy body in 13, explosions in 15, and miscellaneous sounds in 1, per cent.

Miscellaneous Phenomena. In Mersea Island, two small fissures, or rather cracks, were formed by the earthquake. One of them was traced for more than 200 yards along the southern edge of the road which skirts the shore in a westerly direction from West Mersea. The crack, which occurred on a low drift-covered hill, was soon obliterated, but, on the day of the earthquake, it was reported to be several inches wide and more than two yards deep. The water of St Peter's well, close to the crack, was turbid for about two hours after the earthquake. The other crack, also running east and west, occurred in London Clay at Cross Farm, about a mile E.N.E. of West Mersea. From this crack there flowed for a short time two small streamlets of fresh water; they were about two yards apart, an inch in width, and were discoloured by a reddish sand. In a neighbouring well, the water rose about two feet and became turbid.

Within and near the epicentral district, there was a general rise in the level of water in wells. At the Colchester water-works, two wells, sunk to a depth of 420 feet, penetrate some distance into the Chalk. In these wells, the earthquake caused a rise of 7 to 7½ feet, and this level was maintained for about six months. At Bocking, near Braintree, the Messrs Courtauld's well, sunk to a depth of 241 feet and also reaching the Chalk, showed a rise of 19½ inches between Apr. 21 and 28, with a further rise of 25½ inches until June 3, after which the level slowly but gradually declined. The change of level from Mar. 3 to June 16 is represented by the continuous line in Fig. 87. The level during the corresponding period of the preceding year is represented

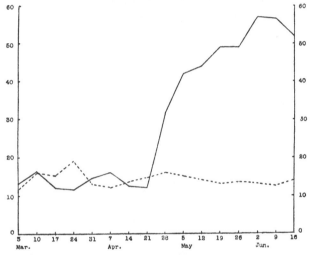

Fig. 87. Colchester Earthquake of 1884 Apr. 22: Level of water in well at Bocking.

by the broken-line. At Great Oakley (12½ miles E. of Colchester), a well that had been empty for months previously was quickly half-filled with water, and continued so until the summer, although water was drawn off daily. At White Colne (8 miles W.N.W. of Colchester), the level of the water in a deep well was lowered, this being the only known exception to the general rise in the level of the underground water.

Two slight after-shocks are reported, both resting on somewhat scanty evidence:

About a fortnight after Apr. 22, a slight shock at Tiptree (near Kelvedon).

June 24, about 5 a.m., the inmates of one house in Colchester were awakened by a slight shock.

344 EARTHQUAKES OF THE SOUTH-EAST OF ENGLAND

Earth-Sounds at East Mersea and Peldon. During 1896 and several years previously, noises, evidently of underground origin, were frequently heard at East Mersea. They resembled that caused by the slamming of a heavy door or the shooting down of a load of coal. Though especially noticeable on quiet days and at night, they were often loud enough to attract the attention of farm-labourers while ploughing in the fields, and were known to them as "earth-grumblings." Similar earth-sounds have also been heard at Peldon, which is close to the S.W. epicentre of the earthquake. In neither place have they been studied in sufficient detail to determine their seismic origin; but it is possible that they may be due to small creeps, the successors of the great movement of 1884 (W. Cole, *Essex Nat.* vol. 9, 1896, pp. 230–231).

CHAPTER XVIII

EARTHQUAKES OF UNKNOWN EPICENTRES IN ENGLAND

1. **974**. Cat. No. 1 (Symeon of Durham, vol. 2, p. 131; Roger de Hoveden, vol. 1, p. 64; *Chron. of Melrose*, p. 101; *Flor. Hist.* vol. 1, p. 513; etc.).

"In this year a great earthquake took place over all England" (Symeon of Durham).

1043, an earthquake in England (Mallet, 1852, p. 19).

2. **1060** JULY 4. Cat. No. 3 (*Anglo-Saxon Chron.* vol. 2, p. 161).

"In this year there was a great earthquake on the Translation of St Martin."

3. **1067**. Cat. No. 4 (William of Malmesbury, p. 125; *London Mag.* 1750, p. 102).

"On the 11th of August, a great earthquake terrified the whole of England by a dreadful marvel, so that all buildings recoiled for some distance and afterwards settled down as before."

4. **1076** MAR. 27. Cat. No. 5 (Matthew Paris, vol. 1, pp. 22–23; B. de Cotton, p. 50; *Flor. Hist.* vol. 2, p. 9; *Annal. Monast.* vol. 3, p. 425; vol. 4, p. 10; etc.).

"The earth trembled, and a general earthquake took place all over England" (*Flor. Hist.*).

1076 Apr. 6, "a terrible Earthquake which shook the Ground very much" (Speed, p. 431, possibly the same as the preceding).

5. **1081** DEC. 25. Cat. No. 7 (*Flor. Hist.* vol. 2, p. 11).

"A great earthquake, accompanied with a terrible subterraneous noise, took place all over England, in a manner contrary to the usual course of nature."

6. **1088**. Cat. No. 8 (*Flor. Hist.* vol. 2, p. 21; Henry of Huntingdon, p. 215; *Brut y Tywysogion*, p. 55; *Annal. Cambriae*, p. 29; etc.).

"A dreadful earthquake in all the island of Britain" (Brut y Tywysogion).

7. **1089** AUG. 11. Cat. No. 9 (Symeon of Durham, vol. 2, p. 217; Roger de Hoveden, vol. 1, p. 142; John de Oxenedes, p. 38; B. de Cotton, p. 53; *Anglo-Saxon Chron.* vol. 2, p. 193; *Annal. Monast.*

vol. 1, pp. 4, 43; vol. 2, pp. 36, 200; vol. 3, p. 427; *Flor. Hist.* vol. 2, p. 21; William of Malmesbury, p. 282; etc.).

"About the third hour of the day, there was a very great earthquake throughout England" (Roger de Hoveden); "all the buildings were lifted up, and then again settled as before" (William of Malmesbury).

1090 Aug. 13, "a great earthquake throughout all England" (*Le Livere de Reis de Engleterre*, p. 161); possibly the same as the preceding.

1091, an earthquake (*Annal. Monast.* vol. 4, p. 12), without the name of any place disturbed.

8. 1099 NOV. 3. Cat. No. 10 (Perrey, p. 118).

An earthquake that terrified all England.

9. 1116 DEC. 13. Cat. No. 12 (Holinshed, vol. 2, p. 66).

"A great earthquake." In the *Annal. Monast.* (vol. 4, p. 376), a strong earthquake is recorded on 1117 Jan. 3, which may be the same as the preceding.

10. 1120 SEPT. 28. Cat. No. 14 (Stow, p. 199; Short, vol. 1, p 113).

"This yeere was a great Earthquake in manie places of Englande... about the thirde houre of the daie" (Stow). Short states that the earthquake occurred in the Vale of Trent, that it overthrew many houses and buried their inhabitants in the ruins, and "gave daily 10, 17 or 20 Shocks."

1121, first week in Advent, an earthquake in England (Short, vol. 2, p. 167).

11. 1129. Cat. No. 16 (*Anglo-Saxon Chron.* vol. 2, p. 227; Short, vol. 1, p. 115).

"On the night of the mass of St Nicholas, a little before day, there was a great earthquake" (*Anglo-Saxon Chron.*).

12. 1132. Cat. No. 17 (*Annal. Monast.* vol. 4, p. 19).

An earthquake in England.

13. 1133 AUG. 4. Cat. No. 18 (Matthew Paris, vol. 1, p. 247; B. de Cotton, p. 62; *Flor. Hist.* vol. 2, p. 56; *Eulog. Hist.* vol. 3, pp. 62–63; Stow, p. 204).

"Earlie in the morning, in manie parts of England an earthquake was felt, so that it was thought that the earth woulde have sunke under the feete of men, with such a terrible sound, as was horrible to heare" (Stow).

1135 Dec. 1, "the Earth was so shaken, than many Edifices fell down" (Short, vol. 1, p. 118).

1136 Aug. 4, "early in the morning, there was a great earthquake in many parts of England" (John of Hexham, p. 5), apparently the same as the earthquake of 1133 Aug. 4.

1140 Apr., an earthquake in England (Short, vol. 1, p. 119).

1145 Jan. 25, at midnight, a great earthquake (*London Mag.* 1750, p. 102).

14. **1158**. Cat. No. 20 (*Flor. Hist.* vol. 1, p. 566; *Annal. Monast.* vol. 1, p. 48; Gervase of Cant. vol. 1, p. 166; etc.).

A great earthquake in many parts of England.

1176, an earthquake in England (Short, vol. 2, p. 168).

15. **1184** JAN. 16. Cat. No. 23 (*Annal. Monast.* vol. 4, pp. 39, 40).

An earthquake in many parts of England, about the middle of the night.

16. **1201** JAN. 8. Cat. No. 26 (Florence of Worcester, p. 311).

An earthquake in England.

1202, an earthquake in different parts of England (Mallet, 1852, p. 30).

1218, winter, a violent earthquake in England (Perrey, p. 120).

1219, an earthquake in England (Short, vol. 1, p. 136).

17. **1228** APR. 23. Cat. No. 27 (*Annal. Monast.* vol. 1, pp. 36, 70).

An earthquake in many parts of England.

1240, "a terrible sound was heard, as if a huge mountain had been thrown forth with great violence, and fallen in the middle of the sea; and this was heard in a great many places at a distance from each other, to the great terror of the multitudes who heard" (Matthew Paris, vol. 4, pp. 2–3; *Annal. Monast.* vol. 1, p. 115).

18. **1246** MAR. 11. Cat. No. 30 (*Annal. Monast.* vol. 1, p. 135; vol. 2, p. 90; vol. 4, pp. 95–96; *Eulog. Hist.* vol. 3, p. 303; John de Oxenedes, p. 177).

A great earthquake throughout all England and in Ireland.

1273, an earthquake in England on the vigil of St Nicholas (W. Rishanger, p. 80).

1278, an earthquake in France and England (Mallet, 1852, p. 35).

1284, an earthquake in England (Mallet, 1852, p. 35).

19. **1298** JAN. 5. Cat. No. 37 (*Annal. Monast.* vol. 4, p. 539; *Flor. Hist.* vol. 3, pp. 105, 297; *Chartularies, etc., St Mary's Abbey, Dublin,* vol. 2, p. 328; etc.).

An earthquake in England towards twilight.

20. **1318** NOV. 14. Cat. No. 38 (T. Walsingham, vol. 1, p. 154; *Hist. Monast. S. August. Cant.* p. 102).

A great earthquake in England.

21. **1319** DEC. 1. Cat. No. 39 (*Le Livere de Reis de Engleterre,* p. 337).

"A general earthquake in England, with great sound and much noise."

1320, an earthquake in England (Mallet, 1852, p. 37).

1333, about, a great earthquake (*Monum. Franc.* vol. 2, p. 153).

1361, an earthquake in England (*John Hardyng's Chron.* p. 330):

> "In the same yere was on sainct Maurys day,
> The great winde and earth quake meruelous,
> That greately gan the people all affraye,
> So dredfull was it then and perelous."

On 1382 May 21, one of the strongest of British earthquakes wrought much injury to churches, etc., in Kent (p. 330).

22. **1382** MAY 24. Cat. No. 42 (Stow, p. 472; Holinshed, vol. 2, pp. 754–755; *Eulog. Hist.* vol. 3, p. 356).

"There followed also another Earthquake...in the morning before the sunne rising, but not so terrible as the first" (Stow, who gives the day as May 23). "Earelie in the morning, chanced an other earthquake, or (as some write) a watershake, being of so vehement and violent a motion, that it made the ships in the hauens to beat one against the other, by reason whereof they were sore brused by such knocking togither" (Holinshed).

23. **1385.** Cat. No. 43 (T. Walsingham, vol. 2, pp. 128, 130; *Chron. Angliae,* pp. 364–365).

Two earthquakes were felt during this year in England, one of them on July 18, the other "ante medium noctis Inventionis Sanctae Crucis."

24. **1508** SEPT. 19. Cat. No. 46 (Holinshed, vol. 5, p. 468).

"A great earthquake in manie places both of England and Scotland."

25. **1575** FEB. 26, *between 5 and 6 p.m.* Cat. No. 52; Intensity 8 (Holinshed, vol. 4, p. 326; Stow, p. 1149; *Gent. Mag.* vol. 20, 1750, p. 56; *London Mag.* 1750, p. 103).

A great earthquake at York, Worcester, Gloucester, Tewkesbury, Bredon, Hereford and Bristol. Part of Ruthen Castle fell, and several brick chimneys. The year is given as 1574 in Stow's *Chronicle* and the *Gentleman's Magazine.*

1627, an earthquake in England (Short, vol. 1, p. 308).

26. **1661.** Cat. No. 64 (*Gent. Mag.* vol. 20, 1750, p. 56).

An earthquake in England.

27. **1683** OCT. 9, *about 11 p.m.* Cat. No. 70 (T. Pigot, *Phil. Trans.* vol. 12, 1683, p. 321; R. Pickering, *Phil. Trans.* vol. 46, 1752, p. 624; R. Plot, *Hist. of Staffordshire*, p. 143; *Life of Sir W. Dugdale*, p. 146).

A strong earthquake, accompanied by noise, felt in the counties of Oxford, Warwick, Stafford and Derby.

28. **1687** MAY 12. Cat. No. 72 (*Evelyn's Diary*, vol. 3, p. 37).

An earthquake felt in several parts of England.

29. **1690** OCT. 7, *about 7.30* —. Cat. No. 74 (*Evelyn's Diary*, vol. 3, pp. 90–91).

An earthquake felt at Althorp (Northamptonshire), and also, it is said, at Barnstaple, Holyhead and Dublin.

30. **1692** SEPT. 7, *about 2 p.m.* Cat. No. 75 (*Gent. Mag.* vol. 51, 1781, p. 378; *Evelyn's Diary*, vol. 3, p. 105; *Notes and Queries*, vol. 1, 1862, p. 94; etc.).

A strong earthquake felt in London, Suffolk, Norfolk, Essex, Kent, Sussex and Hampshire, and also in Flanders and Holland. A chimney fell in Colchester and a crack was opened in the steeple of St Peter's church in that town. The streets of London were filled with panic-stricken crowds.

31. **1727.** Cat. No. 84 (*Gent. Mag.* vol. 20, 1750, p. 56; *London Mag.* 1750, p. 103).

An earthquake in Cheshire and Wales, and almost all along the western coast.

1728 Oct., an earthquake in N. England (Short, vol. 2, p. 170).

1729 Jan. 30, an earthquake in N. England (Short, vol. 2, p. 170).

1753 May, an earthquake felt in some parts of England (Mallet, 1852, p. 157; Perrey, p. 135).

1753 July and Sept., shocks felt in different parts of England (Mallet, 1852, p. 157; Perrey, p. 174).

1767, in the north of England (Prestwich, p. 544).

32. **1786 JUNE 16.** Cat. No. 159 (W. Creech, *Edinburgh Fugitive Pieces*, 1815, p. 122; Lauder, p. 366).

A strong shock felt in Whitehaven, in the S.W. of Scotland, the Isle of Man and Dublin.

33. **1788 JULY 8.** Cat. No. 163 (Lauder, p. 367; W. Creech, *Edinburgh Fugitive Pieces*, 1815, p. 123).

An earthquake in the Isle of Man.

1791 Oct. 28, an earthquake in England (Mallet, 1854, p. 29).

1805 Mar., an earthquake in England (Mallet, 1854, p. 61).

34. **1839 SEPT. 2.** Cat. No. 354 (Perrey, p. 156; Parfitt, I, p. 651).

An earthquake in the west of England, Monmouthshire, Bristol and Cardiff.

1852 Apr. 9; 1853 Feb. 27, Mar. 6, 21, 31; 1854 Feb. 12; slight (reported) earthquakes in England, no details being given (E. J. Lowe, *Brit. Met. Soc. Proc.* vol. 2, 1865, p. 60).

35. **1852 NOV. 9, 4.25 *a.m.*** Cat. No. 709; Intensity 6 or 7; Centre of disturbed area in about lat. 53° 14′ N., long. 4° 7′ W.; No. of records 135 from 105 places; Fig. 88 (Mallet, *Irish Acad. Trans.* vol. 22, 1855, pp. 397–410; *New Monthly Mag.* vol. 96, 1852, pp. 446–455; etc.).

In one respect, this earthquake seems to be unique among British earthquakes, for it was probably felt in all four portions of the United Kingdom. Its disturbed area, so far as it can be traced, is 290 miles long from W.N.W. to E.S.E., 245 miles wide, and contains about 56,000 square miles. It includes the whole of Wales, about half of England, the eastern counties of Ireland, the Isle of Man, and, though there is no undoubted observation from Scotland, the southern portions of the counties of Wigtown and Kirkcudbright. Its boundary, which is an isoseismal of intensity 4, is represented by the continuous line in Fig. 88. Outside this line, the shock may have been felt at Hammersmith (in Middlesex, 64 miles to the S.E.) and Glasgow (60 miles to the N.), but the record from the latter city is uncertain as the time of occurrence is not given and there is a wide intervening tract without observations. The centre of the area coincides approximately with the town of Bangor in North Wales; but, as the intensity was greatest in the east of Co. Dublin, it is probable that the epicentre, or the more important epicentre if there were two, was submarine and not far from the Irish coast.

At two places in Ireland, some slight damage was attributed to the shock; at Kingstown part of a wall, and at Phibsborough a shattered

Fig. 88. Earthquake of 1852 Nov. 9.

chimney, are said to have been thrown down. The intensity was 6 at Dublin and Kilbride (Co. Wicklow). In the mountainous districts of North Wales, the intensity was also considerable; but at no place in England was it greater than 5.

In several places, the shock was observed to consist of two parts, as, for instance, at Liverpool, Dublin, Glaslough (Co. Dublin) and Bray (Co. Wicklow). The mean duration of the shock was 8·2 seconds. The sound was compared to passing waggons, etc., in 22 per cent.

of the records, thunder in 14, wind in 22, the fall of a heavy body in 28, explosions in 7, and miscellaneous sounds in 7, per cent.

At a railway goods store in Dublin, the man on watch heard sparrows fall from their perches in the roof, and in the morning several were picked up dead and others unable to fly.

Mallet remarks (p. 409) that there are grounds for believing that earlier during the night of Nov. 8–9 several slight shocks were felt in and near Dublin, and also during the following night. The following records depend on the evidence of a single observer:

Nov. 9, 8.28 a.m., a very slight shock at Dolserey (2 miles from Dolgelley), followed immediately by a loud rumbling noise as of a heavily laden waggon driven over a stony road.

Nov. 9, 9 a.m., a very slight shock, lasting about 3 seconds, at the same place.

CHAPTER XIX

EARTH-SHAKES IN MINING AND LIMESTONE DISTRICTS

THE term "earth-shake" is applied to local disturbances of the earth which owe their origin, either wholly or in part, to artificial causes. Typical examples are those which are occasionally felt at Sunderland and other places on the magnesian limestone of the Durham coast. These are described in the concluding section of this chapter. Other examples are the shocks that are so frequently felt in mining districts, and especially in Cornwall, in the South Wales coalfield, and in the neighbourhood of Manchester.

EARTH-SHAKES IN MINING DISTRICTS

The characteristic features of these earth-shakes are given later. They enable us to remove a few disturbances from the list of British earthquakes, under which title they have hitherto appeared. The following shocks (nine in number) are probably of semi-artificial origin:

1. 1765 Nov. 27, a violent shock at Long Benton (3 miles N.E. of Newcastle), houses built of stone being disjointed and fields, etc., sinking by more than two feet. In this case, there can be little doubt that the shock and subsidence were caused by the complete withdrawal of coal over an area two miles square (*Gent. Mag.* vol. 35, 1765, p. 587).

2. 1822 Sept. 18, 1.30 a.m., a strong shock, accompanied by a noise like distant thunder, at Dunston (1 mile W. of Gateshead) (*Ann. de Chim. et de Phys.* vol. 21, 1822, p. 396; *Journ. of Sci.* vol. 14, 1823, p. 450; *Phil. Mag.* vol. 60, 1822, p. 234).

3. 1839 June 11, a shock at Chorlton-on-Medlock and in the district to the north of Manchester (Milne, vol. 31, p. 122).

4. 1881 Aug. 26, about noon, a strong shock felt in the mining district of Teversall (3 miles W. of Mansfield), bricks being dislodged from a chimney at Fackley (*Times*, Aug. 30).

5. 1886 Apr. 8, a shock (of intensity 5 or more) in a district (about 10 miles S.E. of Edinburgh) including Gorebridge, Cowdengrange and Newtongrange, accompanied by a low rumbling sound (*Nature*, vol. 33, 1886, p. 611).

6. 1886 Aug. 20, about 9 p.m., a shock, accompanied by a sharp rumbling noise, at Kilsyth in Stirlingshire (*Times*, Aug. 23).

7. 1887 Oct. 12, 6.20 a.m., a shock (of intensity 5 and lasting 2 or 3 seconds) at Merthyr Tydvil and Abermorlais (*Times*, Oct. 13).

8. 1887 Nov. 4, about 1 a.m., tremors (of intensity 4) lasting a few seconds, at Methley and Allerton, near Leeds (*Times*, Nov. 5).

9. 1888 Aug. 4, a shock (of intensity 5 and lasting a few seconds) at Kilsyth (*Daily Express*, Aug. 6).

I will now describe in greater detail some of the more important earth-shakes felt in 1889 and the following years.

EARTH-SHAKES IN CORNWALL

1. **1895** AUG. 25, *about* 0.30 *p.m.* BLISLAND. Intensity 4; Centre of disturbed area in lat. 50° 31·9′ N., long. 4° 40·9′ W.; No. of records 20 from 19 places and negative records from 2 places; Fig. 89 (*Geol. Mag.* vol. 7, 1900, pp. 164–166; vol. 2, 1905, pp. 220–221).

This earth-shake disturbed a small but elongated area, 6 miles long, nearly 2 miles wide, and containing 9 square miles. Its centre is ½ mile

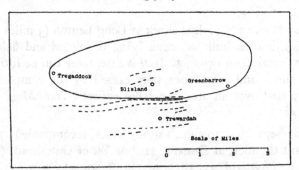

Fig. 89. Blisland Earth-shake of 1895 Aug. 25.

N. of Blisland, and the direction of its longer axis N. 85° E. The sound-area overlaps the disturbed area both to the north and south.

In all parts of the disturbed area, the shock was merely a slight tremor, a shaking like that experienced on a bridge when a heavy weight is passing over it.

The Geological Survey map shows no fault anywhere near the disturbed area. Not far distant, however, are several series of elvan-dykes which are nearly parallel to the longer axis of the disturbed area. The broken-lines in Fig. 89 represent one of these series, and if one of the dykes, especially either of those which coincide nearly with the southern boundary of the area, should run along a fault hading to the north, it would satisfy all the conditions required. The focus must have been about 4 miles long, and, judging from the form and dimensions of the disturbed area, must have been at a slight depth.

2. **1902 JUNE 4, 10.30 p.m. CAMBORNE.** Intensity 5; Centre of disturbed area in lat. 50° 13·4′ N., long. 5° 16·6′ W.; No. of records 22 from 12 places and 30 negative records from 22 places; Fig. 90 (*Geol. Mag.* vol. 2, 1905, pp. 221–222).

The boundary of the disturbed area is 4½ miles long, 3 miles wide, and contains about 10 square miles. Its centre is 1 mile N.E. of Camborne, and the direction of its longer axis N. 24° W.

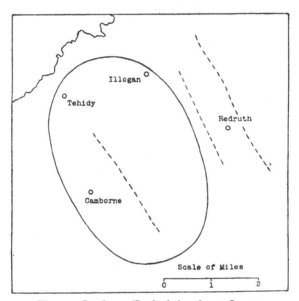

Fig. 90. Camborne Earth-shake of 1902 June 4.

The shock consisted of one prominent vibration, like that produced by the fall of a heavy weight, followed by a tremulous motion lasting for 1 or 2 seconds. The sound was heard by all the observers, and the sound-area must therefore coincide with the disturbed area.

Of the three faults, indicated by the broken-lines in Fig. 90, the more westerly one passes through a point about ¾ mile N.E. of Camborne in the direction N. 33° W. It is the " great cross course " of the Dolcoath Mine, heaving the Dolcoath Lode 140 yards to the south; its hade is practically zero.

3. **1904** SEPT. 5, 1.55 *a.m.* CAMBORNE. Intensity 4; No. of records 7 from 6 places and negative records from 3 places (*Geol. Mag.* vol. 5, 1908, p. 307).

The earth-shake was felt chiefly in the district between Camborne and Redruth. The boundary of the disturbed area is similar to that of the preceding earth-shake (Fig. 90), but falls short of it by about ½ mile towards the north, west and south. It is thus about 3½ miles long, 2½ miles wide, and 7 square miles in area, its longer axis being roughly parallel to that of the earlier shock and consequently to the principal faults of the district. The shock and sound were both of brief duration.

EARTH-SHAKES IN THE RHONDDA VALLEYS (GLAMORGANSHIRE)

1. **1889** JUNE 22, *about* 10.30 *p.m.* LLWYNYPIA. Intensity 5; Centre of disturbed area in lat. 51° 37·8′ N., long. 3° 26·1′ W.; No. of records 6 from 6 places; Fig. 91 (*Geol. Mag.* vol. 8, 1891, p. 371).

The boundary of the disturbed area is indicated by the smaller broken-line in Fig. 91. It is roughly circular in form, 3¼ miles in diameter and 8 square miles in area, with its centre 1 mile E. of Llwynypia.

The shock, which lasted 1 or 2 seconds, consisted of one vibration, like that produced by blasting. It was accompanied by a deep rumbling noise. The shock was felt underground in the Ynyshir Steam Colliery, and in one or two other collieries in the adjoining district.

2. **1894** APR. 11, *about* 2.40 *a.m.* PORTH. Intensity 5; Centre of disturbed area in lat. 51° 36·9′ N., long. 3° 23·3′ W.; No. of records 13 from 12 places; Fig. 91 (*Geol. Mag.* vol. 7, 1900, p. 174).

The boundary of the disturbed area, represented by the continuous line in Fig. 91, is almost exactly circular, about 5 miles in diameter and 20 square miles in area, with its centre ¾ mile E. of Porth. Two loud reports, like those of cannon, were heard in quick succession, and, at the same moments or immediately afterwards, sharp vibrations were felt. They seem to have been almost equally distinct underground, for, in several pits, the miners rushed to the bottom of the shafts, thinking that a violent explosion had occurred.

3. **1894** MAY 2, *about noon.* PORTH.

At the surface, this shock was much slighter than the preceding, but underground it produced similar effects, miners in the same pits again leaving their work to escape from what appeared to be an explosion. The boundary of the disturbed area, though smaller, was probably concentric with that of the earth-shake of Apr. 11.

4. **1896** OCT. 16, *about 11 p.m.* PENTRE. Intensity 5; Centre of disturbed area in lat. 51° 38·9′ N., long. 3° 29·3′ W.; No. of records 9 from 4 places and negative records from 13 places; Fig. 91 (*Geol. Mag.* vol. 7, 1900, pp. 174–175).

The area disturbed by this earth-shake was a very small one, circular and about 1 mile in diameter. Its centre lies about ⅓ of a mile S. of Pentre.

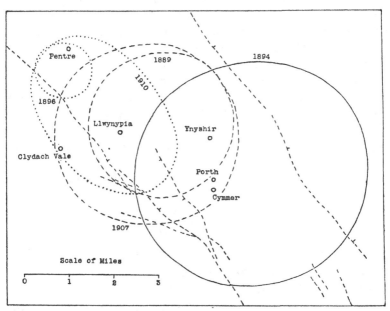

Fig. 91. Rhondda Valleys Earth-shakes of 1889 June 22, 1894 Apr. 11, 1896 Oct. 16, 1907 May 17 and 1910 Feb. 16.

A loud boom, like the muffled sound of blasting, was heard, followed by a brief shaking. The sensation resembled that experienced in a room immediately beneath another in which a heavy article of furniture had fallen. The shock seems to have been felt much more severely in pits than at the surface. In the pit at Gelli, a loud boom was heard, like the discharge of a shot in the rock, followed by a rumbling noise; workmen lying down felt the ground heave, while dust rose in clouds from the floor.

As will be seen from the map, the centre of the disturbed area lies about ⅓ mile on the N.E. or downthrow side of the Dinas fault.

5. **1907** MAY 17, 3.15 *p.m.* LLWYNYPIA. Intensity 5; Centre of disturbed area in lat. 51° 37·7′ N., long. 3° 26·6′ W.; No. of records 22 from 13 places, negative records from 7 places, and 6 records from mines; Fig. 91 (*Geol. Mag.* vol. 5, 1908, pp. 307-308).

Near the centre of the disturbed area, the intensity of the earth-shake was 5, very nearly 6. The boundary of the disturbed area is an isoseismal of intensity 4, and is very nearly a circle, about 4 miles in diameter, and 13 square miles in area. Its centre lies ½ mile E. of Llwynypia.

The shock in all parts of the disturbed area consisted of a single series of vibrations, lasting on an average for 1·5 seconds. Close to the centre of the area, as at Llwynypia, it began with one or two strong vibrations, rapidly decreasing in strength; while near the boundary, as at Gelli and Cymmer, only a tremulous motion was felt. The sound was much louder near the centre of the area than near the boundary; at Llwynypia, it was described as a violent report and a great crash; near the boundary, it was either described as a dull thud or a distant explosion or not heard at all. It was heard by 86 per cent. of the observers.

The earth-shake was observed in pits at Llwynypia, Wattstown, Gelli and Ynyshir, in the first two at a depth of 500 yards. Gelli is close to the boundary of the disturbed area, and Ynyshir about ½ mile from it. The disturbed areas on the surface and underground were thus roughly coincident. In the pit at Llwynypia, the sound was a terrific crash followed by a low rumbling.

6. **1910** FEB. 16, 2.55 *a.m.* TONYPANDY. Intensity 5; Centre of disturbed area in lat. 51° 38·0′ N., long. 3° 28·0′ W.; No. of records 18 from 7 places and 8 negative records from 7 places; Fig. 91.

This is the only known earth-shake in the Rhondda district with an elongated disturbed area. The boundary is represented by the larger dotted curve in Fig. 91. It is 4 miles long, 2½ miles wide, and contains 8 square miles. The centre is 1 mile N.W. of Tonypandy, and the direction of its longer axis is N. 40° W.

The shock consisted of a single series of vibrations lasting for about 3 seconds. The sound was heard by all the observers. The earth-shake was felt underground at the Cambrian Collieries, Clydach Vale. A number of workmen employed in the 7-foot seam hurried to the bottom of the shaft, thinking that an explosion had occurred.

As will be seen from Fig. 91, the longer axis is parallel to the Dinas fault and about ½ mile from it on the N.E. or downthrow side.

PENDLETON EARTH-SHAKES

1. **1899** FEB. 27, 10.1 *p.m.* Intensity 4; Centre of disturbed area in lat. 53° 30·0′ N., long. 2° 17·4′ W.; No. of records 14 from 11 places; Fig. 92 (*Geol. Mag.* vol. 7, 1900, p. 175).

The disturbed area, represented by the broken-line in Fig. 92, is nearly circular in form, 4 miles in diameter, and 13 square miles in area. Its centre lies about 1 mile N. of the centre of Pendleton. A booming sound, like that of a gas-explosion, accompanied the shock.

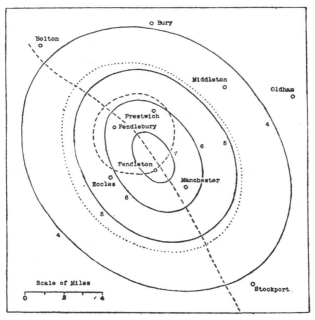

Fig. 92. Pendleton Earth-shakes of 1899 Feb. 27 and 1905 Nov. 25.

In the Pendleton Colliery, some fragments fell from the roof, and a cloud of dust rose, which put out the workmen's lamps. The centre of the disturbed area is close to the Pendleton or Irwell Valley fault, the course of which is represented by the nearly straight broken-line.

2. **1900** APR. 7, 1.17 *a.m.* Intensity 4 or 5 (*Geol. Mag.* vol. 8, 1901, p. 361).

As in the preceding earth-shake, the disturbed area was not more than 4 or 5 miles in diameter, with its centre close to the Irwell Valley fault, but a mile or two farther to the S.S.E. of the other. At

Pendleton, the vibration resembled that felt in a house when a heavy traction-engine passes. In the pits, at a depth of about 1000 yards, the noise was loud and the shock caused dust to rise.

3. 1905 NOV. 25, 3.42 *a.m.* Intensity 7; Centre of isoseismal 7 in lat. 53° 29·6′ N., long. 2° 15·8′ W.; No. of records 139 from 45 places and negative records from 4 places; Fig. 92 (*Geol. Mag.* vol. 3, 1906, pp. 171–176)

At Pendleton, the intensity of the shock was not much below 8, for there was some slight damage done, several chimney-pots and one chimney-stack being thrown down.

On the map in Fig. 92, four isoseismals are represented by continuous lines, corresponding to intensities 7 to 4. The isoseismal 7 is 3 miles long, 1¾ miles wide, and 4 square miles in area. Its centre lies ¾ mile N. of the centre of Pendleton. The next isoseismal, of intensity 6, is 6 miles long, 4½ miles wide, and contains 21 square miles; the isoseismal 5 is 9½ miles long, 7½ miles wide, and 56 square miles in area; while the isoseismal 4, which bounds the disturbed area, is 15¼ miles long, 12 miles wide, and includes an area of 144 square miles. The longer axes of the isoseismal lines are parallel or nearly so, being directed N. 37° W. The distances between the isoseismals are approximately the same on both sides of the longer axes.

The shock was brief in all parts of the disturbed area, the average duration being 2·5 seconds. It consisted of a few prominent vibrations, quick-period tremors having been apparently absent.

The sound-area, the boundary of which is represented by the dotted line in Fig. 92, is 10¾ miles long, 8¼ miles wide, and contains 70 square miles. It includes the whole of the isoseismal 5, but falls short of the isoseismal 4 in all directions. The sound was heard by 75 per cent of the observers, and was compared to passing waggons, etc., in 14 per cent. of the records, thunder in 13, wind in 4, loads of stones falling in 9, the fall of a heavy body in 31, and explosions in 29, per cent.

The Irwell Valley fault has been traced from the neighbourhood of Bolton to that of Poynton in Cheshire, a distance of more than 20 miles. Its downthrow is to the N.E. In the neighbourhood of Pendleton, the mean direction of the fault is N. 34° W., and its position agrees with that required by the evidence of the earth-shake provided the depth of the focus were very small. The approximate equality of the distances between successive pairs of isoseismals on both sides of the longer axes is no doubt due to the small vertical dimensions of the focus.

BARNSLEY EARTH-SHAKE

1903 OCT. 25, 11.5 *p.m.* Intensity 7; Centre of disturbed area in lat. 53° 31·3′ N., long. 1° 28·2′ W.; No. of records 13 from 9 places and 12 negative records from 10 places; Fig. 93 (*Geol. Mag.* vol. 2, 1905, pp. 222–223).

The disturbed area is 5¾ miles long, 3½ miles wide, and contains 16 square miles. Its centre is almost coincident with the village of Worsborough, and the longer axis is directed N. 38° W.

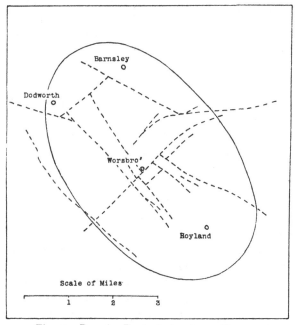

Fig. 93. Barnsley Earth-shake of 1903 Oct. 25.

The intensity of the shock was 7 at Worsborough and Hoyland, 6 at Worsborough Bridge and Barnsley, and from these places it died away rapidly towards the boundary. The shock consisted of a single series of vibrations, lasting 3 to 5 seconds. The sound was also of brief duration.

As will be seen from the map, there are several faults (indicated by broken-lines) which are parallel, or nearly parallel, to the longer axis of the disturbed area. One of them passes on the S.W. side of the centre of the disturbed area, hades to the N.E., and its direction near the centre is N. 41° W.

KILSYTH EARTH-SHAKE

1898 FEB. 16, 1.33 *p.m.* Intensity 4 or 5; No. of records 6 from 1 place and negative records from 10 places (*Geol. Mag.* vol. 7, 1900, p. 175).

Kilsyth lies in the valley of the Kelvin, about 11 miles N.E. of Glasgow, and the same distance from Falkirk and Stirling. The earth-shake was felt and heard by hundreds of people in the town, but there is no evidence that it was noticed in the surrounding country.

The shock consisted of a single movement, which rapidly died away, lasting about 1 second, a dull heavy thud, as if some heavy body had fallen on the ground, shaking all around. Men working in the mines ran to the bottom of the shaft.

CHARACTERISTICS OF EARTH-SHAKES IN MINING DISTRICTS

(i) *Magnitude and Form of Disturbed Area.* In the 14 earth-shakes during the years 1889–1910 described above, the disturbed area ranges from 1 to 144 square miles, but in only one case (the Pendleton earth-shake of 1905) exceeds 20 square miles. Including this earth-shake, the mean disturbed area is 20 square miles; excluding it, the mean area is 10 square miles.

In 7 of these earth-shakes, the boundary of the disturbed area is approximately circular; in the others, it is slightly elongated, the difference between the lengths of the longer and shorter axes ranging from $1\frac{1}{4}$ to 4 miles, the average difference being about 2 miles.

The most remarkable feature of the earth-shakes is their great intensity when the areas disturbed by them are taken into account. Thus, two of the shocks were of intensity 7, six of intensity 5, and the rest of intensity 4 or 4–5. How great is the contrast in this respect between earth-shakes and true earthquakes (1889–1916) is shown in the following table[1]:

INTENSITY	MEAN DISTURBED AREA IN SQ. MILES	
	EARTH-SHAKES	EARTHQUAKES
7	80	24600
5	10	790
4	9	160

(ii) *Nature of the Shock and Sound.* As a rule, the shock consists of one or two strong vibrations, followed sometimes by a brief tremor,

[1] In determining these average areas, after-shocks are omitted.

such as would be produced by the fall of a heavy body on the floor of an adjoining room. The mean duration of the shock is 2 seconds; that of ordinary earthquakes (1889–1916) ranges from 3 to 6 seconds.

The following table contains the percentages of reference to the different types of sound for earth-shakes with elongated and circular disturbed areas, and, as a contrast, the corresponding percentages for strong, moderate and slight earthquakes are added. The figures in the last column are percentages of reference to types of short duration (types 4–6):

	TYPE							
	1	2	3	4	5	6	7	Short
Earth-shakes with elongated dist. areas	16	18	3	12	27	24	0	63
Earth-shakes with circular dist. areas	3	21	0	3	31	45	0	79
Strong earthquakes	45	20	11	4	7	8	5	20
Moderate earthquakes	46	26	4	4	8	7	4	20
Slight earthquakes	31	29	5	6	11	14	4	32

(iii) *Relations of Sound-Area and Disturbed Area.* In the Pendleton earth-shake of 1905, the sound-area falls short of the disturbed area in all directions, the sound-area being about one-half the size of the other. In the much weaker Llwynypia earth-shake of 1907, the sound was very faint near the boundary of the disturbed area or not heard. In the Blisland earth-shake of 1895, the sound-area overlaps the disturbed area laterally; in 11 other cases, the boundaries of the two areas were approximately coincident.

Again, the sound-area shows no sign of displacement with respect to the isoseismal lines, as occurs in some earthquakes, for example, the Helston earthquake of 1898 Apr. 1 (p. 302). So far as our evidence goes, the sound-area and disturbed area are similar and concentric.

(iv) *Relative Effects on the Surface and in Mines.* In the Llwynypia earth-shake of 1907, the shock was felt in mines close to the boundary of the disturbed area on the surface. It is also obvious, from several of the accounts given above, that the shock was stronger and the sound louder in pits than on the surface.

Now, in ordinary earthquakes, the shock is much weaker in pits than on the surface, and the underground disturbed area is much less than the surface disturbed area. For instance, the Hereford earthquake of 1896 was perceptible on the surface to a mean distance of 177 miles, and in pits to a distance of 58 miles. In the Derby earthquake of 1903, the corresponding distances were 62 and 20 miles; in the Swansea earthquake of 1906, 146 and 19 miles from the nearer epicentre.

(v) *Relations between Earth-Shakes and Faults.* The position of the epicentre is known with some accuracy for 14 earth-shakes. In the Blisland earth-shake of 1895, the epicentre is close to several elvan-dykes, and in the Camborne earth-shakes of 1902 and 1904 to the "great cross course" of the Dolcoath Mine. The centres of the Porth earth-shakes of 1894 lie near no mapped fault. In the Llwynypia earth-shake of 1907, the centre lies close to the north end of the Cymmer fault, and in that of 1899 not far distant. In the Pentre earth-shake of 1896 and the Tonypandy earth-shake of 1910, the centre lies from ⅓ to ½ a mile from the Dinas fault and on its downthrow side. In the Pendleton earth-shakes of 1899, 1900 and 1905, the centre is either close to the Irwell Valley fault or on its downthrow side. The centre of the Barnsley earth-shake of 1903 is close to a fault on its downthrow side; that of the Kilsyth earth-shake of 1898 is in a region traversed by many faults.

Again, in the six earth-shakes with an elongated disturbed area (namely, those of Blisland in 1895, Camborne in 1902 and 1904, Tonypandy in 1910, Pendleton in 1905, and Barnsley in 1903), the longer axis of the disturbed area is parallel or very nearly parallel to an elvan-dyke or fault in its immediate neighbourhood.

ORIGIN OF EARTH-SHAKES IN MINING DISTRICTS

The most significant feature of these earth-shakes is the rapid decrease of intensity from the centre to the boundary of the disturbed area. This of course implies that the foci are extremely shallow compared with those of ordinary earthquakes[1], a conclusion which is supported by the observations made in mines. The extent of the underground disturbed area and the great intensity of the shock and sound in pits show that the foci cannot be far above or below the workings in which the earth-shakes are observed.

A second conclusion is that the foci of the earth-shakes are usually of small dimensions. This is clear from the small size of the disturbed areas, their approximate circularity or slight elongation of form, and the marked brevity of the shock and sound.

Lastly, the proximity of the epicentres to known faults, their situation on the downthrow side of the faults, the parallelism of the longer

[1] See my *Manual of Seismology* (Camb. Univ. Press, 1921), pp. 132–133.

axes of the isoseismals to the fault-lines, and the migration of the epicentres along, or parallel to, the Dinas and Irwell Valley faults, all lead to the conclusion that the earth-shakes are caused by small fault-slips; and the fact that the foci are at about the same depth as the pit-workings leads to the further conclusion that the fault-slips are precipitated by the working of the mines. It is important to notice that some coal-seams are worked out right up to the faults referred to. This is certainly the case at Pendleton and Barnsley, probably also at Kilsyth and in the Rhondda Valleys.

Now, by the withdrawal of the coal—or, it may be, by the lowering of the water-level by pumping in other parts of the mine—the rock above is deprived to a great extent of its support, and tends to sink down and close up the worked-out seam. Nowhere can this tendency be greater than where the rock is severed by a fault. Here, the sinking would take place by a series of fault-slips, each of which must give rise to a rather strong shock on the surface of the ground above. But, as the slip would only affect a small region of the fault-surface and would occur at a slight depth, the intensity of the shock would fade away rapidly from the epicentre; the disturbed area would therefore be small and circular or slightly elongated in form; and the shock and sound would be of brief duration. Again, as the focus would be limited approximately by the extent of the worked-out seams, its boundaries would be more or less sharply defined, so that the onset of the shock would be sudden, the sound and shock practically co-terminous, and the sound-area and disturbed area nearly or quite concentric.

If this be the correct explanation of the earth-shakes in mining districts—and it seems to account for all their peculiar features—it follows that the shocks cannot be classed either with true or with spurious earthquakes. They are of natural origin in so far as they are produced by fault-slips, but of artificial origin in that the slips are precipitated by human labour and not by the slow and gradual cooling and warping of the earth[1].

EARTH-SHAKES IN LIMESTONE DISTRICTS

Earth-shakes, not unlike those described above but usually slighter, are felt from time to time in the limestone areas of this country, especially in parts of Sunderland. From 1883 to 1885, they were

[1] *Geol. Mag.* vol. 7, 1900, pp. 174–177; vol. 2, 1905, pp. 219–223; vol. 3, 1906, pp. 171–176.

especially frequent on the south side of that town and along certain lines in that district. They have, however, been observed at other times, both before and since[1]. They are usually sudden shocks, strong enough to make windows and crockery rattle, occasionally even to crack the walls of houses, being often but not always followed by loud noises and dull rumbles, and of disturbed area so limited that persons living only a short distance from the centre have felt and heard nothing. One observer counted 37 such earth-shakes in four months, from 1883 Dec. 7 to 1884 Apr. 7, 14 of them being described as slight and the rest as sharp or severe.

The origin of these earth-shakes is by no means clear. Prof. Lebour, who studied the series of 1883–85, attributed them to rock-falls in underground channels worn or enlarged by the natural passage of water or by pumping for the local water-supply. As the magnesian limestone at and near Sunderland is not more than 400 feet in thickness and the coal-mines there are very deep, this theory is in accordance with the evidently slight depth of the foci. It would also account for the formation of what Prof. Lebour called "breccia-gashes" —fissures or channels filled with angular fragments of limestone now cemented together—of which as many as fifteen are shown in the cliff-section of Marsden Bay (5 miles N. of Sunderland). It is possible, however, that some of the shocks—the stronger ones especially—owe their origin, as in the mining districts, to small fault-slips precipitated by the action of underground water.

There may be some connexion between the occurrence of the Sunderland earth-shakes and the general subsidence of the town. In 1855, a series of levels was carried out by the Ordnance Survey, and repeated in 1895. A comparison of the two series revealed a general subsidence of the town, in some parts by more than 6 feet (G. A. Lebour, *N. of Eng. Inst. of Min. and Mech. Eng. Proc.* vol. 33, 1884, pp. 165–174; *Geol. Mag.* vol. 2, 1885, pp. 513–515. The results of different series of levels are given in the *Newcastle Daily Chronicle* for 1895 Feb. 10).

---

[1] About 1869 or 1870 I heard two on one evening. They were rumbling sounds, without any tremor, each lasting about 2 seconds, and resembling the noise that would be produced by a mass of rock falling down a vertical cavern with uneven sides.

CHAPTER XX

EXTRA-BRITISH EARTHQUAKES

IN this chapter are given lists of earthquakes in three districts which lie outside the area of Great Britain proper. One or two of the Irish earthquakes may have been felt across the channel, and the strong earthquake of 1852 Nov. 9, though the centre of its disturbed area (as drawn, Fig. 88) lies near Bangor, was probably of submarine origin, and connected more closely with Ireland than with Great Britain. To the east of the Channel Islands is a focus in which several strong earthquakes originated, two of them disturbing areas of about 70,000 square miles and being felt along the southern coast of England and as far north as London. Several of the earthquakes felt in the Shetland Islands came from Norway, and it is possible that the majority were of Norwegian, rather than of local, origin. Lastly, the Lisbon earthquake of 1755 Nov. 1 gave rise to seiches or oscillations of the water of numerous ponds, lakes and rivers in this country, and to sea-waves chiefly along our southern coasts.

EARTHQUAKES IN IRELAND

Few countries are so free from earthquakes as Ireland. The following list contains notices of 21 shocks, and it is possible that several of these were really land-slides which in old writings are often ascribed to earthquakes. Taking area into account, for every 1000 British earthquakes, only 43 originated in Ireland.

660, "an earthquake in Hibernia" (*Chron. Scotorum*, p. 99).

681, "a great wind and earthquake in the island of Hibernia (*Chron. Scotorum*, p. 107).

1118, "a very great earthquake in Sliabh-Elpa, which extinguished many cities, and a multitude of people in them" (*Annals of Loch Cé*, vol. 1, p. 111).

1586 Apr. 11, an earthquake in Ireland (Short, vol. 2, p. 169).

1690 Jan. 28, an earthquake at Kingston (Perrey, p. 128).

1690 Oct. 7, a strong earthquake felt in Dublin, Wicklow, Kilkenny, etc. (*Morning Post*, 1852 Nov. 11).

1713 Mar. 10, an earthquake at Clogher (Perrey, p. 129).

1734 Aug., an earthquake in Ireland that destroyed a hundred houses and five churches (*Gent. Mag.* vol. 4, 1734, p. 456).

1756 Jan. 2, an earthquake near Tuam (Perrey, pp. 142–143).

1756 Jan. 27, 4 p.m., an earthquake at Bailyborough (*Gent. Mag.* vol. 26, 1756, p. 91).

1760 Jan. 20, about 7 p.m., a shock, with a rumbling noise, at Wicklow and for several miles round (*Gent. Mag.* vol. 30, 1760, p. 99).

1761 Mar. 31, an earthquake at Cork (*Gent. Mag.* vol. 31, 1761, p. 185).

1762 Mar. 16, an earthquake at Wexford (Perrey, p. 146).

1777 Oct. 1, an earthquake at Kinsale (Mallet, 1852, p. 189).

1820 Apr. 6, between 2 and 3 a.m., a shock, with a rumbling noise, at Cork and the neighbouring towns (*Ann. de Chim. et de Phys.* vol. 15, 1820, p. 423; *Phil. Mag.* vol. 55, 1820, p. 312).

1821 June 25, numerous and violent shocks at Cork (Mallet, 1854, p. 133).

1840 Jan. 8, an earthquake in the northern part of Co. Donegal (Mallet, 1854, p. 292).

1868 Oct. 24, a slight shock at Newton (near Mallow, Co. Cork) (Perrey, *Mém. Cour.* vol. 22, 1872, pp. 104–105).

1869 Jan. 2, two shocks at Finachely (Co. Wicklow) (Perrey, *Mém. Cour.* vol. 24, 1875, p. 5).

1879 Dec. 6, about 11.30 p.m., a rather strong shock, accompanied by noise like that of a building falling, at Stranorlar (Co. Donegal) and many miles round, also at Baron's Court (Co. Tyrone) (*Nature*, vol. 21, 1880, p. 188).

1881 Aug. 27, a shock (of intensity 4 and lasting 2 or 3 seconds), accompanied by a rumbling noise like thunder, at Courtown House, Gorey, Co. Wicklow (*Times*, Sept. 1).

EARTHQUAKES IN THE CHANNEL ISLANDS

1761 Aug. 14, an earthquake felt all over the island of Guernsey (*Gent. Mag.* vol. 31, 1761, p. 378).

1773 Apr. 15, 2.15 p.m., two shocks felt in Guernsey and France (Milne, vol. 31, p. 105).

1773 Apr. 16, shocks felt in Jersey (at 1 and 2 p.m.), Guernsey and France (at 4 p.m.), and on the coast of Dorset (Milne, vol. 31, p. 105).

1773 Apr. 23, shocks in Jersey at about noon and 11.30 p.m. (Milne, vol. 31, p. 105).

1799 Feb. 6, several houses in Guernsey rent (*Gent. Mag.* vol. 69, 1799, p. 245).

1843 Mar. 10, about 1 a.m., a shock, accompanied by a noise like the rumbling of a carriage, in Jersey and Guernsey (*Times*, Mar. 18).

1853 Apr. 1, 10.45 p.m., a shock, accompanied by a noise like that of an artillery waggon rolling quickly over a newly macadamised road, in Jersey and Guernsey, and felt along the south coast of England from Plymouth to Brighton (*Times*, Apr. 5 and 13).

1868 Apr. 4, between 1 and 2 a.m., two distinct shocks, separated by a brief interval and accompanied by a rumbling noise, felt in Jersey and at Brest (*Times*, Apr. 8; *Brit. Rainfall*, 1868, p. 66).

1872 Jan. 3, an earthquake in Jersey (Fuchs, p. 137).

1878 Jan. 28, about 11.55 a.m., intensity 6. This shock was felt along and near the south coast of England from Bovey Tracy on the west to St Leonards on the east, and inland as far as London, Kew and Greenwich. To the south, it was felt at Havre, Rouen and Paris. The boundary of the disturbed area is about 320 miles long from E. to W., 270 miles wide, and contains about 68,000 square miles. Its centre lies probably a short distance to the east of Jersey. At several places (London, Bovey Tracy, Netley, Brighton and St Leonards) the shock consisted of two distinct parts separated by a very brief interval (*Times*, Jan. 29—Feb. 2).

1884 Aug. 26, a strong shock in Jersey (*Times*, Aug. 27).

1887 Apr. 20, a few minutes past 3 a.m., a shock (of intensity about 5), in Jersey and Guernsey, felt slightly at Lymington (*Times*, Apr. 22 and 27).

1889 May 30, 8.21 p.m., a shock of intensity 7 (nearly 8) and lasting about 4 seconds, was felt over an area nearly the same as that disturbed by the earthquake of 1878 Jan. 28, being strongest in Guernsey and near Cherbourg. It was felt by many persons in London, and in France as far south as Paris. The shock consisted of two distinct parts, separated by an interval of about 3 seconds, in Guernsey, Sark and London, and at Bognor, Bournemouth, Surbiton and Wareham (*Nature*, vol. 40, 1889, pp. 141–142; *Times*, May 31 and June 1).

EARTHQUAKES IN THE SHETLAND ISLANDS

1755 Oct. 20, a shock felt at Scalloway, near Lerwick (Milne, vol. 31, p. 101).

1866 Mar. 9, 1.20 a.m., at the Flugga Rock lighthouse, which is built on a rock rising 180 feet above the sea about $1\frac{1}{2}$ miles from the north shore of Unst, the tower was severely shaken by a shock which lasted about 30 seconds. There was no wind or sea at the time to cause the shaking, which was almost certainly due to the Norwegian earthquake of that day. The times given for this earthquake vary from $1.45\frac{1}{2}$ a.m. at Christiansand to 2.52 a.m. at Soderhamn (Christiansand time) or 1.14 to $2.20\frac{1}{2}$ a.m. (Greenwich mean time). The disturbed area on the mainland was 593 miles long from Bodö to Langesund and 390 miles wide from Bergen to Soderhamn. The Flugga Rock is 214 miles from Bergen, and it was probably owing to the great height of the lighthouse above the sea that the earthquake was felt there (*Times*, Apr. 6; T. Ch. Thomassen, "Berichte uber die, wesentlich seit 1834, in Norwegen eingetroffenen Erdbeben," 1888, pp. 29–31).

1871 Apr. 10, a shock, attended by three distinct reports like the firing of a heavy gun, at Lerwick and along the range of hills above the town. The sounds are said to have been heard at two places 30 miles apart. It is uncertain whether the disturbance was due to an earthquake or to gun-firing (*Times*, Apr. 14).

1876 Nov. 28, 6 p.m., a slight shock (intensity 4, duration 2 or 3 seconds) at Sumburgh Head lighthouse at the southern point of the Shetland Islands (*Nature*, vol. 36, 1887, p. 325).

1879 Jan. 4, 1.5 p.m., a smart shock, lasting about 4 seconds, at North Unst lighthouse, which may be connected with the Norwegian earthquake felt at 1.30 — at Flesje, Sogn (*Nature*, vol. 36, 1887, p. 325; T. Ch. Thomassen, p. 37).

1880 July 18, 0.20 a.m., a smart shock, lasting 34 or 35 seconds, at North Unst lighthouse, followed by a second less strong and of shorter duration (*Nature*, vol. 36, 1887, p. 325).

1885 Sept. 26, 10 p.m., the North Unst lighthouse tower was shaken suddenly, there being no heavy sea at the time and the wind light (*Nature*, vol. 36, 1887, p. 325).

1886 Oct. 4, 11 p.m., a shock (intensity 4, duration several seconds) at Baltasno, Unst (*Times*, Oct. 6; *Nature*, vol. 34, 1886, p. 553).

1886 Oct. 5, 1 a.m., the same.

THE LISBON EARTHQUAKE OF 1755 Nov. 1

Two phenomena of this remarkable earthquake were certainly observed in Great Britain—(i) the seiches, or rhythmical oscillations of the water, in many pools, lakes and rivers, and (ii) the seismic sea-waves which followed several hours later. It has been asserted that the earthquake was both felt and heard in this country, and the statement cannot be discredited on account of the distance from Lisbon, for several great earthquakes (for example, the Charleston earthquake of 1886, the Assam earthquake of 1897 and the Kangra earthquake of 1905) have been felt at places more than 800 miles from the origin. The evidence under this heading is given in the next section.

The interest excited by the phenomena observed in this country was not less than that produced by the remarkable series of earthquakes five years before. Many letters on the subject were communicated to the Royal Society, and these were collected together and published in the *Philosophical Transactions* (vol. 49, 1756, pp. 351–398). In 1757, a collection of papers on earthquakes was issued by an anonymous editor (a Member of the Royal Academy of Berlin)[1] under the title *The History and Philosophy of Earthquakes, etc.* The last chapter (pp. 280–334) deals with the phenomena of the Lisbon earthquake in various parts of the globe, and is founded mainly, though containing many additions, on the letters printed by the Royal Society. A few local historians also describe the phenomena which occurred in the districts with which they are concerned.

The position of the epicentre is given by D. Milne (*Edin. New Phil. Journ.* vol. 31, p. 263) as in about lat. 39° N., long. 11° W., and the distances of the various places of observation are measured from this point. Milne also estimates that the time of occurrence was 9.23 a.m. (Lisbon time) or 9.40 a.m. (Greenwich mean time)[2]. Adopting this as the correct time, it follows that, if the seiches were caused by the primary or compressional waves, they should have been first observed at Dartmouth at about 9.44 a.m. and in Loch Ness about 9.45 a.m.; if by the long or surface waves at 9.46 and 9.49 a.m. respectively.

---

[1] Thomas Young (*Lectures on Natural Philosophy*, vol. 2, p. 492) gives the name of the author as Bevis. John Bevis (1693–1771), a well-known astronomer, was elected a member of the Berlin Academy of Sciences in 1750. He "is said to have, from modesty, concealed his authorship of several creditable works" (*Dict. Nat. Biog.* vol. 4, pp. 451–452).

[2] Mallet (*Brit. Ass. Rep.* 1852, p. 163) gives the time of occurrence as between 9.30 and 9.40 a.m.

REPORTED OBSERVATIONS OF THE EARTHQUAKE

Scilly Isles (804 miles). The shock was so strong that several people ran out of their houses, being afraid that they would fall upon them (Edmonds, p. 108).

Helston (830 miles). A sound, like that of a carriage passing, was heard (Edmonds, p. 108).

Cork (918 miles). "At 36 minutes after 9 [Dublin time?] two shocks of an earthquake were felt at about half a minute's interval" (*Hist. and Phil.* p. 283).

Reading, near (993 miles). About 11 a.m., a gardener, standing by a fish-pond, felt a violent trembling of the earth, which lasted more than 50 seconds, and immediately afterwards saw the water in the pond oscillating (*Phil. Trans.* p. 366).

Eyam Edge (Derbyshire, 1080 miles). About 11 a.m., an observer on the surface felt one shock, which raised him sensibly in his chair and caused pieces of plaster to fall from the sides of the room. A fissure, 6 inches wide, 1 foot deep, and nearly 150 yards long, opened at the same time in a neighbouring field. In the same place, two miners were at work in a lead mine in a drift about 50 yards long and 120 yards deep. A violent shock was felt, followed soon after by a still stronger shock, and this again by three slighter shocks, the interval between the first and fifth shocks being about 20 minutes. Each shock was followed by a loud rumbling sound. As they went along the drifts, the miners noticed that fragments had fallen from the sides and roof (*Phil. Trans.* pp. 398–402).

Leadhills and Wanlockhead (near Dumfries, 1180 miles). The earthquake was very distinctly felt (*Edin. Adver.* 1820).

Drymen and Kilmaronock (near Glasgow, 1220 miles). The shock was particularly severe around Drymen. "On that day, Whitefield, the great English divine, was preaching in the adjoining parish of Kilmaronock.... The speaker and his hearers occupied the face of an eminence. Instantly the earth heaved, and the people were bent forwards as if by a wave" (W. Nimmo, *Hist. of Stirlingshire*, 3rd edit. 1880, vol. 2, p. 207).

Although it is difficult, in the absence of necessary information, to account for most of the above disturbances, it seems clear, for the following reasons (in addition to the general omission of the time of occurrence), that none of them was due to the Lisbon earthquake:

(i) Of the writers who describe the seiches in letters communicated to the Royal Society, only one (near Reading) refers to a movement

—a violent trembling of the earth. Another observer, in the same place, makes no reference to any shock. A few, near Durham and in the Lake District, state that no motion or tremor was perceived (*Phil. Trans.* pp. 367, 382, 384 and 386).

(ii) At places between 800 and 1200 miles from the epicentre, the movement, if observed, would have consisted of slow gentle oscillations, and not of a trembling or as a series of violent shocks; it could hardly have been felt in mines; and no sound at all would have been heard at distances so great.

(iii) The phenomena observed at Eyam Edge closely resemble those of the earth-shakes observed in mining districts, which, as urged in the last chapter, are probably due to fault-slips precipitated by mining operations.

Thus, while it is not impossible that the earthquake might have been felt in the south of England, it would seem that we have no evidence that the disturbed area extended so far northwards[1].

SEICHES IN RIVERS, LAKES, POOLS, ETC.

Seiches were observed in the sea at Poole (937 miles from the epicentre) and in Loch Long (1220 miles); in harbours at Portsmouth (950 miles) and Yarmouth (1014 miles); in rivers and streams at Dartmouth (870 miles), Midhurst (973 miles), Guildford (978 miles), Tonbridge (1004 miles), Rotherhithe (1007 miles), Gainsborough (1127 miles), Hull (1139 miles) and Queensferry (1270 miles), and near Fort Augustus (R. Tarff, 1303 miles, and R. Oich, 1306 miles); in a canal at Busbridge (near Godalming, 973 miles); in the lakes of Coniston (1137 miles), Esthwaite (1139 miles), Windermere (1139 miles), Closeburn (near Dumfries, 1177 miles), Lomond (1220 miles), Katrine (1230 miles), Coulter (near Stirling, 1240 miles), and Ness (1320 miles); and in pools at or near Midhurst (973 miles), Lee (Surrey, 973 miles), Guildford (978 miles), Earley Court (near Reading, 987 miles), Cobham (Surrey, 988 miles), Shirburn Castle (Oxfordshire, 991 miles), Cranbrook (Kent, 994 miles), Eaton Bridge (Kent, 1000 miles), London (1007 miles), Tenterden (Kent, 1009 miles), Luton (1014 miles), Rochford (Essex, 1035 miles), Thaxted (Essex, 1037 miles), Albury (Hertfordshire, 1048 miles), Wickham Hall (near Bishop Stortford, 1048 miles), Bardfield and Bocking (Essex, 1054

[1] Mr R. D. Oldham, in his report on the Assam earthquake of 1897 June 12 (*India Geol. Surv. Mem.* vol. 29, 1899, pp. 371–376) also concludes that the observations at Eyam Edge refer to a local disturbance and that there is no evidence to show that the Lisbon earthquake was felt in this country.

miles), Finchingfield (Essex, 1056 miles), Royston (1058 miles), Toppesfield (Essex, 1066 miles), Dunstall (near Bury St Edmunds, 1075 miles), Barlborough (Derbyshire, 1091 miles), Framlingham (1093 miles), and Durham (1182 miles)

The following accounts of seiches observed in a canal, lake and pool, respectively, give a general idea of the nature of the movements:

Busbridge (near Godalming, 973 miles). The observers were near the east end of a canal, about 700 feet long from E. to W., about 58 feet wide, the water being usually 2 to 4 feet deep at the east end, increasing to about 10 feet at the west end. On the north side the path is about 14 inches above the water-level, and on the south side about 10 inches. The greatest height of the bank over which the water flowed was about 20 inches. At half-past 10 a.m., an unusual noise was heard in the water at the east end, and the water in great agitation was raised in a ridge extending lengthwise about 30 yards and about 2 or 3 feet above the usual level of the water, and flowed about 8 feet over the grass walk on the north side. The water then returned into the canal, was raised in a ridge in the middle, and flowed over the grass walk on the south side. During this latter motion, the bottom of the canal on the north side was bare of water over a width of several feet. The water having returned to the canal, the vibrations grew less, but several times the water flowed over the south bank. In about a quarter of an hour from the first movement, the water became quiet and smooth as before. The motion of the water during the whole time was attended by a great perturbation of the sand at the bottom of the canal and by a great noise like that of water turning a mill (*Phil. Trans.* pp. 354–356).

Loch Lomond (1220 miles). Loch Lomond, suddenly and without the least gust of wind, rose against its banks with great rapidity, but immediately retired, and in five minutes' time subsided, till it was as low in appearance as anybody then present had ever seen it in the time of the greatest summer drought; and then it instantly returned towards the shore, and in five minutes' time rose again as high as it was before. The agitation continued at this rate from $9\frac{1}{2}$ to $10\frac{1}{4}$ a.m., taking five minutes to rise and as many to subside, and from $10\frac{1}{4}$ to 11 a.m. every rise came somewhat short in height of the one immediately preceding, taking five minutes to flow and five to ebb, until the water settled as it was before the agitation. The height to which the loch rose was measured and found to be 2 feet 4 inches (*Phil. Trans.* p. 390).

Durham, near (1182 miles). The observer was near a pond about 40 yards long and 10 yards wide. At about 10½ a.m., a sudden rushing noise was heard like the fall of water. The surface of the pond gradually rose, without any fluctuating motion, until it reached a grating which stood some inches above the usual water-level. Through this, the water was discharged for a few seconds; it then subsided as much below the mark it rose from as it was above it in its greatest elevation; and continued thus rising and falling for about six or seven minutes, making four or five returns each minute. The ebb and flow were each about half a foot in the vertical direction (*Phil. Trans.* pp. 385–386).

Though the measurements in these and other observations are only approximate, it would seem that: (i) the oscillations occurred in the N.–S., rather than in the E.–W., direction; (ii) the water rose and fell each way from about 4½ inches to about a foot in pools and 2 feet 4 inches in Loch Lomond; (iii) the period of the seiches varied from about a quarter of a minute in pools to about 10 minutes in Loch Lomond; and (iv) the duration of the whole movement from its first observation ranged from 2 or 4 minutes in pools to an hour and a half in Loch Lomond. The dependence of the duration of the movement on the length of the body of water was noted by the Rev. J. Harrison in the case of the neighbouring lakes of Windermere, Coniston and Esthwaite. These lakes are respectively about 10¼, 5¾ and 2 miles in length, and the movements in them lasted about 8 or 10 minutes, above 5 minutes, and less than 5 minutes. "And as to the differences of the time the agitations lasted," he adds, "may they not be thought proportionate to the different dimensions of the lakes, as the vibrations of pendulums, after the impelling powers are taken away, may be found to continue in proportion to their lengths?" (*Phil. Trans.* p. 384).

SEISMIC SEA-WAVES

While the seiches were usually observed between 10 and 11 a.m., the seismic sea-waves did not reach our southern coasts until after 2 p.m. Shortly after this hour, they were seen in Mount's Bay. In Penzance Bay, they were noticed first at 2.45 p.m. and attained their greatest rise (of 8 feet) at about 3 p.m. At Kinsale (on the south coast of Ireland, about 15 miles S. of Cork), they were observed between 2 and 3 p.m. (Dublin time?), and at and near Plymouth at about 4 p.m. Though the times are not given, there can be no doubt that they swept up the shore at Hayle and St Ives, the water at the latter place rising between 8 and 9 feet and floating two vessels that before were dry, and even at Hunstanton on the north coast of Norfolk. At Kinsale, the waves were observed a second time between 6 and 7 p.m., and it may have been this renewal that was experienced at Swansea at 6.45 p.m.

The following account of the sea-waves at Mount's Bay (by the Rev. W. Borlase, F.R.S.) is typical of others: "A little after two in the afternoon, the weather fair and calm,...there happened here, and the parts adjacent, the most uncommon and violent agitation of the sea....About half an hour after ebb, the sea was observed at the Mount-pier to advance suddenly from the eastward. It continued to swell and rise for the space of ten minutes; when it began to retire, running to the west and south-west, with a rapidity equal to that of a mill-stream descending to an undershot wheel. It ran so for ten minutes, till the water was six feet lower than when it began to retire. The sea then began to return, and in ten minutes it was at the before-mentioned extraordinary height. In ten minutes more it was sunk as before, and so it continued alternately to rise and fall between five and six feet, in the same space of time. The first and second fluxes and refluxes were not so violent at the Mount-pier as the third and fourth, when the sea was rapid beyond expression, and the altera-tions continued in their full fury for two hours. They then grew fainter gradually, and the whole commotion ceased about low water, five and a half hours after it began" (*Phil. Trans.* pp. 373–374).

CHAPTER XXI

SOUND-PHENOMENA OF BRITISH EARTHQUAKES

IN the accounts of many recent earthquakes in the above pages, reference is made to the sound-phenomena—especially to the type of sound, the percentage of audibility and the extent of the sound-area. In the present chapter, the phenomena are considered generally for the different classes into which British earthquakes may be divided, in addition to other phenomena which it seemed unnecessary to describe in detail for individual earthquakes. The classes are four in number—(i) Strong earthquakes in which the area within the iso-seismal 4 exceeds 5000 square miles; (ii) Moderate earthquakes in which the area within the isoseismal 4 lies between 5000 and 1000 square miles; (iii) Slight earthquakes with a long focus, the area being less than 1000 square miles; and (iv) Slight earthquakes with a short focus[1]. The Menstrie earthquakes (all of which belong to Class (iv)) are divided into three classes: (i) The principal earthquakes, which disturbed areas of from 60 to 1000 square miles; (ii) The stronger after-shocks, which disturbed areas from 74 to 150 square miles; and (iii) The remaining earthquakes, the disturbed areas of which were less, and usually much less, than 60 square miles[2].

NATURE OF THE EARTHQUAKE-SOUND

The following table shows the frequency of reference to the different types of sound (p. 9), the figures being percentages of records under each type for each class of earthquake considered:

CLASS OF EARTHQUAKES	TYPE						
	1	2	3	4	5	6	7
Strong	45	20	11	4	7	8	5
Moderate	46	26	4	4	8	7	4
Slight (long focus)	40	37	4	1	4	10	4
Slight (short focus)	28	26	6	8	13	16	3
Menstrie (i)	30	22	4	5	22	16	1
,, (ii)	20	11	11	6	30	14	8
,, (iii)	7	9	3	4	49	27	1

[1] The average lengths of focus in earthquakes of these four classes are, respectively, 12·8, 12·3, 12·4 and 3·9 miles (see Chapter XXIII).

[2] In Class (i) are included the earthquakes of Comrie 1839, Kendal 1843 and 1871, Unknown centre 1852, Hereford 1863 and 1896, Colchester 1884, Inverness 1890 and 1901, Pembroke 1892, Carmarthen 1893, Derby 1903 and 1904, Carnarvon

Thus, while there is considerable variety in the type of sound referred to in different earthquakes, this table shows that there is a close general resemblance between strong earthquakes, moderate earthquakes and slight earthquakes with a long focus. In slight earthquakes with a short focus, there is a marked tendency to more frequent reference to types of short duration. This is shown more clearly if we omit the seventh type, which includes sounds of various durations, and group the first three types together as of long duration, and the next three as of short duration. The percentages of reference to types of long duration for the four classes of earthquakes are respectively 80, 79, 84 and 62, and for the three classes of Menstrie earthquakes 57, 46 and 19.

INAUDIBILITY OF THE SOUND TO SOME OBSERVERS

The inaudibility of the sound may be either total, temporary or partial, the latter consisting in the suppression of some vibrations only, so that observers in the same place may refer the sound to various types. For the four classes of ordinary earthquakes, the percentages of audibility within the sound-area are respectively 82, 92, 98 and 96; and for the three classes of Menstrie earthquakes 89, 91 and 97. Classifying the earthquakes by intensity only, the average percentages of audibility for earthquakes of intensities 7–6, 5, 4 and 3 are respectively 92, 95, 98 and 100. Thus, the percentage of audibility increases as the strength of the earthquake decreases, the reason for this being that the sound in slight earthquakes is usually a more prominent feature than the shock.

The fact that deep explosive crashes are heard by some observers in the midst of the rumbling sound and not by all points to the unequal partial suppression of the sound-vibrations. In other cases, the sound presents itself differently to different persons. In a few strong earthquakes, there are a large number of records from within certain limited

---

1903, Doncaster 1905, Swansea 1906, and Stafford 1916; in Class (ii) the earthquakes of Bolton 1889, Leicester 1893, Carlisle 1901 (No. 3), Strontian 1902, Derby 1906, and Oban 1907; in Class (iii) the earthquakes of Ullapool 1892, Barnstaple 1894, Carlisle 1901 (No. 5), Doncaster 1902, Dunoon 1904, St Agnes 1905, and Swansea 1907; and in Class (iv) the earthquakes of Helston 1892 and 1898, Wells 1893, Launceston 1896, Oakham 1898, Menstrie 1900 (Nos. 12 and 13), 1905 (Nos. 16 and 19), 1908 (No. 70) and 1912 (No. 169), Penzance 1904, Leicester 1904, Carnarvon 1906, Malvern 1907, Dunoon 1908, Glasgow 1910 and Grasmere 1911. In Class (i) of the Menstrie earthquakes are included Nos. 1, 2, 5, 6, 8, 19, 70 and 169; in Class (ii) Nos. 12, 16, 17, 31, 38, 42, 71, 121, 128, and 171; and in Class (iii) the remaining earthquakes.

districts. The next table gives the percentages of reference to the different types of sound:

EARTHQUAKE	PLACE	TYPE						
		1	2	3	4	5	6	7
Hereford 1896	Birmingham	35	18	17	6	7	11	6
Inverness 1901	Inverness	26	34	13	16	0	8	3
Derby 1903	Derby	61	11	6	5	11	3	2
,,	Nottingham	46	12	4	0	12	21	4

An explanation often given of the inaudibility of the sound is that it is due to the inattention of the observer. In the case of earthquakes occurring in the middle of the night, the audibility of the fore-sound is a decisive test of this explanation; for, if the observers are asleep, they can only be awakened before the shock begins if they are capable of hearing the sound. The following table gives the percentage of audibility of the fore-sound in the case of three strong earthquakes to observers who were respectively awake and asleep when the earthquake began:

EARTHQUAKE	TIME OF OCCURRENCE A.M.	PERCENTAGE OF AUDIBILITY	
		observers awake	observers asleep
Hereford 1896	5.32	72	74
Inverness 1901	1.24	72	72
Doncaster 1905	1.37	78	67
Average		74	71

It follows, therefore, that the inaudibility of the sound is rarely, if ever, due to inattention.

VARIATION THROUGHOUT THE SOUND-AREA

The average percentage of audibility for strong earthquakes is 97 within the central isoseismal, and 94, 87, 68 and 55 within the zones bounded by successive isoseismals. The rate of decline is thus at first slow, but it afterwards becomes more rapid, especially near the boundary of the sound-area. In three earthquakes (Hereford 1896 and Derby 1903 and 1904), the variation in audibility is represented by means of isacoustic lines (Figs. 51, 56 and 59). All three earthquakes were twins, and in each the isacoustic lines are strongly distorted in the neighbourhood of the synkinetic band.

If we may judge from the frequency with which the word "heavy" occurs in descriptions of the sound, there seems to be but little variation in the lowness of the sound, the average percentages of descriptions in which the word occurs in the successive zones of strong earthquakes being respectively 28, 29, 34, 27 and 22.

In the next table are given the average percentages of reference to various types of sound in the successive zones (A–E, A being the central zone) of strong earthquakes:

ZONE	TYPE							TYPES OF LONG DURATION
	1	2	3	4	5	6	7	
A	38	27	6	8	5	13	4	73
B	44	23	9	6	6	8	5	79
C	52	19	10	4	5	6	4	84
D	47	20	14	3	5	5	4	86
E	47	12	23	3	1	4	1	91

Thus, as the distance from the origin increases, there is a steady decrease in the frequency of reference to types 2, 4 and 6, a steady increase to type 3, on the whole an increase to type 1, and a steady increase to types of long duration. The percentages of comparison to distant thunder out of the total number of references to type 2 in successive zones are 10, 43, 53, 57 and 100—this continued increase suggesting a greater monotony in the sound, a closer approach to uniformity both in intensity and tone, as the distance from the epicentre increases.

RELATIONS BETWEEN THE SOUND-AREA AND
DISTURBED AREA

In all strong British earthquakes, the disturbed area extends beyond the sound-area in every direction. The magnitude of the sound-area ranges from 7800 square miles for the Derby earthquake of 1903 to 70,000 square miles for the Hereford earthquake of 1896; and the ratio of the sound-area to the disturbed area from 35 per cent. for the Hereford earthquake of 1863 to 77 per cent. for the Kendal earthquake of 1871, the average value of this ratio being 58 per cent. In earthquakes of moderate strength and in some slight earthquakes, the sound-area and disturbed area approximately coincide, the areas in such cases ranging from about 400 to 2000 square miles. In many slight earthquakes (for example, the Comrie earthquakes of 1895 and 1898, the Helston earthquake of 1898 and the Oakham earthquake of 1898), the sound-area overlaps the disturbed area either on one side only or in every direction.

In strong earthquakes, the divergence of the boundary of the sound-area from an isoseismal line is not clearly manifested. In two moderate earthquakes (the Edinburgh and Bolton earthquakes of 1889), the sound-area was excentric with regard to the isoseismal lines. It is in slight earthquakes, however, that the displacement of the sound-area is most easily seen, as, for example, in the Leicester earthquake of 1904 and the Helston earthquake of 1898 (Figs. 63 and 73), in both of which the sound-area, relatively to the isoseismal lines, is shifted towards the line of the originating fault.

TIME-RELATIONS OF THE SOUND AND SHOCK

In the following tables, the figures in the columns headed p, c and f denote the number of records per cent. for each epoch in which that epoch of the sound preceded, coincided with, or followed, the corresponding epoch of the shock.

EARTHQUAKES	BEGINNING			EPOCH OF MAX. INTENSITY			END		
	p	c	f	p	c	f	p	c	f
Strong	66	26	8	20	61	9	16	41	43
Moderate	66	29	5	24	73	3	12	43	45
Slight (long focus)	71	26	3	33	67	0	25	36	39
Slight (short focus)	49	43	8	33	67	0	14	53	33
Menstrie (i)	51	43	6	—	—	—	17	51	32
,, (ii)	35	52	12	—	—	—	18	54	28
,, (iii)	34	43	23	—	—	—	9	50	41

There is thus a close resemblance, as regards time-relations, between strong earthquakes, moderate earthquakes, and slight earthquakes with a long focus. In slight earthquakes with a short focus, there is a tendency to coincide in both terminal epochs.

In the next table are given the average time-relations for the zones between successive pairs of isoseismal lines in strong earthquakes:

EPOCH	ZONE											
	A			B			C			D		
	p	c	f	p	c	f	p	c	f	p	c	f
Beginning	65	30	5	69	24	7	69	22	9	62	23	15
Epoch of Max. Intensity	15	78	6	20	68	12	23	59	17	30	56	14
End	18	46	36	18	40	42	16	46	38	21	43	36

This last table shows that the sound-vibrations and the vibrations which constitute the shock must travel with approximately the same velocity; for, if, as is generally supposed, the sound-vibrations were to travel the more rapidly, the beginning of the sound should precede that of the shock by an interval increasing with the distance from the epicentre; while the end of the sound should be heard less frequently after the shock. The figures in the table show no tendency whatever to vary in either of these directions.

CHAPTER XXII

DISTRIBUTION OF BRITISH EARTHQUAKES IN SPACE AND TIME

RELATIONS BETWEEN THE EARTHQUAKES OF ENGLAND, SCOTLAND AND WALES

THE earthquakes of Scotland differ in some respects from those of England and Wales:

(i) Of the total number (1190) of earthquakes, 310 originated in England, 822 in Scotland, and 54 in Wales, while 4 are of unknown origin; or, for every 100 earthquakes felt within a given area in England, 452 are felt in Scotland and 120 in Wales.

(ii) The number of earthquakes that have caused marked injury to property in this country is comparatively small. As already mentioned (p. 6), Messrs Meldola and White estimate the number between 103 and 1881 at 74; but 18 of these are recorded by Dr Short alone, 3 originated outside the area of Great Britain, while at least 7 —and probably a much larger number—though strong enough perhaps to detach a few bricks from chimneys, were really of an intensity below 8. The number of known earthquakes that certainly attained this or a higher intensity between 974 and 1921 is, I believe, not more than 22, of which 13 occurred in England, 7 in Scotland and 2 in Wales[1]. These figures do not, however, represent correctly the distribution of destructive earthquakes in this country, for the earliest shock of this strength occurred in 1180 in England, in 1248 in Wales, and in 1769 in Scotland. From 1750, when our record of strong earthquakes becomes comparatively full, until 1921, 4 destructive earthquakes occurred in England, 5 in Scotland and 1 in Wales. If we include all earthquakes from the year 1750 of intensities 4 and upwards, the relative numbers for equal areas in England, Scotland and Wales are 100, 249 and 109.

(iii) The number of well-defined earthquake centres is 57 in England, 28 in Scotland and 8 in Wales, the relative numbers for equal areas being 100, 72 and 97. Again, the numbers of earthquakes assigned to these particular centres are 216 in England, 797 in Scotland and 37 in Wales, giving an average number of earthquakes per centre

---

[1] If we include earthquakes which have caused slight damage but were of intensity less than 8, the numbers would be 26 in England, 9 in Scotland and 8 in Wales, or altogether 43.

of 3·8 in England, 28·4 in Scotland and 4·6 in Wales. The large average number for Scotland is chiefly due to the remarkable series of Comrie, Menstrie and Invergarry earthquakes. If these three districts be deducted, the average number per centre is still higher (namely, 4·9), though not much higher, than in the other divisions of the country.

(iv) There is a marked difference in average disturbed area for each degree of intensity. In this respect, the comparison can only be made for English and Scottish earthquakes, and for those of intensities 8, 7, 5 and 4, the numbers in other cases being too small to give satisfactory results. For earthquakes of intensity 8, the average disturbed areas are 71,000 square miles for England and 27,650 for Scotland; for those of intensity 7, 16,400 and 4900; for those of intensity 5, 1600 and 460; and for those of intensity 4, 560 and 130[1]. Thus, the area for each degree is much greater in England than in Scotland, the ratios for the different degrees of intensity being 2·6, 3·3, 3·5 and 4·3, the ratio thus increasing as the strength diminishes. The explanation of this inequality is, no doubt, that the seismic foci are situated at a less depth in Scotland than in England; and, in this connexion, it is worthy of notice how much more frequently earthquakes can be assigned to known faults in Scotland than in England[2].

(v) Lastly, there are differences in the nature of the earthquakes in the three divisions of the country. It is only in Scotland that we meet with long series of frequent earthquakes, such as those of Comrie (421 in number) and Menstrie (200). Again, twin earthquakes, so far as we know, are entirely absent from Scotland, in Wales we have only twin earthquakes of the second class (see Chapter XXIII), in England we have twin earthquakes of both classes. As will be seen later, this difference seems to be connected, to some extent, with the different depths at which the earthquakes originate.

---

[1] In these estimates, after-shocks are excluded.

[2] In part, this greater frequency is due to the relatively larger surface-area occupied by the older rocks in Scotland than in England. Though our earthquakes are probably not confined to fault-movements in such rocks, they are much more numerous in them than elsewhere. It is remarkable that some of the strongest of British earthquakes (for example, the Canterbury earthquake of 1382 and the Colchester earthquake of 1884) occurred in the south-east of England. But the coating of newer rocks in this district is comparatively thin, and there can be little doubt that the earthquakes referred to originated in movements in the older rocks below, and that the high intensity of these shocks was partly due to their being felt on soft or earthy formations.

DISTRIBUTION OF BRITISH EARTHQUAKES
IN SPACE

The distribution of British earthquakes is represented in Fig. 94, the map being constructed as follows:

(i) The positions of well-defined centres are indicated by dots. Though the names of neighbouring towns are as a rule affixed to them, it should be remembered that the towns themselves are usually a few miles away.

(ii) The lines drawn through the dots are parallel to the average direction of the longer axes of the isoseismal lines of earthquakes belonging to those centres. When no such lines are drawn, the disturbed areas are either circular or the directions of the longer axes are unknown. The omissions in such cases are for the most part of little consequence, except in the important centre of Invergarry (50 earthquakes). The earthquakes of the Bolton centre are no doubt connected with the Irwell Valley fault, the direction of which in that region is N. 50° W.

(iii) The lengths of the lines indicate roughly the numbers of earthquakes belonging to the centres, different lengths being used for the following numbers of earthquakes, namely, 1–5, 6–10, 11–25, 26–50, 200–421. The lines might have been drawn proportional to the numbers of earthquakes, but, as these numbers vary from 1 to 421, this would have involved a map on a much larger scale than that of Fig. 94.

The map is thus intended to show the positions and directions of the faults to the growth of which our earthquakes are due, the portions of those faults which are now growing, and, very roughly no doubt, the rates at which they are growing.

It will be seen from this map that there are tracts of country, five in number, that are nearly free from earthquakes. These are: (i) North-west Scotland, comprising all the district on the north-west side of the Caledonian canal, about 8700 square miles in area and containing only one earthquake centre, that of Ullapool with 3 slight earthquakes; (ii) North-east Scotland, including the countries of Nairn, Elgin, Banff, Aberdeen, Forfar, Kincardine, and part of Inverness, about 6500 square miles in area; (iii) South Scotland and North-east England, including the countries of Ayr, Wigtown, Kirkcudbright, Dumfries, Lanark, Linlithgow, Edinburgh, Haddington, Peebles, Selkirk, Roxburgh, Berwick, Northumberland, Durham and the North Riding

Fig. 94. Distribution of British Earthquakes in Space.

of Yorkshire—an area of about 12,200 square miles, with only four centres visited by slight earthquakes, namely, Edinburgh (2 earthquakes), Leadhills (8 earthquakes), Penpont (3 earthquakes) and Ecclefechan (4 earthquakes); (iv) Central Wales and West England, comprising the counties of Cardigan, Brecon, Radnor, Montgomery, Flint, Shropshire and Cheshire—an area of 5200 square miles with no important centre; and (v) a wide band including the counties of Dorset, Wiltshire, Hampshire, Berkshire, Oxford, Buckingham, Hertford, Bedford, Northampton (the greater part), Huntingdon, Cambridge and Norfolk—an area of 11,600 square miles and containing only two well-defined centres, namely, those of Banbury (2 earthquakes) and Oxford (4 earthquakes). Altogether, these five aseismic districts contain about 44,200 square miles—about half the area of Great Britain—and are represented by 26 earthquakes. The remainder of the country contains about 43,800 square miles and has been visited by 1024 earthquakes. The relative earthquake-frequency of the aseismic and seismic districts of the country is thus nearly as 1 to 40. If intensity were taken into account, the disproportion would be still more marked.

The map shows also how very varied are the directions of the isoseismal axes. Lines of a given direction are not confined to a particular district, and, in the same district, the directions are diverse and even at right angles. This shows that the underground forces which are now concerned in modifying our surface-features act, not in one prevailing direction, but in at least two directions.

Notwithstanding the general diversity in direction, there are lines that prevail in certain districts. (i) In Scotland, the directions, with two exceptions, lie between N. 30° E. and N. 70° E., the average of eight directions being N. 46° E. or very nearly N.E. The exceptional lines are those of the Ullapool centre (N. 30° W.) and the Glasgow centre (N. 83° E.). (ii) In the North-west of England, the principal directions are nearly N. (Carlisle centre) and nearly E. (Grasmere, Kendal and Rochdale centres), with one exception—about N.W.—in the minor centre of Clitheroe. (iii) In South Wales, the principal directions are N. and E. in the Pembroke centre, N. 70° E. in the Carmarthen centre, and N. 85° E. in the Swansea centre. (iv) In South-west England, though one direction is N., and others diverge from an easterly direction by as much as 30, 31 and 45 degrees, the prevailing direction is E., the average of 13 directions being 3° N. of

E. (v) In South-east England, the direction is E. in the Maidstone centre and between N. 28° E. and N. 67° E. (with an average of N. 47° E.) in the Chichester, Canterbury and Colchester centres. (vi) Lastly, in the twin earthquakes of the Midland Counties of England, the directions lie between N. 42° W. and N. 68° W. and between N. 28° E. and N. 65° E.; the average directions being N. 57° W. and N. 47° E. The directions in this district will be considered more fully in the next chapter.

It will thus be seen that there are four prevailing directions, which are approximately N.E. (Caledonian), N.W. (Charnian), N. (Malvernian) and E. (Armorican). Of the four, the north-easterly direction is the most persistent. It ranges over the whole country, from the north of Scotland to the south-east of England. It includes fractures which are responsible for such important earthquakes as those of Inverness, Comrie, Carnarvon and Colchester. The north-westerly direction is more limited in range. Except for the isolated centre of Ullapool, it is confined to Lancashire, Herefordshire and the Midland Counties of England. The easterly direction prevails from Glasgow to the south coast of England, being especially prominent in Lancashire, South Wales and the south-west of England. Of all the four directions, the northerly occurs most rarely, being found only in the Carlisle, Mansfield, Malvern, Pembroke and Camelford centres.

The importance of the Caledonian movements becomes more evident when we take into account the intensity of the earthquakes. Since the year 1750, there have been 10 earthquakes of intensity 8 or 9. In seven of these (the Derbyshire earthquake of 1795, the Inverness earthquakes of 1769, 1816 and 1901, the Comrie earthquakes of 1839 and 1841, and the Colchester earthquake of 1884), the isoseismal axes were directed approximately N.E.; in two others (the Hereford earthquakes of 1863 and 1896) N.W.; and in one (the Swansea earthquake of 1906) nearly E. For the intensity 7, there were 10 earthquakes with their axes N.E., 1 N.W., 3 N. and 4 E. For the intensity 6, the corresponding numbers are 4, 1, 0 and 3; for the intensity 5, 26, 5, 3 and 7; and for the intensity 4, 24, 7, 4 and 21. Altogether, 71 earthquakes have their axes N.E., 16 N.W., 10 N. and 36 E. For every intensity, and especially for the highest, the N.E. direction predominates, though, for the intensity 4, the number in this direction is closely approached by the number with an easterly trend.

DISTRIBUTION OF BRITISH EARTHQUAKES IN TIME

Our record of British earthquakes extends over a period of about 950 years, but it is only during the last 170 years that we can claim for it any approach to fulness and little, if any, to completeness. The latter period is too short for us to detect any secular change in earth-quake-frequency. There are, however, years or groups of years that are marked by distinct increases of frequency and intensity. For instance, 1750 was such a year and still deserves to be called "the year of earthquakes." From 1839 to 1841, 217 shocks (including two of intensity 8) were recorded, all but 10 in the Comrie centre. During the four years 1903–1906, four earthquakes of intensity 7 and one of intensity 8 occurred in England and Wales. In 1912, 74 slight shocks were felt, all in the Menstrie district, this being the year of greatest known earthquake-frequency, with the exception of 1841, when 81 shocks occurred at Comrie. In Wales, more than half the known earthquakes since 1750 occurred during the sixteen years 1892 to 1907, the numbers being 23 from 1750 to 1891 and 26 afterwards; in other words, for every earthquake felt during a given interval in the earlier period, two were felt during an equal interval in the later. Thus, earthquake-frequency in Great Britain, as elsewhere, is subject to marked fluctuations rather than to any general increase or decrease.

To some extent, these fluctuations appear subject to law. For instance, neglecting after-shocks, important earthquakes occurred in the Inverness centre in 1769, 1816, 1888, 1890 and 1901, the mean interval between successive shocks being 33 years. In other Scottish centres, in which three or more earthquakes occurred, the mean intervals are 12 years in the Fort William and Ullapool centres, 22 years in the Oban centre, 36 years in the Glasgow centre, and 47 years in the Dunoon centre, the average interval in all these centres being 30 years. In England, the mean intervals are 17 years in the Hereford centre, 22 years in the Grasmere centre, 23 years in the Rochdale centre, 27 years in the Taunton and Barnstaple centre, 28 years in the Helston and Kendal centres, and 49 years in the Penzance centre, the average interval in all these centres being 27 years, or nearly the same as in the Scottish centres. In addition to the above centres, there are three centres (Comrie, Menstrie and Chichester) in which series of earthquakes occurred, and twelve others with only two principal earthquakes in each, namely, those of Ardvoirlich, Dunkeld and Perth in Scotland, and those of Altarnon, Banbury, Exmouth, Falmouth, Grantham, Lincoln, Liskeard, Northampton and Reigate in England. At Comrie, the principal earthquakes occurred in 1801, 1839 and 1841, the average interval being 20 years; in the Menstrie district, in 1736, 1767, 1802, 1809, 1872, 1881, 1905, 1908 and 1912, the average interval being 22 years; at Chichester in 1707, 1734, 1750, 1811, 1824 and 1834, the average interval being 25 years. The average of the 15 intervals in these three districts is 23 years. In the twelve centres with two principal earthquakes each, the average interval is just under 50 years. Thus, the more frequent the earthquakes are in a district, the less is the average interval between successive earthquakes.

PERIODICITY OF BRITISH EARTHQUAKES

Annual Period. In the following table are given the numbers of earthquakes during successive half-months from 1750 to 1916, the upper figure being the number during the first half of the month (days 1–15, in February 1–14) and the lower figure that during the second half:

Jan.	Feb.	Mar.	Apr.	May	June	July	Aug.	Sept.	Oct.	Nov.	Dec.
50	36	38	41	41	22	42	23	36	55	44	42
59	39	60	54	44	32	29	38	59	79	50	59

The curve in Fig. 95 represents the monthly variation in earthquake-frequency from 1750 to 1916.

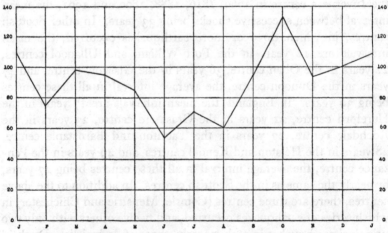

Fig. 95. Monthly variation in Frequency of British Earthquakes.

If the approximate method of harmonic analysis known as the method of overlapping means[1] be applied to these figures, the maximum-epochs and amplitudes of the annual and semi-annual periods may be determined. The results of the analysis are given in the next table and represented graphically in Fig. 96.

PERIOD	EPOCH	AMPLITUDE
Annual	Dec. (mid.)	·22
Semi-annual	Apr. and Oct. (mid.)	·19

They agree closely with those obtained for other countries in the northern hemisphere and especially for various districts in the empire of Japan[2].

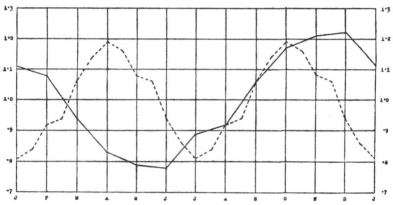

Fig. 96. Annual Periodicity of British Earthquakes.

Diurnal Period. The numbers of earthquakes felt during successive hours for the period from 1750 to 1916 are as follows:

	0–1	1–2	2–3	3–4	4–5	5–6	6–7	7–8	8–9	9–10	10–11	11–12
a.m.	33	74	58	51	42	42	35	26	31	38	24	27
p.m.	19	43	33	33	44	33	25	31	35	36	35	59

The curve in Fig. 97 represents the hourly variation in earthquake-frequency during the period 1750–1916. The application of the method of overlapping means gives the following results for the epochs and

---

[1] *Manual of Seismology*, pp. 185–190.
[2] *Phil. Trans.* 1893 A, pp. 1107–1169; *Phil. Mag.* vol. 41, 1921, pp. 908–916.

amplitudes of the diurnal and semi-diurnal periods, the results being represented graphically in Fig. 98.

PERIOD	EPOCH	AMPLITUDE
Diurnal	1 a.m.	·27
Semi-diurnal	2–4 a.m. and p.m.	·22

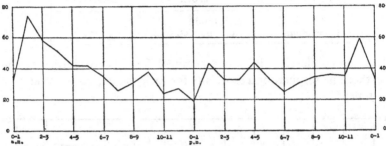

Fig. 97. Hourly variation in Frequency of British Earthquakes.

When earthquake records, as in this case, are founded on personal observations only, the diurnal variation in recorded earthquake-frequency is the resultant of two components, one the actual variation in frequency such as would be determined by seismographic records

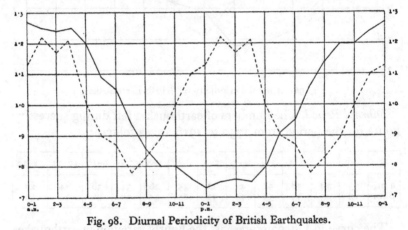

Fig. 98. Diurnal Periodicity of British Earthquakes.

only, the other an apparent variation due to changes in the conditions of observation. The curve in Fig. 97 resembles nearly all other curves based on personal observations in showing two maxima, one shortly before midnight, the other between 1 and 2 a.m.; and there can be little doubt that both maxima are to a great extent apparent only and due to unusually favourable conditions of observation, many persons being awake and resting at these quiet times. In the case of a recent series of earthquakes, those of Menstrie, a third, though less marked, apparent maximum occurs during the hour 4–5 p.m., which is probably due to the quiet of the tea-hour especially on Sunday afternoons.

Owing to what may be called the personal equation of the observers, the true diurnal period can only be determined by seismographic records. The discussion of those obtained at various stations in Japan shows that, for ordinary earthquakes, the epoch of the diurnal period occurs shortly before or about noon; for after-shocks it occurs about midnight. In the semi-diurnal period, the maximum-epoch occurs about 8 or 9 a.m. and p.m. for ordinary earthquakes; for after-shocks it is usually earlier at first, returning after the lapse of some time to about 8 or 9 a.m. and p.m. The apparent diurnal periodicity of British earthquakes may thus be due in part to the large number of after-shocks which have occurred in this country and especially in the Comrie and Menstrie districts.

CHAPTER XXIII

ORIGIN OF BRITISH EARTHQUAKES

BRITISH earthquakes are divisible into two classes—simple and twin earthquakes. In the former the shock consists of a single series of vibrations, in the latter of two series separated by an interval of 2 or 3 seconds. They differ, also, in other respects, of which the following are the most important:

(i) The disturbed area is greater in twin than in simple earthquakes for shocks of the same intensity. Taking earthquakes of intensities 8 and 7 from the year 1890, the disturbed area for simple earthquakes ranges from 7500 to 33,000 square miles, the average being 21,800 square miles[1]. For twin earthquakes, the area ranges from 12,000 to 98,000 square miles, the average being 47,200 square miles. For earthquakes of intensity 5, the corresponding averages are 400 and 2950 square miles.

(ii) For simple earthquakes, the disturbed areas of the after-shocks range from 35 to 2200 square miles, the average being 440 square miles. For twin earthquakes, they range from 112 to about 29,000 square miles, the average being 4200 square miles. Some of the disturbed areas of the minor shocks of twin earthquakes are remarkable considering their low intensity. For instance, one after-shock of the Pembroke earthquake of 1892 and two of the fore-shocks of the Hereford earthquake of 1896 were of intensity 4 or about 4, yet they were felt over areas of about 4800, 6300 and 6400 square miles, respectively; another after-shock of the Pembroke earthquake of intensity 5 disturbed an area of about 29,000 square miles. The average disturbed areas of earthquakes of intensities 4 and 5 are respectively 400 and 1080 square miles.

(iii) Simple earthquakes are much more frequent than twin earthquakes. Leaving out of account the slight tremors of Comrie and Invergarry, of the principal earthquakes since the beginning of 1889, 42 have been simple and 11 twin. In other words, 21 per cent., or about one in five, have been twin earthquakes.

(iv) After-shocks are more frequent with simple than with twin earthquakes. In simple earthquakes of intensities 7 and 8, the numbers range from 6 to 17, the average being 11·0. In twin earthquakes of the same intensities, they range from 0 to 7, the average number being

---

[1] In this estimate, I have omitted the Menstrie earthquakes of 1908 and 1912. Had they been included, the lower limit would have been 605 square miles and the average 13,400 square miles.

2·4 or 18 per cent. (nearly one in five) of the total number of after-shocks.

(v) Lastly, in the arrangement of their foci, there is considerable difference between simple and twin earthquakes. Compare, for instance, the orderly linear arrangement of the foci of simple earthquakes in central Scotland, along both the Great Glen fault and the Highland border fault, with the complicated arrangement of twin earthquake foci in the midland counties of England.

ORIGIN OF SIMPLE EARTHQUAKES

It is clear, from the disturbed areas of both principal earthquakes and their after-shocks, that twin earthquakes originate at a greater depth than simple earthquakes. And this is no doubt the reason why it is usually possible to associate simple earthquakes with known faults, and why it is so rarely the case that twin earthquakes can be so connected.

Another feature of simple earthquakes that has an important bearing on their origin is that the longer axes of their isoseismals are parallel to strike-faults, or to the lines of superficial crust-folds. Twin earthquakes, on the other hand, as will be seen later, are probably due to movements along faults which intersect transversely the more deeply-seated crust-folds.

In Great Britain, the growth of strike-faults is effected by slips which rarely exceed 18 or 20 miles in length and, on an average, are about 11 miles. In the slightest shocks, the slips are mere creeps over an area which may be a mile or less in length. When the focus is on the larger scale (11 miles in length), we may imagine the stresses in its neighbourhood to have been gradually increasing. The stresses might not be uniform throughout the whole focus; at any rate, there would be variations in the resistance to slipping, and the first action of the growing stresses would be to remove these obstructions by small slips, giving rise to fore-shocks with nearly circular disturbed areas. The effective stress is thus rendered more nearly uniform throughout the focus, and this results in a general movement over the whole area —greatest in the central region and gradually decreasing towards the margins—that gives rise to the principal earthquake of the series.

Whether the fault concerned be normal or reversed, whether the relative movement be downward or upward, the first effect of such a slip would be a rapid increase of stress along both lateral and upper margins of the focus. The increased stress in the lateral margins would cause slips within and beyond them, thus gradually extending the area of displacement in both directions; that in the upper margin would cause slips in the central region but on the whole continually approaching the surface.

The series of Inverness earthquakes in 1901 may be referred to in illustration of these remarks. The sequence of events has already been described (pp. 43–56)—the slight fore-shock of Sept. 16, followed by the principal earthquake of Sept. 18, and this again by a number of after-shocks, six being of some consequence, of which one was due to a movement that extended the principal focus about half a mile to the north-east, and the other five to movements that extended the focus 6 miles or more to the south-west. In all these after-shocks, there was also a continuous decrease in the depth of the focus, as shown by the gradual approach of the epicentres towards the fault-line.

Westerly Migration of Foci. The tendency of the Inverness after-shocks of 1901 to shift to the south-west and as far as and even below Loch Ness is characteristic of other earthquakes of the same important centre. The strongest known earthquake of this district occurred on 1816 Aug. 13, and the after-shocks lasted for more than two years. The 15 recorded shocks are probably only the strongest of the series, but, while two-thirds of them originated in the central district, the remaining five had epicentres to the south-west in the neighbourhood of Dores and even below Loch Ness. The earthquake of 1888 Feb. 2 had no recorded after-shocks. That of 1890 Nov. 15 was followed by 10 after-shocks, of which 8 originated in the central district, one (on Nov. 16) near the north-east end of Loch Ness and another (on Nov. 18) near Drumnadrochit, 5 miles farther to the south-west.

A similar tendency to migrate in a westerly direction is shown by the earthquake-foci along the Great Glen fault. The principal centres in its neighbourhood are those of Inverness, Fort William and Oban. The strong Inverness earthquake of 1816 was followed after a long interval—54 years, which, however, in the history of a great fault, is but as one day—by the Fort William earthquake of 1870, and later still by the Oban earthquakes of 1877 and 1880. After the Inverness earthquake of 1901, occurred the Fort William earthquake of 1906 and the Oban earthquake of 1907.

Since the beginning of the nineteenth century, the Menstrie earthquakes originated within a portion of the fault (about 9 miles in length) ranging from a mile or two east of Tillicoultry to a short distance west of Bridge of Allan. Dividing this portion of the fault into four regions —east, east-central, west-central and west—we find that the principal earthquake of 1802 originated in the east region, those of 1872 and 1905 in the east-central, that of 1908 in the east and that of 1912 in the west-central region. Of the slighter earthquakes since 1900, activity predominated until 1912 May 3 in the east-central region, and after that date in the west region, no minor earthquakes during the whole period of the series (17 years) being confined to the east region.

In England, we have only one clear case of migration of foci in simple earthquakes, that of the Taunton earthquakes of 1868 and 1885 and the Barnstaple earthquakes of 1894 and 1920. In the Taunton earthquake of 1868, the centre of the disturbed area was 6 miles S. 78° E. of Taunton and the length of the focus about 12 miles. In that of 1885, the centre was $\frac{1}{2}$ mile N.E. of Wiveliscombe, or 14 miles west of the preceding centre, and the focus was about 10 miles long. The centre of the Barnstaple earthquake of 1894 was $5\frac{3}{4}$ miles west of Exford or 22 miles west of that of the earthquake of 1885 and the length of its focus about $10\frac{1}{2}$ miles. In the Barnstaple earthquake of 1920, the centre was $\frac{1}{2}$ mile N. of East Downe or 10 miles west of the preceding and the length of the focus about 9 miles. There was thus a continual westerly migration of the epicentre in these four earthquakes. But, while the foci of the Taunton earthquakes and those of the Barnstaple earthquakes just about overlapped, there was an interval of 11 miles between the foci of the Taunton earthquake of 1885 and of the Barnstaple earthquake of 1894[1]. The total distance of about 56 miles which separated the terminal regions of the foci of 1868 and 1920 may thus be divided into five nearly equal portions, of which the first and second (counting from the east end) suffered displacement in 1868 and 1885 respectively, the third has apparently remained undisturbed, while the fourth and fifth were displaced in 1894 and 1920 respectively.

Stability of the Comrie Foci. While, in some earthquakes, the aftershock foci oscillate from end to end of the principal focus, and while, in certain districts, there is on the whole a westerly migration of activity, we are confronted at Comrie with migration within such very narrow limits that in this one district—that in which earthquakes are

---

[1] This interval may be due to the connexion of the Taunton earthquakes with the southern boundary fault of the Morte Slates and of the Barnstaple earthquakes with the northern boundary fault (pp. 316, 319).

more numerous than in any other British centre—the foci have yet been nearly stationary for a century and more.

Both phenomena—extraordinary frequency and practical constancy of site—probably depend on the branching of the great fault in the neighbourhood of Comrie. The rock there being cut by a series of faults, it is evident that a movement along any one fault must involve changes of stress along at least one other, and that these increased stresses, when relieved, may similarly affect the stresses along the former or a third fault.

NATURE OF TWIN EARTHQUAKES

(i) *Wide Area of Observation.* The wide area over which the twin-shock is felt is perhaps the strongest evidence that it is not of local origin. In the Pembroke earthquake of 1892, the twin-shock was observed throughout the disturbed area, from Rhyl to the Scilly Isles and from Worcester to Tullow (Co. Carlow). In the Carmarthen earthquake of 1893, it was noticed in nearly all parts of the disturbed area, and at places so near its boundary as Derby, Ashley and Bournemouth. In the Hereford earthquake of 1896, the twin-shock was felt over the district bounded by the isoseismal 5, or over more than 40,000 square miles; towards the north-west, it was also perceptible in Westmorland, in the Isle of Man, and in Ireland, or very nearly to the boundary of the disturbed area. The twin-shock of the Derby earthquake of 1903 was recorded by 68 per cent. of the observers, and was perceptible over the whole disturbed area of about 12,000 square miles. That of the following year was felt over an area of about 8000 square miles or nearly one-third of the disturbed area. In the Doncaster earthquake of 1905, the twin-shock was recognised throughout a district overlapping the isoseismal 5 by a few miles in every direction, and to some observers as far as, and even beyond, the isoseismal 4. In the Swansea earthquake of 1906, the weaker part of the twin-shock usually escaped notice, but it was felt over an area of 13,500 square miles or slightly more than one-fifth of the disturbed area. Lastly, in the Stafford earthquake of 1916, both parts of the shock were felt over the greater part of the disturbed area and in some directions as far as the isoseismal 3, which formed its boundary.

(ii) *Coalescence of the Two Parts of the Shock.* Though, as a general rule, the twin-shock is perceptible over most of the disturbed area, there may exist within that area a narrow band along which the two parts of the shock are no longer distinctly separated but coalesce and form a single continuous series of vibrations. As the two movements are felt together in this band, I have called it the *synkinetic band*. The band has been traced in only four British twin earthquakes—namely, the Hereford earthquake of 1896, the Derby earthquakes of 1903 and 1904, and the Stafford earthquake of 1916—that is, in nearly half the twin earthquakes of the last thirty years. The band is invariably only a few miles in width and crosses the central curves at right angles to their longer axes. In the Derby earthquake of 1903, the synkinetic band is straight, and in the other three earthquakes curved. In every case, it is worthy of notice that two maxima of intensity, connected by weaker tremulous motion, were felt by observers within, but close to the boundaries of, the synkinetic band.

(iii) *Mean Duration of the Interval between the Two Parts of the Shock.* The duration of the interval between the two parts rarely exceeds a few seconds. Taking, first, the whole disturbed area, the mean duration of the interval was 2·1 seconds in the Derby earthquake of 1904, 2·2 seconds in the Stafford earthquake of 1916, 2·3 seconds in the Carmarthen earthquake of 1893, 2·5 seconds in the Leicester earthquake of 1893, 3·0 seconds in the Pembroke earthquake of 1892, the Carlisle earthquake of 1901 and the Derby earthquake of 1903; 3·5 seconds in the Doncaster earthquake of 1905, 3·6 seconds in the Hereford earthquake of 1896, and 3·7 seconds in the Swansea earthquake of 1906, the average of all these estimates being 2·9 seconds.

Again, the mean duration of the interval varies but little with the distance from the epicentres. In the following table, the mean duration in seconds is given for successive zones bounded by isoseismal lines:

EARTHQUAKE	ZONE				
	A	B	C	D	E
Hereford 1896	2·4	2·4	2·2	3·0	3·2
Swansea 1906	2·2	2·1	2·0	2·2	—
Stafford 1916	2·0	2·1	2·4	2·4	1·1
Average	2·2	2·2	2·2	2·5	2·1

(iv) *Relative Nature of the Two Parts of the Shock.* As a general rule, the observations under this heading refer to the relative intensity of the two parts of the shock. The only exception is in the case of the Hereford earthquake of 1896. In this earthquake, the longer axes of the isoseismal lines were directed N.W. In the north-western half of the disturbed area, the first part of the shock was the stronger and of greater duration; in the south-eastern half, the same features characterised the second part of the shock; but the boundary between these two portions of the disturbed area was not straight, but concave towards the south-east. In the Derby earthquake of 1903, the two parts of the shock were nearly equal in strength; all over the disturbed area, 61 per cent. of the observers stated that the first part was the stronger and 39 per cent. the second, and this proportion was nearly the same on both sides of the minor axis of the isoseismal lines. The first part of the Derby earthquake of 1904 was also generally regarded as the stronger, but the difference in intensity between the two parts was clearly greater in the south-western, than in the north-eastern, half of the disturbed area. In the Stafford earthquake of 1916, the two parts were also nearly equal in strength, the first part being slightly the stronger in the eastern half of the disturbed area and the second part in the western half.

In the remaining twin earthquakes, the distribution of intensity is somewhat different. In the Carmarthen earthquake of 1893, the second series was the stronger near the western end of the isoseismal 6 and farther west in Pembrokeshire, and the first in other parts of the disturbed area. In the Doncaster earthquake of 1905, the first part of the shock was the stronger over most of the disturbed area; while the second part was the stronger within a small and nearly circular area about 20 miles in diameter and including the centre of the N.E. portion of the isoseismal 7. In the Swansea earthquake of 1906, the first part was the stronger within a nearly circular area about 22 miles in diameter near the eastern end of the isoseismal 8; outside this area, the second part was much the stronger.

TWIN EARTHQUAKES CONNECTED WITH A TWIN-FOCUS

The two series of vibrations which form a twin earthquake may be due to (i) a single impulse duplicated by reflection or refraction at the bounding surfaces of different strata, (ii) a simple impulse, the condensational and distortional waves becoming separated with increasing distance from the origin, (iii) a repetition of the impulse within the same or an overlapping focus, and (iv) the occurrence of simultaneous or nearly simultaneous impulses in two detached foci.

I will now give the reasons for believing that the last-named theory offers the best interpretation of the evidence.

(i) In the first place, twin earthquakes cannot be generally due to the reflection or refraction of the earth-waves, although here and there the shock may be duplicated in this manner. The wide area over which the twin-shock is almost uniformly felt shows that it is not a local phenomenon. The existence of a synkinetic band in some twin earthquakes, the nearly constant duration of the interval between the two parts of the shock, both in different earthquakes and at different distances from the origin in any one earthquake, and the definite law of variation in their relative intensity, etc., are equally opposed to a haphazard origin. Moreover, on this theory, the first part of the shock should always be the stronger, for energy is lost by the reflection or refraction of a wave. And, again, of the earthquakes originating in a given district, a few are twins and the majority simple. Yet, according to this explanation, the earth-waves should undergo deflection at the same surfaces on every occasion.

(ii) Nor can the two parts of a twin earthquake be respectively waves of condensational and distortional vibrations; for, if they were, no synkinetic band would ever cross the central district, and the relative nature of the two parts would be uniform throughout the disturbed area. Also, the interval between the two series does not increase, as it should do, with the distance from the origin. The earthquake-sound accompanies both series in precisely the same manner; if the second series consisted of distortional vibrations only, it would be unattended by sound. In submarine twin earthquakes of other lands, both parts have been felt at sea, and distortional vibrations cannot be propagated in a liquid. Lastly, if the second series consisted of distortional vibrations, every earthquake would be a twin earthquake.

(iii) Again, twin earthquakes cannot be due to a repetition of the impulse within the same or an overlapping focus; for, if so, the order of relative intensity would not vary in a definite manner throughout the disturbed area, the two parts of the shock would never coalesce, and the mean duration of the interval would not generally be confined within the limits of 2 or 3 seconds.

(iv) Thus, as twin earthquakes are not due to the separation of the waves arising from a single impulse, nor to repeated impulses in the same focus or in overlapping foci, it follows that they must be caused by nearly simultaneous impulses in two detached, or practically detached, foci. On this view of their origin, the phenomena of twin earthquakes may be readily explained. The twin character of the shock will be perceptible as far as the weaker of the two parts can be felt; and the fact that both parts are often observed nearly all over the disturbed area shows that, in such cases, the impulses were of nearly equal strength. Again, in the neighbourhood of the epicentre corresponding to the weaker impulse, the vibrations from that focus may be of greater intensity than those from the more distant focus; the stronger impulse does not necessarily occur first, and thus the order of relative intensity may vary in different earthquakes and in different parts of the disturbed area of the same earthquake. Lastly, if the interval between the two impulses be less than the time required to traverse the distance between the two foci, the two series of vibrations must coalesce along a band passing between the two epicentres and crossing the line joining them approximately at right angles.

CLASSIFICATION OF TWIN EARTHQUAKES

The twin earthquakes described in Chapters VIII, X, and XIII belong to two distinct classes, one including the Hereford earthquake of 1896, the Derby earthquakes of 1903 and 1904, and the Stafford earthquake of 1916, the other including the Northampton earthquake of 1750, the Leicester earthquake of 1893, the Carlisle earthquake of 1901, the Doncaster earthquake of 1905 and the Swansea earthquake of 1906. The earthquakes of the two classes differ in the following respects:

(i) The chief point in which they differ is in the distance between the epicentres. On the one hand, we have such distances as 9 miles in the Hereford earthquake, and 8 or 9 miles in the Derby and Stafford earthquakes; on the other 17 miles in the Leicester and Doncaster earthquakes, $22\frac{1}{2}$ miles in the Swansea earthquake, and 23 miles in the Carlisle earthquake. In the twin earthquakes of the Midland Counties, the distance between the epicentres in one class ($8\frac{1}{2}$ miles) is just half that in those of the other class (17 miles).

(ii) Another important point of difference is the occurrence of a synkinetic band in earthquakes of the first class and its absence in those of the second. In other words, the two impulses in earthquakes of the first class occur either simultaneously or so closely together that the second focus is in action *before* the waves from the first focus can reach it, so that the second impulse is not a consequence of the first; in earthquakes of the second class, the impulses also occur closely together, but the second focus comes into action just *after* the instant when the waves from the first focus reach it, so that the second impulse may be, but is not necessarily, a consequence of the first.

(iii) In earthquakes of the first class, the interval between the two parts of the shock is one of absolute rest and quiet; it is only near the margins of the synkinetic band that two maxima are felt connected by weaker motion. In earthquakes of the second class, tremulous motion nearly always joins two maxima of intensity at places near the central region; though, at some distance from the epicentre, the tremulous motion ceases to be felt and the shock then consists of two distinct parts separated by an interval of rest and quiet. The small percentage of observers who notice the twin-character of the shock is also characteristic of earthquakes of the second class. In the Swansea earthquake of 1906, only 14 per cent. of the observers within the isoseismal 8, and only 16 per cent. of those within the isoseismal 5, detected the duplication of the shock.

(iv) In every pair of foci in earthquakes of the second class, an impulse occasionally occurs in one focus only. Compare, for example, the Leicester earthquakes of 1893 and 1904, the Doncaster earthquakes of 1902 and 1905, and the Swansea earthquakes of 1906 and 1907.

(v) Earthquakes of the first class are followed after a short interval by a slight shock originating in the interfocal region. This was also the case with the Carlisle earthquake of 1901, but not with the other earthquakes of the second class.

Earthquakes of the first class are twin earthquakes proper. They are due to distinct impulses in two detached foci initiated by a single generative effort. Those of the second class may be similar in origin, but they are due to successive efforts, though coming into being at a single birth.

ORIGIN OF TWIN EARTHQUAKES

In a simple earthquake, the immediate consequence of the parent fault-slip is a change of stress within and near the focus, especially an increase of stress along its margins. The after-slips in consequence take place either in the focal region or just beyond it. Thus, the foci of successive earthquakes are not as a rule detached, but are either coincident or overlapping. A twin earthquake is clearly of a different and more complex origin. Through a single generative effort, movements occur almost simultaneously in two distinct regions of the fault which are separated by a portion in which there is little or no displacement. A double movement such as this cannot always be due to an interrupted slip; for this would involve an interval of time between the component slips long enough for the increased stress resulting from the first slip to take effect in the second focus, and therefore longer than the time of transit between the two foci—a supposition which is negatived by the existence in some earthquakes of a synkinetic band. At the same time, the movements cannot be independent, for the chances against two movements occurring within 2 seconds or less, and that in a district like Staffordshire where none has been known for a thousand years, are almost infinite. Even higher must they be in a district such as Derbyshire, in which practically simultaneous movements occurred in the same two detached foci in 1903, 1904 and 1906, though previously and since unknown there.

The only movement that can produce such nearly simultaneous displacements is one of rotation, and it is difficult to conceive of

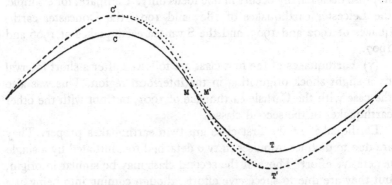

Fig. 99. Origin of Twin Earthquakes.

rotation taking place within the crust except in the formation of a fold. In Fig. 99, the curve CMT is supposed to represent a section of a crust-fold along a transverse fault, the crest C and the trough T of the fold being separated by a distance of 8 or 9 miles. If a small step were to take place in the growth of the fold, that is, if the crest and trough were to become more pronounced, simultaneous movements of the fold—the crest C to C' and the trough T to T'—would occur, and these movements would be accompanied by a rotation of the median limb, the central portion M of which would undergo no displacement. Thus, we should have two foci, CC' and TT', entirely separated by the interfocal region about M. Moreover, the forces which give rise to an earthquake of this kind must act at right angles to the direction of the fold, that is, parallel to the direction of the fault.

A step in the growth of a crust-fold such as that here considered would leave the median limb subjected at both ends to increased stresses, which should afterwards be relieved by a slip in the interfocal region. As these interfocal slips are simple in character—the shock showing no sign of duplication—it follows that the movement of the median limb is then one of translation; as if the growth of the arches were followed by a smaller bodily advance of the fold. If the initial impulses in the two foci be nearly equal in strength, the interval that elapses before the occurrence of the after-slip may be considerable —40 days in the case of the Derby earthquake of 1903. If they differ slightly in strength, the interval may be brief—10 minutes in the case of the Hereford earthquake of 1896, 8 hours in that of the Derby earthquake of 1904 and 15 hours in that of the Stafford earthquake of 1916.

In twin earthquakes of the second class, the distance between the two epicentres is 17 miles in the Midland Counties or just double the distance between the epicentres in twin earthquakes of the first class. This relation suggests that the origin of such earthquakes may be similar to that of twin earthquakes of the first class, but that both foci coincide either with the crests or with the troughs or with corresponding median limbs of successive folds. As, however, there is in these earthquakes no synkinetic band, it is evident that the second impulse, though no doubt nearly on the point of occurrence, is precipitated by the waves from the focus first in action. The existence near the epicentre of a continuous tremor between the two maxima of intensity shows that the two foci are not completely detached and that some slight movement takes place throughout the whole interfocal region. With such an origin, we should expect that the two parts of the shock would differ in strength and that movements might sometimes occur in one focus without any disturbance in the other. The effect of the first slip must be an immediate increase of stress in the interfocal region, relieved, however, soon afterwards by the slip in the second focus. If the two impulses be of nearly equal strength, as in the Doncaster earthquake of 1905, the increased stress in the interfocal region has little, if any, immediate result, though an after-slip might be expected after the lapse of many years. But if, as in the Carlisle earthquake of 1901, one slip be much stronger than the other, the increase of stress in the interfocal region may be enough to precipitate an after-slip of strength inferior to that of the twin earthquake. The slips of the original earthquake—as in the Leicester earthquake of 1893—would probably occur near the crests of the successive folds, those being the regions in which there is the least resistance to motion.

Effects of the Growth of Deep-Seated Folds on that of the Surface-Crust. Deep-seated movements, such as those described above, could hardly occur without in some way affecting the structure of the surface-crust. The effect may not be immediately evident in every case. A few twin earthquakes have, however, been followed by a movement —apparently a superficial movement—along a fault in a nearly perpendicular direction. The interval varies considerably in length. For instance, the Pembroke earthquake of 1892, which originated along a deep north-and-south fault, was succeeded after less than four hours by an earthquake originating in an east-and-west fault, and this again by another four days later. The Carlisle earthquake of 1901, due to movements along a deep north-and-south fault, was followed after ten years by a movement in a nearly east-and-west fault, giving rise to the Grasmere earthquake of 1911. The Hereford earthquake of 1853 Mar. 27, which preceded the strong twin earthquake of 1863 by 10½ years, was no doubt the consequence of some long-past and forgotten movement in the Hereford twin-foci[1]. But, while the axes of the Pembroke and Grasmere earthquakes were crossed centrally by the longer axes of the preceding twin earthquakes—earthquakes of the second class—those of the Hereford earthquakes of 1853 and 1924 lay respectively on the south-west and north-east sides of the longer axis of the related twin earthquakes of the first class. The superficial character of these transverse movements is shown by the association of the Pembroke earthquake with a known fault and by the elongated form and closeness of the isoseismal lines of all of them.

Stability of Twin Earthquake Foci. Little, if any, perceptible migration takes place in the foci of twin earthquakes, at any rate within the limited period of our records. In each of two districts three twin earthquakes have occurred during the last century. The Hereford earthquakes of 1863 and 1896 evidently, and the earthquake of 1868 probably, originated in the same foci, one near Hereford and the other near Ross. In 1863 and 1896, the Hereford focus was the first in action; in 1863 and 1868, the impulse in the Ross focus was the stronger, and

---

[1] The Hereford earthquake of 1871 Mar. 20 may have originated in the same fault.

in 1896 that in the Hereford focus. The Derby earthquakes of 1903, 1904 and 1906 originated in the same foci, one near Ashbourne and the other near Wirksworth. In 1903, the impulses were simultaneous, in 1904 the Wirksworth focus, and in 1906 the Ashbourne focus, was the first in action. In all three earthquakes, the impulse in the Ashbourne focus was the stronger.

Twin Earthquakes and the Structure of the Earth's Crust. How uniform the structure of the crust is at the depth of a few miles is shown by the twin earthquakes of the Midlands, namely, the Northampton earthquake of 1750, the Leicester earthquake of 1893, the Derby earthquakes of 1903, 1904 and 1906, and the Stafford earthquake of 1916.

Beginning at the western end of the area, we have first the Stafford earthquake with two foci 8 or 9, say $8\frac{1}{2}$, miles apart, the Derby earthquakes with their foci separated by the same distance, the Leicester earthquake with its foci distant 17 miles, and lastly the Northampton earthquake with its foci badly defined but about the same distance apart. The lines joining the epicentres are represented in Fig. 100 by the thick lines *SS*, *DD*, *LL* and *NN*.

Now, if the axes of the Stafford and Derby earthquakes be produced, they meet in a point *A*, which is 9 miles from the eastern epicentre of the Stafford earthquake and 18 miles from the southern epicentre of the Derby earthquakes. In both districts, the distance between the epicentres is that between the crest and trough of a crust-fold, or half a wave-length. Thus, the point of intersection *A* is half a wave-length from the nearer epicentre of the Stafford earthquake and a whole wave-length from the nearer epicentre of the Derby earthquakes. Again, if the axes of the Derby earthquakes and the Leicester earthquake be produced, they meet in a point *B* which is about 10 miles from the point of intersection *A* and about 8 miles from the southern epicentre of the Derby earthquakes, or about half a wave-length from either. The same point *B* is about 26 miles from the north-western epicentre of the Leicester earthquake and 43 miles from the south-eastern epicentre, that is, about three and five half-wave-lengths from the two epicentres. Lastly, the south-western epicentre of the Northampton earthquake lies on the axis of the Leicester earthquake and at about 8 miles, or half a wave-length from the south-eastern epicentre of that earthquake.

If we may assume that the earthquake-faults are approximately at right angles to the folds to the growth of which the earthquakes are due, it would seem that the crust at a depth of a few miles below the counties of Stafford, Derby and Leicester is corrugated in two systems of perpendicular folds. In Fig. 100, the broken-lines and dotted lines (drawn parallel in each series for simplicity) represent either the anticlines or synclines. As the north-western focus of the Leicester earthquake coincides with an anticline, it is probable that the broken-lines represent anticlines and the dotted lines synclines. The arrow-heads denote the directions of the forces that produced the twin

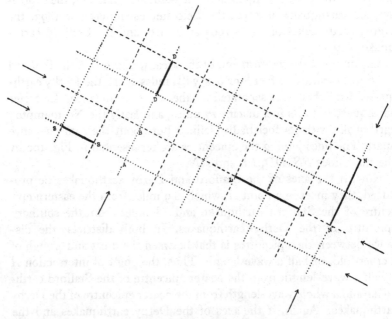

Fig. 100. Connexion between Twin Earthquakes and the
Structure of the Earth's Crust.

earthquakes. Thus, the Stafford and Leicester earthquakes were due to movements along transverse faults intersecting Caledonian folds; the Derby and Northampton earthquakes to movements along transverse faults intersecting Charnian folds. It will be seen that twin earthquakes of both classes originate in the same system of folds, but that, if we may judge from such a limited number of examples, the two classes occupy different ends of the area in question.

One other point with regard to the directions of these crust-folds is worthy of notice. The directions of the folds are not quite parallel, but exhibit a fan-shaped arrangement. The direction of the folds is about N. 25° E. through the Stafford foci, N. 28° E. through the Derby foci, N. 30° E. through the north-western Leicester focus and N. 42° through the other, and N.E. through the Northampton foci. In other words, successive anticlines are inclined at an average angle of 5°.

Confining our attention to the twin earthquakes, it is interesting to notice the westerly progression of the series—the Northampton earthquake in 1750, the Leicester earthquake in 1893, the Derby earthquakes in 1903–6, and the Stafford earthquake in 1916.

The existence of this deep-seated double system of corrugations may perhaps throw light on another peculiarity of British earthquakes. Dividing them into three classes of strong, moderate and slight earthquakes according as the areas embraced by the isoseismal 4 are greater than 5000, between 5000 and 1000, or less than 1000, square miles, it appears that the average length of focus for strong earthquakes is 12·8 miles and for moderate earthquakes 12·3 miles. Slight earthquakes are divisible into two groups, in one of which the focus is 9 miles or more in length, in the other 7 miles or less. The average length of focus for the former is 12·4 miles, and for the latter (omitting a very large number of very slight shocks) 3·9 miles. If we leave out of account the second division of slight shocks, which are of the nature of local creeps, it follows that the average length of focus in all three classes is very nearly the same, lying between 12·3 and 12·8 miles.

INDEX

ABER-DINLLE fault and Carnarvon earthquakes, 179
Aberfeldy earthquakes, 159
After-shocks of British earthquakes, 37, 40, 46, 74, 81, 128, 135, 145, 176, 186, 194, 201, 208, 213, 215, 218, 256, 262, 269
Airy, G. B., 244, 246
Altarnon earthquakes, 308
Amulree earthquake, 159
Anglo-Saxon Chronicle, 229, 321, 345–6
Animals, effects of earthquakes on, 70, 255, 335
Annales Cambriae, 203, 345
Annales Monastici, 13, 229, 291, 293, 319, 321–2, 330, 333, 345–8
Annals of Loch Cé, 367
Ardvoirlich earthquakes, 159

BAKER, D. E., 226
Baker, R., 224, 231, 235, 323, 332
Bala earthquake, 181
Ballachulish earthquake, 58
Banbury earthquakes, 241
Barlow, W., 317
Barmouth earthquakes, 180
Barnsley earth-shake, 361
Barnstaple earthquakes, 314; origin of, 316
Barograph, Earthquake registered by, 342
Barrel, E., 332
Becket, Materials for the History of Thomas, 293
Beddgelert earthquake, 180
Beeston, Earthquakes registered at, 235, 246
Belgium, British earthquakes felt in, 333, 341, 349
Bevis, J., 371
Bidston, Earthquakes registered at, 174, 199, 267
Birds, Effects of earthquakes on, 67, 327, 336, 352; killed by earthquakes, 256, 352
Birmingham earthquakes, 238
Birmingham, Earthquakes registered at, 174, 199, 267, 288
Blisland earth-shake, 354
Blomefield, F., 293–4
Bolton earthquakes, 219; earthquake of 1889 Feb. 10, 219; origin of, 220
Borlase, W., 3, 296, 376
Boswell, J., 217n.
Boyle, R., 242
Bray, Mrs, 312
Breccia-gashes of Durham coast, possible relations with earth-shakes, 366
British Association Committees on British earthquakes, 3, 5, 64–5
Brut y Tywysogion, 203, 333, 345
Bryce, J., 5, 64–5
Brydone, P., 204

Buckland, W., 64
Bungay earthquake, 294
Burrow, J., 331

CALDERWOOD, D., 168–9
Camborne earth-shakes, 355
Cambridgeshire earthquakes, 293
Camden, W., 231, 333
Camelford earthquakes, 307
Campbell, R., 118
Canterbury earthquakes, 330; earthquake of 1382, 384n.
Carlisle earthquakes, 204; earthquakes of 1901 July 9, 206; origin of, 209
Carmarthen earthquakes, 191; earthquake of 1893 Nov. 2, 192
Carnarvon earthquakes, 172; earthquake of 1903 June 19, 173; origin of, 179
Catalogues of British earthquakes: Fuchs, C. W. C., 6; *Gentleman's Magazine*, 2; Grey, Z., 2; Lauder, T. D., 3; *London Magazine*, 2; Lowe, E. J., 4; Meldola, R., and W. White, 6; Milne, D., 4; O'Reilly, J. P., 5; Perrey, A., 4; Roper, W., 6; Walford, C., 5
Chambers, R., 160, 169
Channel Islands, Earthquakes in, 368; felt in England, 368–9
Charnwood fault and Leicester earthquakes, 281
Cheshire earthquakes, 227
Chichester earthquakes, 325; earthquake of 1750 Mar. 29, 1; earthquakes of 1833, 3, 327
Chronica Monasterii de Melsa, 226
Chronicles of Melrose, 229, 345
Chronicon Angliae, 330, 348
Chronicon Scotorum, 367
Chronology of older annalists, Inaccurate, 12
Classification of British earthquakes, 377
Clitheroe earthquakes, 222
Clustering of British earthquakes, 389; of Comrie earthquakes, 115
Colchester earthquakes, 337; earthquake of 1884 Apr. 22, 6, 338, 384n.
Cole, W., 247, 344
Comrie earthquakes, 3, 5, 62; earthquake of 1789 Nov. 5, 66, 68; of 1801 Sept. 7, 73; of 1839 Oct. 23, 3, 65, 79; of 1841 July 30, 94; characteristics of, 113; clustering of, 115; connexion with Highland border fault, 118; distribution in space, 117; intensity, 114; nature of shock and sound, 116; number, 113; origin of, 118; periodicity of, 116; recorded by seismometers, 65, 90, 92, 94–5, 99, 101–3; registers of, 62
Comrie foci, Stability of, 397; its cause, 398

Comrie, Seismometers at and near, 64–5
Coningsby earthquakes, 293
Cornwall, Earthquakes of unknown epicentres in, 311
Coseismal lines of Hereford earthquake of 1916 Dec. 17, 254
Cotton, Bartholomew de, 13, 293, 324, 333, 345
Cox, J. C., 317
Creech, W., 68, 161, 169, 350
Crieff, Doubtful earthquakes at, 111
Cunningham, 9, 333

DALTON-IN-FURNESS earthquakes, 221
Dartmouth earthquake, 313
Depth of seismic foci: earthquakes with a shallow focus, 150, 190, 212, 231, 298, 307, 316; earthquake with a focus of moderate depth, 221; earthquakes with a deep-seated focus, 61, 190, 195, 209; earthquakes with successive foci approaching the surface, 54, 56, 304
Derby earthquakes, 263; earthquake of 1903 Mar. 24, 263; isacoustic lines, 268; observations in mines, 268; seismographic records, 266; velocity of earth-wave, 266; earthquake of 1904 July 3, 271; isacoustic lines, 273; earthquake of 1906 Aug. 27, 275; origin of, 276
Derbyshire earthquakes, 233; earthquake of 1795 Nov. 18, 3, 234
Destructive earthquakes in England, 234–5, 244, 249, 291, 293, 322, 330, 333, 338, 349; Scotland, 36, 44, 79, 94, 123; Wales, 196, 203; earthquakes that caused slight damage in England, 204, 216, 231, 236, 263, 283, 333, 335, 346, 349, 350; Scotland, 39, 60, 73, 77, 123, 169; Wales, 173
Devon, Earthquakes of unknown epicentres in, 317
Diceto, Ralph de, 291
Dinas fault and Rhondda Valleys earth-shakes, 357
Direction of shock, 252, 273
Distribution of British earthquakes in space, 283, 385; aseismic areas in Great Britain, 385; construction of map, 385; direction of isoseismal axes, 387; prevalence of Caledonian movements, 388; distribution of Comrie earthquakes, 117
Distribution of British earthquakes in time, 389; fluctuations in frequency, 389; intervals between successive earthquakes in particular districts, 389
Doncaster earthquakes, 284; earthquake of 1905 Apr. 23, 285; seismographic record, 288; origin of, 289
Dorsetshire earthquakes, 323
Drummond, J., 63–4, 76–8, 81–98, 114

Dugdale, W., 349
Dumfriesshire earthquakes, 165
Dunkeld earthquakes, 62
Dunning earthquakes, 160
Dunoon earthquakes, 118–9; earthquake of 1904 Sept. 18, 119

EARTH-SHAKES in limestone districts, 365; origin of, 366
Earth-shakes in mining districts, 353; Barnsley earth-shake, 361; Blisland earth-shake, 354; Camborne earth-shakes, 355; Kilsyth earth-shake, 362; Pendleton earth-shakes, 359; Rhondda Valleys earth-shakes, 356; characteristics of, 362; disturbed area, its magnitude and form, 362; earth-shakes and faults, relations between, 364; origin of, 364; nature of shock and sound, 362; sound-area and disturbed area, relations between, 363; relative effects on the surface and in mines, 363
Earth-sounds after Colchester earthquake of 1884 Apr. 22, 344
Ecclefechan earthquakes, 166
Edinburgh earthquakes, 163; origin of, 164
Edmonds, R., 221, 297, 305–6, 311, 331–2, 372
Ellman, E. B., 329
England, Earthquakes of the east of, 290; midland counties, 233; north-east, 224; north-west, 204; south, 323; south-east, 325; south-west, 296; west, 227; earthquakes of unknown epicentres in, 223, 345
England, Scotland and Wales, Relations between earthquakes of, 383; relations between average disturbed areas and intensities, 384; relative frequency, 383; of destructive earthquakes, 383; relative numbers of earthquake-centres and of earthquakes per centre, 383; variations in nature of earthquakes, 384
England, Welsh earthquakes felt in, 173, 183, 187, 192, 196
English earthquakes felt in Belgium, 333, 341, 349; France, 331, 333, 341; Holland, 349; Ireland, 212, 244, 249, 350; Scotland, 205–6, 212, 214, 348, 350; Wales, 223, 231–2, 244, 247–9, 260, 263, 271, 350
Eulogium (Historiarum), 322, 333, 346–8
Evelyn, J., 349
Everton (Liverpool) earthquake, 222
Ewing seismographs, Earthquakes registered by, 57–8
Excentricity of sound-area with reference to isoseismal lines, 48, 153, 163, 193, 199, 209, 219, 230, 238, 253, 280
Exmouth earthquakes, 314

Fᴀʙʏᴀɴ, R., 228, 241, 325, 330
Falmouth earthquake, 305
Fasciculi Zizaniorum, etc., 330
Faults, Determination of fault from
seismic evidence, 8; relations between
British earthquakes and faults, 7;
Aber-Dinlle fault and Carnarvon
earthquakes, 179; Bala fault and Bala
earthquake, 181; Charnwood fault and
Leicester earthquakes, 281; Dinas fault
and Rhondda Valleys earth-shakes,
357; Glen App fault and Leadhills
earthquakes, 165; Great Glen fault and
earthquakes of Ballachulish, 58; Dhu
Heartach, 61; Fort William, 56; Inverness, 36; Oban, 58; Phladda, 61;
Torosay, 61; Highland border fault
and earthquakes of Comrie, 62, 118;
Dunkeld, 62; Dunoon, 119; Kintyre,
121; Rothesay, 121; Irwell Valley
fault and Bolton earthquake, 219; and
Pendleton earth-shakes, 359; Malvern
fault and Malvern earthquake, 229;
Morte Slates faults and Barnstaple
earthquakes, 315; and Taunton earthquakes, 319; Ochil fault and Menstrie
earthquakes, 123, 150; Pentland fault
and Edinburgh earthquake, 164;
Queenzieburn fault and Glasgow
earthquake, 163
First committee on British earthquakes,
3; map of distribution of British
earthquakes, 5; scientific study of a
British earthquake, 3; trustworthy
record of a British earthquake, 12
Fissures in ground formed by British
earthquakes, 46, 342
Fletcher, J., 228
Flores Historiarum, 13, 291, 293, 322,
324, 333, 336, 345-8
Focus, Length of, 8, 55-6, 149, 179,
202, 221, 252, 304, 307, 317, 377*n*.,
395, 397, 409
Forbes, J. D., 65; earthquakes registered
by his seismometer, 90, 92, 94-5, 99,
101-3; seismometer, 64
Fordham, H. G., 247
Fore-shocks of British earthquakes, 43,
76, 124, 173, 183, 244, 248, 270, 274,
279, 285
Forster, J., 317
Fort William earthquakes, 56; origin of,
36; registration of earthquakes, 57-8
Framlingham earthquakes, 295
France, British earthquakes felt in, 331,
333, 341
Fuchs, C. W. C., 6, 227, 235, 247, 294,
369
Furley, R., 330, 333

Gᴀɪɴsʙᴏʀᴏᴜɢʜ earthquake, 290
Geikie, A., 164
Germany, British earthquake registered
in, 267

Gervase of Canterbury, 293, 347
Giles, J. A., 337
Gilfillan, S., 3, 4, 65
Glasgow earthquakes, 160; origin of, 163
Glastonbury, Church of St Michael
damaged by earthquake of 1275, 12,
296, 322
Glen App fault and Leadhills earth-
quakes, 165
Gloucestershire earthquakes, 231
Glover, S., 233
Godwin-Austin, T., 244, 322
Göttingen, Earthquake registered at, 267
Graham, D., 81
Grantham earthquakes, 292
Grasmere earthquakes, 210
Gray, E. W., 3, 234
Great Glen fault and earthquakes of In-
verness, etc., 35-6, 53, 57
Greenly, E., 247
Greenwich, Earthquake registered at,
342
Grey, Z., 2
Gruggen, J. P., 3, 327
Gun-firing, Spurious earthquakes due
to, 10

Hᴀʟᴇs, S., 1, 332
Halifax earthquakes, 222
Hampshire earthquakes, 324
Hardyng, John, 330, 348
Harrison, J., 375
Helston earthquakes, 299; earthquake
of 1898 Apr. 1, 302; origin of, 304
Henry of Huntingdon, 345
Henry of Knyghton, 322
Henry, T., 216
Henwood, W. J., 299, 309, 311
Hereford earthquakes, 244; earthquake
of 1863 Oct. 6, 4, 244; of 1868 Oct.
30, 247; of 1896 Dec. 17, 249; isa-
coustic lines, 253; observations in
mines, 254; origin of, 257
Herefordshire earthquakes, 231
Hicks, H., 316
Highland border fault and earthquakes
of Comrie, etc., 62, 118
Holinshed, R., 239, 241, 291-2, 324,
330, 346, 348-9
Holland, British earthquake felt in, 349
Hopkinson, J., 337
Horne, J., 151
Hoveden, Roger de, 229, 291, 345-6
Howard, L., 157, 236, 314
Huntingdonshire earthquakes, 241
Hutton, W., 238

Iɴᴛᴇɴsɪᴛʏ, Scales of, 9, 64
Invergarry earthquakes, 153
Inverness earthquakes, 36; earthquake
of 1816 Aug. 13, 3, 36, 54; of 1888
Feb. 2, 38, 54; of 1890 Nov. 15, 39,
54; of 1901 Sept. 18, 44, 55; origin of,
53

Investigation, Methods of, 7
Ireland, Earthquakes in, 367
Ireland, English earthquakes felt in, 212, 244, 249, 350; Scottish, 59; Welsh, 173, 183, 192, 196
Irwell Valley fault and Bolton earthquake, 219; and Pendleton earthshakes, 359
Isacoustic lines, 253, 268, 273
Isbell, E. J., 245
Ixworth earthquake, 295

Jehu, T. J., 118
Jewitt, Ll., 317
John of Brompton, 229
John of Hexham, 291, 347
Johnson, S., 217 n.
Judd, J. W., 238

Kendal earthquakes, 212; earthquake of 1871 Mar. 17, 213
Kew, Earthquake registered at, 342
Kilsyth earth-shake, 362
Kinahan, G. H., 337
Kinloch Rannoch earthquake, 159
Kintyre earthquakes, 118, 121

Lancashire earthquake of 1750 Apr. 13, 1, 223
Langford, J. A., 238
Lauder, T. D., 3, 36–7, 66, 68, 72–3, 159, 161, 168–9, 204, 350
Launceston earthquakes, 309
Leadhills earthquakes, 165
Lebour, G. A., 366
Leeds, Earthquake registered at, 342
Leicester earthquakes, 278; earthquake of 1893 Aug. 4, 278; of 1904 June 21, 280; origin of, 281
Leicestershire earthquakes, 237
Le Livere de Reis de Engleterre, 346, 348
Lewes earthquakes, 329
Life, Loss of, by British earthquake, 333
Lincoln cathedral, Damage to, 292
Lincoln earthquakes, 291
Lisbon earthquake of 1755 Nov. 1, 371; epicentre and time at epicentre, 371; reported observations of shock and sound in England, 372; seiches in rivers, lakes and pools, 373; seismic sea-waves, 375
Liskeard earthquakes, 309
Lloyd, J., 172
Llwynypia earth-shakes, 356, 358
Lomas, J., 247
London earthquakes, 332; earthquake of 1750 Feb. 19, 1, 333; of 1750 Mar. 19, 1, 335
Lowe, E. J., 4, 235, 245–6, 290, 323, 333, 350
Lowe pendulum, Earthquakes registered by, 235, 246

Macfarlane, P., 5, 63–5, 77–8, 81–3, 94, 99, 100, 108, 114
Magnetographs, Earthquake registered by, 342
Maidstone earthquake, 332
Mallet columns at Dunearn (Comrie), 66
Mallet, R., 4, 196, 216, 223–4, 227–8, 239, 241, 243, 245, 291, 325, 330–1, 336–7, 347–8, 350, 368, 371
Malvern earthquake, 229
Malvern fault and Malvern earthquake, 229
Mansfield earthquakes, 236
Mayhall, J., 212
Meldola, R., 1, 6, 337
Melvill, J., 168
Menstrie earthquakes, 123; earthquake of 1905 Sept. 21, 127; of 1908 Oct. 20, 134; of 1912 May 3, 143; characteristics of, 147; intensity, 147; length of focus, 149; nature of shock and sound, 148; origin of, 149; periodicity of, 148
Meteorites, Spurious earthquakes due to bursting of, 10, 11
Midland counties of England, Twin earthquakes of, 244
Milne, D., 3, 5, 36–8, 58–9, 61–79, 81–103, 117, 123, 157, 159, 161, 165–6, 169, 170–2, 180–1, 191, 204, 206, 212, 216–7, 219, 222–3, 228, 231, 238, 241, 284, 290, 292–3, 307, 323, 325–6, 329, 331, 333, 353, 368–71
Milne, J., 1
Milne seismograph, Earthquakes registered by, 174, 199, 267
Mines, Observations of British earthquakes in: Bolton earthquake of 1889 Feb. 10, 219; Derby earthquake of 1903 Mar. 24, 268; Derbyshire earthquake of 1795 Nov. 18, 234; Helston earthquake of 1842 Feb. 17, 300; Hereford earthquake of 1896 Dec. 17, 254; Penzance earthquake of 1757 July 15, 297; Swansea earthquake of 1906 June 27, 200; Truro earthquakes of 1859 Oct. 21, 305; of 1860 Jan. 13, 306
Montessus de Ballore, F. de, 6
Monumenta Franciscana, 348
Moore, H. C., 247
Morse, S., 204
Morte Slates faults and Barnstaple earthquakes, 315; and Taunton earthquakes, 319

Names of British earthquakes, 35
Nausea, Feeling of, during earthquake, 255
New style of chronology, Introduction of, 13
Nicholson, C., 212
Nimmo, W., 372
Norfolk earthquakes, 293

Northampton earthquakes, 283; earthquake of 1750 Oct. 11, 1, 283
Northamptonshire earthquakes, 239
Norwegian earthquakes felt in Shetland Islands, 370
Nottinghamshire earthquakes, 235

OAKHAM earthquake, 237
Oban earthquakes, 58; origin of, 61
Ochil fault and Menstrie earthquakes, 123, 150
Okehampton earthquakes, 312
Oldham, R. D., 373
Omori horizontal pendulum, Earthquakes registered by, 174, 199, 267, 288
O'Reilly, J. P., 5, 233, 240
Origin of British earthquakes: Barnstaple earthquakes, 316; Bolton earthquakes, 220; Carlisle earthquakes, 209; Carmarthen earthquakes, 195; Carnarvon earthquakes, 179; Comrie earthquakes, 118; Derby earthquakes, 276; Doncaster earthquakes, 289; Edinburgh earthquakes, 164; Glasgow earthquakes, 163; Helston earthquakes, 304; Hereford earthquakes, 257; Inverness earthquakes, 53; Leicester earthquakes, 281; Malvern earthquake, 230; Menstrie earthquakes, 149; Nottinghamshire earthquakes, 235; Pembroke earthquakes, 189; Stafford earthquakes, 263; Swansea earthquakes, 202; Taunton earthquakes, 319; Wales, South, 203; origin of simple earthquakes, 395; of twin earthquakes, 403
Ormerod, G. W., 312
Oxenedes, John de, 322, 345, 347
Oxford earthquakes, 242
Oxfordshire earthquakes, 241

PARFITT, E., 247, 307, 309, 311, 313-14, 317, 350
Paris, Matthew, 13, 291, 293, 319-20, 324, 333, 336-7, 345-7
Parsons, J., 335
Pembroke earthquakes, 183; earthquake of 1892 Aug. 18, 183; origin of, 189
Pendleton earth-shakes, 359
Penpont earthquakes, 166
Pentland fault and Edinburgh earthquakes, 164
Pentre earth-shake, 357
Penzance earthquakes, 296; earthquake of 1757 July 15, 2
Periodicity of British earthquakes, 390; annual periodicity, 390; diurnal periodicity, 391; of Comrie earthquakes, 116
Perrey, A., 4-5, 56-7, 65, 99, 106-11, 161, 165, 170-1, 206, 215, 217, 221-3, 231, 233, 235, 244-5, 247, 294, 297, 305-6, 308, 311, 314, 322-4, 327, 329, 331, 333, 337, 346-7, 350, 367-8

Perth earthquakes, 160
Perthshire earthquakes, Miscellaneous, 159
Pheasants, Effects of earthquakes on, 256, 295, 327
Pickering, R., 290, 349
Pigot, T., 349
Pitlochry earthquakes, 159
Plot, R., 233, 349
Porth earth-shakes, 356
Prestwich, J., 243

QUEENZIEBURN fault and Glasgow earthquake, 163

RAMSAY, A. C., 179
Reid, C., 298
Reigate earthquakes, 331
Rhondda Valleys earth-shakes, 356
Rhyl earthquake, 181
Richardson, R., 163
Rishanger, W., 322, 347
Rivers and lakes, Effects of British earthquakes on, 77, 186-7, 327, 335
Rochdale earthquakes, 216
Roger of Wendover, 13, 324
Roper, W., 1, 6, 171, 215, 224, 226, 294, 311, 323, 336
Rossi-Forel scale of intensity, 9
Rotation of bodies by London earthquake of 1750 Mar. 19, 335
Rothesay earthquakes, 118, 121
Russell, M., 335

SCOTLAND, Earthquakes of central, 62, 123, 157, 159; of the north of, 35, 152-3; of the south of, 161, 163, 165; earthquakes of unknown epicentres in, 168
Scotland, English earthquakes felt in, 205-6, 212, 214, 348, 350
Scottish earthquake felt in Ireland, 59
Sea, Effects of British earthquakes at, 186, 348
Seismographs, etc., British earthquakes registered by: Carnarvon earthquake of 1903 June 19, 174; Colchester earthquake of 1884 Apr. 22, 342; Comrie earthquakes, 65, 90, 92, 94-5, 99, 101-3; Derby earthquake of 1903 Mar. 24, 267; Doncaster earthquake of 1905 Apr. 23, 288; Fort William earthquakes, 57-8; Hereford earthquake of 1863 Oct. 6, 246; Nottinghamshire earthquakes, 235; Swansea earthquake of 1906 June 27, 199
Seismometer, Forbes', 65; origin of term, 64
Settle earthquake, 221
Shap earthquake, 221
Sheahan, J. J., 226
Shetland Islands, Earthquakes in, 370; connexion with Norwegian earthquakes, 370

Shide, Earthquake registered at, 199
Short, T., 1, 6, 204, 233, 235, 237, 241, 284, 290–1, 323–4, 346–7, 349, 367
Shropshire earthquakes, 228
Simple earthquakes, Origin of, 395; extension of foci, 395; westerly migration of foci, along Great Glen fault, 396; in Menstrie earthquakes, 396; in Taunton and Barnstaple earthquakes, 397; differences between simple and twin earthquakes, 394
Somerset earthquake of 1275 Sept. 11, 12, 322
Somerset, Earthquakes of unknown epicentres, in, 321
Sound-phenomena of British earthquakes, 377; Inaudibility of the sound to some observers, 378; nature of the sound, 377; relations between sound-area and disturbed area, 380; time-relations of sound and shock, 381; variation in audibility throughout sound-area, 379; variation in nature of sound throughout sound-area, 380; excentricity of sound-area with reference to isoseismal lines, 300, 303, 307, 311; sound-phenomena of Inverness earthquakes, 40, 46, 48, 51–2
Sound, Types of earthquake-, 9
Spurious earthquakes, Discrimination of, 10
Spurrell, W., 191, 196, 247
St Agnes earthquake, 306
St Albans earthquake, 336
S. Augustini Cantuariensis, Historia Monasterii, 330, 348
St Austell earthquake, 307
St Mary's Abbey, Dublin, Chartularies, etc., 348
Stability of Comrie foci, 397; its cause, 398
Stafford earthquakes, 259; earthquake of 1916 Jan. 14, 260; origin of, 263
Staffordshire earthquakes, 223
Stamford earthquakes, 239
Stevenson, C. A., 38, 59, 61, 152–3
Stevenson, D., 59
Stow, J., 2, 228, 290, 293, 319, 323, 330–1, 333, 346, 348–9
Strontian earthquakes, 157
Strype, J., 331
Stukeley, W., 1, 333
Suffolk earthquakes, 294
Sunderland earth-shakes, 365
Swansea earthquakes, 196; earthquake of 1906 June 27, 196; origin of, 202
Symeon of Durham, 2, 228–9, 291, 345
Symons, G. J., 247, 337

Taunton earthquakes, 318; origin of, 319
Taylor, R., 3, 66–7, 113–4
Thomassen, T. Ch., 370
Thoresby, W., 290

Thorn, W., 330
Tonypandy earth-shakes, 358
Topley, W., 337
Triveti, N., 322
Truro earthquakes, 305
Turnor, E., 293
Twin earthquakes: Carlisle earthquake of 1901 July 9, 206; Carmarthen earthquake of 1893 Nov. 2, 192; Colchester earthquake of 1884 Apr. 22, 338; Derby earthquake of 1903 Mar. 24, 263; of 1904 July 3, 271; of 1906 Aug. 27, 275; Doncaster earthquake of 1905 Apr. 23, 285; Hereford earthquake of 1863 Oct. 6, 244; of 1868 Oct. 30, 247; of 1916 Dec. 17, 249; Kendal earthquake of 1841 Mar. 17, 212; of 1871 Mar. 17, 214; Leicester earthquake of 1893 Aug. 4, 278; of 1904 June 21, 279; Northampton earthquake of 1750 Oct. 11, 283; Pembroke earthquake of 1892 Aug. 18, 183; Stafford earthquake of 1916 Jan. 14, 260; Swansea earthquake of 1906 June 27, 196; classification of, 402; connexion with a twin-focus, 400; investigation of, 8; nature of, 398; coalescence of two parts of shock along the synkinetic band, 398; mean duration of interval between the two parts of the shock, 399; relative nature of the two parts, 399; wide area of observation, 398; origin of twin earthquakes, 257, 263, 276, 281, 403; effects of the growth of deep-seated folds on that of the surface-crust, 406; stability of twin-foci, 406; twin earthquakes and the structure of the earth's crust, 407; synkinetic band, 8, 251, 262, 265, 273; differences between simple and twin earthquakes, 394

Ullapool earthquakes, 152
Underground water, Effects of British earthquakes on, 186, 342
Unknown epicentres, Earthquakes of, in Cornwall, 311; Devon, 317; England, 345; England, north-west, 223; Scotland, 168; Somerset, 321; Wales, north, 181; Wales, south, 203; Yorkshire, 226

Velocity of earth-wave of Derby earthquake of 1903 Mar. 24, 266; of Hereford earthquake of 1896 Dec. 17, 254

Wales, North, Earthquakes of, 172; earthquakes of unknown epicentres in, 181
Wales, South, Earthquakes of, 183; origin of earthquakes in, 203; earthquakes of unknown epicentres in, 203

Walford, C., 5
Wallis, J., 242
Walpole, H., 333, 335
Walsingham, T., 330, 348
Watts, W. W., 282
Wells cathedral, Damage to, 296, 319
Wells earthquakes, 319
Welsh earthquakes felt in England, 173, 183, 187, 192, 196; in Ireland, 173, 183, 192, 196
Wesley, C., 335
Wesley, J., 222, 226, 296, 333, 335
Wetherby earthquakes, 226

White, W., 6, 337
Whyte, R., 161
Wiechert seismograph, Earthquake registered by, 267
William of Malmesbury, 345-6
Wiltshire earthquakes, 324
Winwood, H. H., 247
Woodward, H. B., 337
Worcestershire earthquakes, 229

York earthquakes, 224
Yorkshire, Earthquakes of unknown epicentres in, 226
Young, T., 371